# THE HUMAN VOICE

# THE HUMAN VOICE

## HOW THIS EXTRAORDINARY INSTRUMENT
## REVEALS ESSENTIAL CLUES ABOUT WHO WE ARE

### ANNE KARPF

BLOOMSBURY

Copyright © 2006 by Anne Karpf

All rights reserved. No part of this book may be used or reproduced in any
manner whatsoever without written permission from the publisher except in the
case of brief quotations embodied in critical articles or reviews. For information
address Bloomsbury Publishing, 175 Fifth Avenue, New York, NY 10010.

Published by Bloomsbury Publishing, New York and London
Distributed to the trade by Holtzbrinck Publishers

All papers used by Bloomsbury Publishing are natural, recyclable products made
from wood grown in well-managed forests. The manufacturing processes conform
to the environmental regulations of the country of origin.

Library of Congress Cataloging-in-Publication Data has been applied for.

ISBN 1-58234-299-7
ISBN-13 978-1-58234-299-3

First U.S. Edition 2006

1 3 5 7 9 10 8 6 4 2

Typeset by Hewer Text UK Ltd, Edinburgh
Printed in the United States of America by Quebecor World Fairfield

For Bianca and Lola, delicious voices both

# CONTENTS

PART THREE

# Acknowledgements

The idea that a book is created by a single individual working alone seems to me a giant fiction, especially in the case of non-fiction. For this effort, I've had the benefit of a vast amount of support, encouragement and help.

Peter Lunt arranged what turned into three years as an Honorary Research Fellow in the Department of Psychology at University College, London, allowing me the run of its wonderful libraries. The librarians at the Institute of Education Library and the British Library were also very helpful. Once I'd started writing and no longer had the time to visit these libraries in person, I discovered how well a local library can support specialist research. Colin Carsten of Camden Libraries was an outstanding help, calling in dozens of papers for me, chasing up others, and even coming up with suggestions of his own.

Thanks to Brian Klug for helping set up some of my American interviews, and to Joyce Lorinstein for transcribing them. Sue Woodman and Mike Matthewson were typically generous with accommodation in New York, and friendship. A conversation with Ned Temko first got me excited about the brain (although this might just have been the reckless speed at which he was driving at the time). Stephanie Martin, John Rubin, Lolly Tyler, and Carlos

Fishman all made time in busy schedules to read chapters and comment on them. I thank them for their valuable suggestions, even if I haven't acted on all of them.

I'm also grateful to Dan Rather, Jon Snow, Rory Bremner, and Tim Bell for agreeing to be interviewed. I've tried to preserve the anonymity of all my other interviewees, and thank them for their openness and honesty, as well as interest in the project.

Thanks too to participants in the Sounding Out conference at the University of Nottingham in 2004, and to the Pan-European Voice Conference, PEVOC 6, in London in 2005, for their positive response to the papers I gave there and useful comments.

When I was first thinking about this book, I benefited enormously from discussions with Lennie Goodings, a dear friend who allowed me to take advantage of her editorial expertise and judgement. My agent, Natasha Fairweather of A.P. Watt, has been a constantly encouraging presence: her enthusiastic response as my first reader was tremendously helpful. Thanks to Linda Shaughnessy, also of A.P. Watt, for so energetically selling foreign rights, to my editor at Bloomsbury, Bill Swainson, for helping me excavate this book from its mountainous first draft, and to Mary Tomlinson for her elegant copy-editing. The warm enthusiasm of Karen Rinaldi and Colin Dickerman, of Bloomsbury USA, helped quicken my own passion for the subject.

I'm very grateful to Natalia Karpf for her support. Eve Lowen's unflagging enthusiasm for this project has buoyed me up, and Mark Lowen's interest in it has been touching. Lola Karpf came up with a stream of interesting suggestions.

Thanks to Mary Levens for an illuminating friendship and to Caroline Pick for her love and support. Barbara Rosenbaum's insight and love have been enriching and inspiring. Without Gianna Williams this book would never have been completed. I'm profoundly grateful for her transforming generosity and wisdom.

Five months before I finished the book, I had a fall that resulted in a triple fracture of the leg. In four months of despair the love and support of those I've already mentioned (especially Barbara Rosenbaum, who accompanied me into hospital, and was there to welcome me home), as well as Janine Turkie, Corinne Pearlman, and Mary-Ann Smillie, was indispensable. Members of JFJHR were also terrifically supportive, while fellow parents Teng and Richard Sayes and Kari Holtung were generous with help. Thanks to Sue Summers for recommending the excellent Pippa Worrell, who, along with my other brilliant physiotherapist Nikki Jackson, helped me to walk again.

Cyril Connolly famously said that the pram in the hall was the enemy of good art. He got it wrong – it's the other way round: creative and intellectual endeavour is the enemy of the pram. Although my daughters are well beyond the buggy stage (and I don't lay claim to artistic status), they've suffered from the inevitable preoccupation and long hours involved in my writing this book, only for it to be followed by my accident and enforced immobility. During the successive final sprints to complete the work all sorts of gratification had to be repeatedly deferred. I thank them for their good grace, only occasional outbursts of exasperation about 'the wretched book', love and exuberance.

Peter Lewis has shouldered more domestic burdens than any partner should have to with an extraordinary generosity of spirit. When I couldn't walk he effectively became a single parent, despite his own pressing work commitments. His encouragement and fishcakes, support and love have been utterly sustaining.

A.K.

# *Preface*

EVERY DAY, in countless different ways, we employ sublime skills of which we're only faintly, fleetingly, conscious. We use them when we buy a newspaper as much as when we comfort a bereaved friend, when we ask for a rise as much as when we recite poetry. Without them, our language instincts would count for nothing: we simply wouldn't be able to understand one another. For not only are they the channel through which thoughts become speech, but also they give our words meaning, and tint them with feelings. However hesitant we might think we are, most of us, from the very moment of birth, spontaneously demonstrate the use of this most glorious instrument – our human voice.

Throughout our lives we make decisions, often unwittingly, on the basis of the sound of a person's voice: lovers as well as political candidates get selected for vocal reasons. Our lilt, twang, or tremor are eloquent often beyond words. The voice can also make sentences do somersaults. 'I don't think so' might be an innocent expression of uncertainty or an example of withering sarcasm – the voice tells us which. 'Not bad' – given the right tone – can glow with praise. Yet signal oddly with your voice – by transgressing the normal codes of volume, pause, and pitch – and you can entirely sabotage conversation, turning sense into nonsense.

But the voice isn't just a conduit for language, information, and mood: it's our personal and social glue, helping to create bonds between individuals and groups. One Christmas Eve, for example, a woman in the kitchen of a friend's house accidentally slices open her finger. Her husband, who is outside, comes running in, bandages the finger, and takes her to hospital. But their friends are astonished: she'd said, 'Ow,' softly, over the din of chatting friends, yet her husband had not only heard her, but also could identify something quite serious. 'I know an "*Ow*" versus an "Ow",' he said.[1]

Or consider this. A 63-year-old man phones his 91-year-old mother, who lives some seventy-five miles away, every two or three days. 'I can tell, less than a second into the call, how my mother is. I know every one of her ailments from her voice as she answers. I know that my voice has an emotionally cheering and uplifting effect on her. If she's down, at the end of the call I can hear her very picked up. It doesn't matter what I've said – I can use my voice to pick her up from her 90-something doldrums.' People who have to rely on the phone to maintain close relationships tend to hone those skills. A 50-year-old American woman living in London says, 'I don't ring my mother if I'm feeling distressed or drunk because I know that she's fantastically good at hearing what's wrong.'[2]

Then there's the 12-year-old boy, six weeks into his new school, who has worked out which of his teachers like him mainly through their tone of voice. As a result he speaks in a more friendly way to those teachers, who respond more warmly back to him, creating a loop of approval.[3]

None of these people possesses exceptional vocal abilities: they're simply using the acoustic capacities with which they were born. Of course there are some individuals who are aurally gifted, like the 56-year-old woman who told me, 'Somebody can say hello to me on the phone and I know if I can trust them.'[4] And others with highly developed vocal talents. A 65-year-old GP with an authoritative but mellow voice (and an enthusiastic member of his local choir) is the most popular doctor in his group practice. His sister-in-law is certain that his voice is responsible: it makes patients feel that they can safely entrust themselves to his care.[5]

Her hunch seems less far-fetched when you learn that a recent American study found that, from just forty seconds of surgeon–patient consultations from which the words had been filtered out, leaving tone of voice alone, listeners could tell which doctors had been sued for malpractice and which hadn't. The degree of dominance or concern in the surgeons' voice was a giveaway.[6]

Despite all this, we have very little collective sense in Western societies of the importance of the voice, and almost no shared language with which to talk about it. We persist instead with the idea that the move from a primarily oral to a mainly literate society has made the voice much less important than the image and the written word, as if the voice belonged at the periphery of human experience, rather than at its centre. In all the last few decades' excited debate over the role of language, speech, and conversation, the voice is often no more than an afterthought.

In reality, as I hope to show, the voice lies at the heart of what it is to be human. It plays a crucial role in helping babies establish secure emotional ties, acquire language, develop empathy and social skills. Adults milk it for information in their intimate relationships and professional lives. And now, when radio and television are so prominent and telephone and digital media inescapable, there's a renewed interest in sound[7] that's beginning to stoke fresh fascination with the role of the voice.

In the pages that follow, I hope to convince you of the extraordinary properties of what are essentially vibrations of air. I want you to be bewitched, as you sit on a bus or in a restaurant, by the way that different parts of the body – lungs, abdomen, throat, lips, teeth, tongue, palate, and jaw – unite to make a voice, and how tiny changes in one of them can entirely alter mood and meaning. We can do with our voices what typographers do with print – italicise, put into bold or inverted commas. Some people make their words purr, others use their voice as a bayonet. Within the space of a few minutes, we can become siren or screamer, patron or soother, just through labial flexibility and the reshaping of our internal cavities.

The voice is one of our most powerful instruments, lying at the heart of the communication process. It belongs to both the body and the mind. It's shaped by our earliest infant experience and by powerful social conventions. It bridges our internal and external worlds, travelling from our most private recesses into the public domain, revealing not only our deepest sense of who we are, but also who we wish we weren't. It's a superb guide to fear and power, anxiety and subservience, to another person's vitality and authenticity as well as our own.

You can't really know a person until you have heard them speak. Most of us have hidden under the duvet, remembering how our voice unintentionally betrayed some emotion that we'd thought – and hoped – was securely padlocked away. There can hardly be a person who hasn't bristled over a conversation with someone whose words were unexceptional but whose tone of voice delivered an entirely different message. The voice can invite or discourage intimacy, without our having to be verbally explicit,[8] or even conscious of what we're doing: a lonely person might be unwittingly creating subtle vocal cues that keep other people at a distance, whilst considering themselves a victim, because unable to take responsibility for the negative feelings they elicit in listeners.[9] We use our voice to repel and attract, encourage or undermine. As animals with smell, so are humans with voices.

And yet we're often shockingly indifferent to this instrument – an indifference this book sets out to challenge. Part 1 shows how the voice is implicated in almost every area of our personal and social lives, explains how we modulate it, and explores its origins in human development and experience. Part 2 tracks the psychological, gender and cultural aspects of the voice, using original interviews to examine our relationship with our own voice and those of the people we're close to. Part 3 looks at how the voice has changed and is changing, partly because of technology. This journey round the voice visits the realms of psychology, anatomy, linguistics, anthropology, as well as history, child development, gender and cultural studies. I hope it will play some part in helping us attune to the human voice in a new way, to develop fresh ears and awe for one of our most stunning abilities.

PART ONE

# I

# *What the Voice Can Tell Us*

ONE BALMY FRIDAY evening in Silicon Valley, Amy and Bruno Smart fantasised about inventing a new kind of machine. The Smartacom would reveal everything you'd want to know about another person's education, status, self-confidence, and even state of sexual arousal – in under a minute, even from six feet away. The excited Smarts set about creating and testing a prototype – and it worked! There was just one problem. The Smartacom violated every single privacy law in existence, and had the Smarts been audacious enough to manufacture it, the Federal Trade Commission would have made sure that their business dissolved faster than the Wicked Witch of the West.

The Smarts were actually very dumb. They hadn't realised that an instrument like the one they'd invented already existed. What's more, it breaks no laws and is free. It goes by the name of the human voice.

In an era so preoccupied with privacy and its infringement, with hackers and cookies and data protection, we're remarkably breezy about the personal information that seeps from our voices. Employers award fat contracts to psychometric companies promising to uncover their staffs' hidden flaws and skills, while ignoring what those same staff are freely divulging through their voices. We take this fabulously rich resource for granted, yet if the human voice were a new technology we would be hymning it loudly, and extolling its special properties.

For the moment we open our mouths and start to speak, even if it's only to read out regulations for the disposal of sewage, our voice is doing something terrifyingly intimate – leaking information about our biological, psychological, and social status. Through it, our size, height, weight, physique, sex,[1] age and occupation,[2] often even sexual orientation,[3] can be detected.

The voice is a stethoscope, and transmits information not only about anatomical abnormalities but even illnesses.[4] Our risk of coronary heart disease can be predicted, it's been claimed, on the basis of voice characteristics like volume and speed alone.[5] Doctors have even maintained that picking up on changes to the sound of the voice can be life-saving in cases of throat cancer[6] and people contemplating suicide.[7]

The voice can function as a breathalyser, a rough guide to intoxication.[8] Joseph Hazelwood, captain of the *Exxon Valdez* oil tanker, was acquitted by an Alaskan jury of being drunk when the tanker ran aground in 1989, causing one of the worst accidental oil spills in history. Yet an analysis of his voice from tape recordings of conversations between him and the coast guard before, during, and after the accident found that he didn't only misarticulate certain sounds, in the way that stage drunks do – at one point he called his ship the '*Ekshon Valdez*' – but also took 50 per cent longer to say its name at the time of the accident than the day before (alcohol slows down speech).[9]

Tiredness, too, is audible (and this matters because in certain occupations tiredness can cause death). The voice is such a reliable measure of fatigue[10] that a couple of Japanese researchers have devised a drowsiness-predictor for pilots and air-traffic controllers based entirely on readings from their voice: they hope to use it to improve flight safety by informing exhausted pilots when they should be relieved by other crew members.[11] (Airline pilots also now receive training on how to pitch frightening announcements –

perhaps one reason why they all sound the same.[12]) And the voice is a census: through it we can detect social class, race, and education, sometimes even the number of years that a speaker has spent in school.[13]

Alter a person's voice and you can totally change the way that others react to them. When they wanted to reduce the impact of Sinn Fein leader Gerry Adams, the Conservative government under Margaret Thatcher banned neither his words nor his face but only his actual voice. When the airline captain intones, 'Good evening, ladies and gentlemen, this is your captain speaking,' in a slow baritone, he immediately instils confidence. Imagine the same words shrill and fast.[14] Or a lisping Macbeth, or a camp news-reader. Part of scientist Stephen Hawking's fame comes surely from the contrast between his disabled body and synthesised, American-accented voice. It is as though his mind and voice (unlike his body) were somehow beyond human frailty.

Although cinema is usually characterised as a visual medium, when face quarrels with voice the results can be disastrous. Films dubbed into another language, no matter how expertly, lose some essential synergy between body and sound. And when the dubbed voice belongs to a famous icon, the effect is almost always – at least to those familiar with the original – shockingly comic. Humphrey Bogart *has* to sound like Humphrey Bogart – he is his rasp. A dubbed voice often sounds disembodied: the more it resembles the original actor's voice, the more its difference becomes apparent. It's as if the real film is unfolding tantalisingly somewhere behind or just out of reach of the dubbed version. Even where you don't understand the words' meaning, you take in something important, if non-verbal, about a film through its original voices.

When we travel abroad, we become like dubbed actors. We're infantilised, because even if we have the language we almost invariably lack the music and rhythm particular to each language, however good our accent. (The poet Tom Paulin maintains that W.H. Auden, when he emigrated to the United States, lost 'that sense of the skip and kick of the language and he writes a very perfect but rather chewing-gum standard English, without the deeper rhythms of the language'.[15])

Abroad, we also can't read the locals' voices the way we do at home, and so become deaf to nuances of class, background, and status. (This opens the door to gaffes, but can also free us from the prison of prejudice based on accent and pronunciation.) Yet until recently features like pitch and volume have been almost entirely neglected in foreign-language teaching. A survey of twenty conversation textbooks found not a single reference to voice. The author suggested that language teachers, as well as covering grammar and pronunciation, should also teach sarcasm and reluctance.[16]

A change of voice can destabilise a whole sentence. The playwright Bernard Kops turned up at a rehearsal of one of his plays only to find that its whole centre of gravity had changed. It took him a while to work out what was wrong. The actress playing the main role, instead of saying, 'I can't sleep with anyone,' with the accent on 'any', i.e., any Tom Dick or Harry, had put the stress on the whole word 'anyone' and was playing the character as frigid.[17] There's the joke about a board outside a church bearing the message, 'The Same Yesterday. The Same Today. The Same Tomorrow. Jesus Christ.' A Jew comes along and reads it in an exasperated tone and Jewish accent: Jesus now sounds like an expletive, and the rest like a complaint.

## NO AUDIO

And yet, despite all this, we've developed very little consciousness of the channel on which our ability to communicate so crucially depends. In the process of writing this book I've had a recurring experience. Again and again I've met people who, hearing what I'm writing about, have suddenly gone silent (and not just through self-consciousness). 'I've never given the subject the slightest thought,' they say, once the power of speech returns. And then, almost invariably, they proceed to regale me for ten minutes with anecdotes about voices that they love or hate and why. It's as if we both know and don't know about the voice.

In the fifty interviews with people about their own voices and those of their friends and relatives that I've carried out for this book, I've come upon this chasm between feeling and articulating many times. A lot of people seem to be enormously irritated by certain types of voices – say, nasal, high-pitched, or loud. They can give examples of those they know, or hear on radio and television, whose voices needle them like a dentist's drill. And yet they're almost entirely unable to identify why or how those voices manage to penetrate them with such malign effect. It's as though they don't know much about voices but they know what they dislike.

We're saturated with other people's cadences, most of the time without any inkling of how they work on us or shape our comprehension. As a culture, until quite recently, we've been barely voice-aware – a society that has blocked its ears. In Western cultures' hierarchy of the senses, sound is often placed below sight. We suffer from what Coleridge called 'the despotism of the eye'.[18] Indeed you could say that we often despise sound since we live in a culture where to love the sound of one's own voice is a term of abuse.

The last couple of decades have seen an extraordinary eruption of interest in language – how we acquire it, the skills and rules governing how we use it. Conversation, dialect, the very language instinct have been analysed and the psychology of talk probed. And yet, remarkably, for much of the time this fascination with what we say and how we say it has continued to marginalise the voice. Most linguistic studies on conversation neglect the medium through which it's conducted. Voice and speech are treated as almost identical, and speech as little more than spoken language. Language is thought to be the primary carrier of meaning, as if the voice were only the vehicle for words, the real force governing the direction and speed of a sentence. We raid speech for its semantic meaning, and then discard the voice like leftovers, detritus.

So an entire book on Ronald Reagan's oratory included just one single reference to his voice. The study was called 'Reagan Speaks' but, given its focus on the verbal at the expense of the vocal, it should have been called 'Reagan Reads'.[19] Its omission is all the more extraordinary when you consider how important Reagan's voice was in creating his folksy image. The voice is also usually missing from discussions about film. Where sound is referred to at all, it is in terms of 'the soundtrack'.[20] And radio still gets far less money, status, and critical attention compared to television.

It sometimes seems as if we only pay attention to the voice when it goes wrong. There's a whole literature on teachers' voice disorders going back over thirty years,[21] yet the role of the teacher's voice in enthusing a class has been barely researched.[22] A 7-year-old girl I interviewed, however, had no hesitation in identifying the role it played in her education. 'You can hear if a teacher is interested in what they're saying – their voice goes high. If it's got expression then we learn better, because it's not boring and down – it's up and we'll take notice and start listening.'[23] This neglect of the role of the voice in the classroom is all the more disturbing since the pitch, volume, and tempo of children's voices have even been found unconsciously to affect teachers' opinions of their intelligence.[24]

And yet we have no shared public language through which to speak about the voice or sound, in contrast to the wide vocabulary that we've developed for visual images. Sounds are still part of the great unnamed.[25] Back in 1833 the American physician, James Rush, tried to identify different kinds of voices – whispering, natural, falsetto, orotund, harsh, rough, smooth, full, thin, slender.[26] By the 1970s phoneticians hadn't moved much beyond Rush in naming different types of voice. The terms they had come up with – like whispery voice, harsh voice, creaky voice, tense or lax voice[27] – were never taken up by the public. Neither was more specialist terminology, like vocal fry, jitter, or shimmer, words which anyway have no agreed definition. We're in a state of terminological disarray, and few of us are able to describe the voice in words that aren't either impressionistic or ambiguous.[28]

Perhaps this is because voices are such an inescapable aspect of daily life, girdling us at work and home, suffusing us even when we deliberately tune them out. There's no way to stop sound[29] and we have no 'earlids'.[30] Voices are the audio equivalent of air. For most of us, our voice is something that's simply there when we open our mouths, and other people's voices are givens, as unchangeable and taken-for-granted as their faces.[31] Voices just are.

What's more, the moment you try to describe them, voices seem to evanesce. Made out of breath, they're equally vital and insubstantial. Unlike visual images, the voice exists only in time and can't be frozen. Our voice only comes into being as it passes out of existence.[32] It's gone as soon as it's been produced;[33] in its beginning is its end.[34]

Or maybe our neglect of the voice is connected with age. Because children are dependent on adults and their language skills are limited, they're particularly sensitive to register and cadence: they learn early what inflections are expected in what roles. As we grow and acquire language, though, we begin to neglect the non-linguistic aspects of speech, transferring our attention almost entirely from voices to words. Yet when an anthropologist, to test the importance of the acoustic in people's lives, gave British students a questionnaire asking them to identify two or three important sounds from their childhood, he uncovered entire individual symphonies with deep emotional and personal associations. The scraping of a father's razor, his clearing his throat when concentrating, a mother singing in the kitchen while cooking, all produced a remembered sense of security, as powerful as any gnawed old piece of blanket or other transitional object connected with touch, smell, or sight.[35] Proust's madeleine could have been something aural.

## TALK ABOUT TALK

To see just how thin our thinking about the voice really is, meet the speakers of the Tzeltal language in Tenejapa, Mexico. Not only do they talk a great deal, but they also spend a large part of their time judging, commenting on, or mocking the way the other speaks. The word '*k'op*' is a central feature of their metalinguistic lexicon: combined with other words, it's used to describe more than 400 separate speech situations and characteristics. Tzeltal has words that refer to the personality of the speaker, their physical, mental, emotional, and postural condition, their location, social identity, and volubility. It has others for 'talking with a nice, mellow, singing voice', 'talking very slowly, as if sad', and 'talking with a high voice, not quite falsetto, but almost singing'. Tzeltal can identify 'high, scratchy, cracking voice – characteristic of adolescents', 'speech that is poor and indistinct in which the speaker's head is turned away from the listener', 'pouted whining talk from someone with a wounded ego', and 'speech that is excessively self-assertive, that is loud and forceful coming out with great confidence (negatively valued)'.

Tzeltal-speakers are plain speakers. They have different words to describe 'speech cut off midstream during a conversation so that the speaker can go outside to urinate or defecate', 'speech that trails off into nothing as the speaker falls asleep (especially apt for describing drunk persons)', and 'a kidding-around voice, when someone says, "I'm going now," and doesn't mean it'.[36] The idea that the Inuit have dozens of words for snow, it's now clear, is apocryphal – nothing more than an urban legend.[37] Perhaps it should be replaced in the popular imagination by the number of Tzeltal words for talk.

You could argue that the English dictionary would also, if thoroughly thumbed, come up with 400-plus words to do with speech. But this isn't analogous, for the '*k'op*' words are the equivalent of English words like 'sweet talk', 'baby talk', 'shop talk', etc., and even a demon Scrabble player would be hard pushed to come up with more than a dozen. There's no denying it: we don't do much detailed talking about talk.

## SIMULTANEOUS INTERPRETERS

Yet despite all this, my interviews reveal the enormous amount of skilful voice-reading that we engage in every day, often unconsciously. Interpreting other people's inflections and modifying our own is one of our most important interactive tasks. We voice-read to confirm what our other senses have told us, and sometimes use the voice to express feelings and moods that, if put into words, might leave a trail of embarrassment or shame.

For example: A mother down the years hired nine nannies to look after her children purely by interview over the phone because she felt that she could get a sense of them more accurately from their voice alone. She was disappointed by only one of them (the kleptomaniac daughter of a policeman).[38]

A 56-year-old South African man says of his long-term partner, 'I can tell in her voice when we're going to have sex, long before we go to bed. I know she's up for bed, even if she doesn't mention it – there comes an extra softness in her voice.'[39]

A 14-year-old girl, when her parents complain angrily about her untidy room, failure to do her homework, and other standard parent–teenage grievances, makes a point of responding to them in a calm voice. 'If you don't get loud I think it makes them feel a bit weird for shouting, so they stop. It's very easy to do.' Does she do it often? 'Yes. I'm really aware of doing this, like a plan.'[40]

A 12-year-old boy, in regular conflict with his parents, enjoys the moment when they come and kiss him goodnight 'and talk quietly – there are no other voices to compete with'.[41] He recognises the special property of the intimate voice – that it seems to speak just to us.

A 46-year-old woman says of the loving voice her husband sometimes uses, 'I suppose that's what makes the marriage. It's the private tone between us – I haven't heard him use that voice to anyone else, except maybe slightly to the children.'[42]

## RAISING THE VOICE

So, if most of us are reasonably fluent voice-users and proficient voice-readers, why do we need to understand its physiology, psychology, and acoustic properties? Do we have to learn the Latin name of a plant to enjoy its scent, or the ingredients of a dish to savour its taste? Does knowing about joints and muscles help us walk better?

There are three reasons for exploring the voice. Firstly, the differences and similarities in the way people speak are fascinating. What's more, voice is a distinctive human feature. Other creatures are also vocally skilled. Birds, for instance, can distinguish the voices of their near relatives from thousands of others,[43] and the chaffinch can even recognise a male rival by the individuality of his voice.[44] Monkeys express anger, fear, submissiveness, and dominance through similar vocal cues to humans.[45] Yet while apes and monkeys have a repertoire of about thirteen calls, the human vocal system has a far larger number of sounds.[46] No vocal learning by imitation takes place in mammals below humans; apart from some birds, only humans have voluntary control over the acoustic nature of their vocal utterances, can learn vocal patterns by imitation, and even invent new ones.[47] This means that there's something quintessentially human about the voice, and understanding it enables us to peer more deeply into the unique, complex properties of our own species. So in some important sense, an investigation into the voice becomes an exploration of our humanness.

Finally, the voice is central to our communicative abilities. Women's vocal folds (sometimes called vocal cords) perform more than one million oscillatory cycles a day. Men's accomplish around half a million in the course of a day.[48] Astonishingly, our vocal folds often travel more than two kilometres a day.[49] Our relative ignorance about a channel on which we're so dependent, and that's so critical to debate and negotiation, rumour and argument, is staggering.

From teachers to receptionists to lawyers, around a quarter of the total labour force is in a vocally demanding profession, or uses their voice as their primary tool of trade.[50] Yet even professional voice-users can be remarkably casual about the health of their instrument.[51] And you only have to look at the occupational-safety limits placed on hand movement to find a chastening comparison: if the distance travelled by hands, exposed to vibrations, exceeds 520 metres in one day, it's considered an industrial hazard, but the vocal folds can travel this distance in less than forty-five minutes of continuous speech. We're all familiar with Repetitive Strain Injury, but Repetitive Vocal Injury as a concept doesn't exist, and there's no health-and-safety legislation applying to voice-users.[52]

Although interpreting the human voice is one of our most important daily social activities, the way we speak and hear remains almost entirely hidden, and unexplored. Those who develop a richer understanding of the voice are less likely to mishear their friends, lovers, and rivals, and more able to intuit when people are being authentic or dishonest. And with vocal awareness comes a greater chance of achieving real communication, of expressing oneself clearly, of really hearing another person, of two voices connecting. In an atomised society where isolation and loneliness are rife, this dimension of the voice has never been more essential.

## NO IMPROVEMENT

This isn't a self-help book. There are no tips to be found here on how to develop a rich brown voice, of the kind that sells limousines or instant coffee, or can snare a partner. Plenty of such books already exist, few of them effective. As one speech and language therapist put it, 'If you just give people a voice, it's a bit like a script, they become a cardboard cut-out. And in a moment of stress and anxiety they revert to the voice they had before.'[53] Or they risk losing the natural dynamism of their voice, with all that this implies. Opening one's mouth only to hear another person's voice coming out isn't the same as developing a greater sensitivity to other people's voices and one's own – it's ventriloquism.

This book isn't about the singing voice either, even if there are more similarities between the singing and speaking voice than differences. The same part of the brain may produce both singing and speaking – we have one voice, and not two – but the speaking voice is enough for one volume.[54]

Again, when people speak of the voice, it's often the metaphorical voice they have in mind, and not the embodied one. Today 'having a voice' has lost a lot of its original meaning. Increasingly it's used not literally to mean the sound we make with our mouths but more abstractly, in the sense of having the right to be heard socially and politically. 'Voice' has also become a common term for narrative authority and literary self-expression.[55]

Is this an example of how we've taken the body out of the voice, and the voice out of the body? Or does using 'having a voice' as a metaphor only prove once again the importance of the voice in human experience? The fact that 'finding one's voice', a phrase that on the literal level means being able to speak, has lent itself almost exclusively today to the figurative sense of speaking out testifies to the voice's fabulous ability to express a person's deepest sense of self.

Finally – especially in Britain – if you talk about the voice it's assumed that you're talking about accent. George Bernard Shaw summed up the relationship between accent and class with his famous quip: 'It is impossible for an Englishman to open his mouth without making some other Englishman despise him.'[56] It's rather dispiriting that, all these years later, debate about accent still dominates our thinking about the voice, as if dialect and pronunciation made up the whole of the subject, and not just a small corner of it, albeit a fascinating one. On the other hand, if we've become so sensitive to accent that a native speaker can distinguish instantly between an Atlantic City accent and an Atlanta one, between a Bristolian and a Brummie, or between Hochdeutsch ('High German') and Plattdeutsch ('Low German'), then we have the potential to develop our sensitivity to the many other astonishing dimensions of this vital medium.

## 2

# *How the Voice Achieves its Range and Power*

ONE SEPTEMBER DAY in 1854, Manuel Garcia was strolling through the garden of the Palais Royal in Paris when he had a Eureka moment. The Spanish-born Garcia, teacher of singing at the Royal Academy of Music in London, desperately wanted to find a way of glimpsing the tiny movements that take place inside the human body when we sing. That day, noticing the sun flashing on the window panes of a nearby house, he thought of a method. What he needed was two mirrors: he could then try reflecting the light from one on to the other.

The excited Garcia rushed straight to Monsieur Charrière, a surgical-instrument maker, who had just the thing – a small dentist's mirror with a long handle that had failed to catch on at the London Exhibition of 1851. Garcia bought it for six francs, and hurried home to experiment. Placing it with one hand at an angle against the uvula (the muscle at the back of the mouth), he flashed a ray of sunlight on to it from another mirror held in his other hand. Success! To his great joy he saw, wide open before him, his glottis, so fully exposed that he could even see a part of the trachea (windpipe) beneath. When his excitement had subsided a little, Garcia – with an impressive ability to refrain from gagging – began to observe himself speaking. 'The manner in which the glottis silently opened and shut, and moved in the act of phonation, filled me with wonder.'[1] Garcia had invented the laryngoscope, and the human gaze had penetrated into the very speech organs themselves.[2]

Though there's much that we still don't know, the past thirty years have yielded an extraordinary increase in knowledge about the voice, partly because of the collaboration of so many different disciplines – from engineering to linguistics, neurology to computer science, through aerodynamic studies, acoustic studies, even intonation studies.[3]

What's clear is that the voice isn't produced by any single organ; rather many different body parts combine into a sequence or chain. As the French physician Alfred Tomatis put it, 'We were given a digestive apparatus and a respiratory apparatus, but no specific oral-language apparatus. What ingenious adaptation and unlikely combination was necessary to attain that goal!'[4] That adaptation is so subtle but at the same time so effortlessly achieved that each sullen, inarticulate teenager is, in their own way, as sophisticated a shaper of vocal sounds as someone reciting Shakespeare. Again, in the words of Tomatis, 'The ability to cry, to call out, to listen, to wilfully make sounds for one's own benefit as one's own audience and window on life is one of the most extraordinary humanising mechanisms ever observed in the evolution of language.'[5]

For the production of the human voice is not just a story of physics, biology, and neurology, but also a linguistic, phonetic and acoustic event that consists of three parts.[6] Beginning by conceptualising a preverbal message, we then encode it in grammar and phonetics, only lastly articulating it through the co-ordination of muscles that leads to speech.

We don't need to be neuroscientists or linguists to appreciate this awesome process, but grasping the rudiments of where the voice comes from is a help. This requires no scientific aptitude at all, just a willingness to make a short journey around the anatomy of the human vocal apparatus and the ear.

## THE HUMAN VOICE: WHAT A PRODUCTION

It's obvious that the lips and mouth help create sounds, and that the larynx is also involved, but so too is three-quarters of the torso. We speak with our body: almost every part of it is called upon to make a voice, including the back. The voice can even be affected by a sprained ankle – changing the posture can hurt the abdominal muscles, which in turn can lead to hoarseness.[7]

Remarkably, the voice is produced, as Tomatis noted, by a system designed biologically not for speech but for eating and breathing. Indeed, practically every one of the vocal organs – the lungs, trachea, larynx (and vocal folds), pharynx, nose, jaw, and mouth (including the soft palate, hard palate, teeth, tongue, and lips) – is an impressive multi-tasker, carrying out some other crucial job in the body in addition to producing sounds.

For example, although the larynx is also known as the voice-box, its chief function has nothing to do with the voice: it's to act as a sphincter to prevent anything but air entering the lungs. Alexander Graham Bell called the larynx 'the guardhouse of the lungs'.[8] (Hence the advice not to talk while eating: it's unreasonable to expect the larynx to have to open up to allow air through at the same time as closing off any straying particles of food.) The larynx also helps us lift heavy weights: try lifting the chair you're sitting on and see what happens. The vocal folds immediately close up, helping stabilise the body to allow it to push, lift, or pull.[9] As for the mouth itself, this serves the triple purpose of breathing, chewing, and speaking. The tongue in particular, as Aristotle observed, 'is used both for tasting and articulating'.[10]

The voice is basically audible air. (Without air, we can't make a sound.) In other words, the process by which we breathe in order to live also, with minor changes, provides the energy for speech.[11]

One important change is that we automatically alter our breathing pattern when we talk. Instead of breathing in and out rhythmically, or breathing in more slowly than out, we breathe in quickly and exhale gradually, for that breath must last us through an entire phrase or sentence. So, although our natural frequency of breathing is about fifteen times a minute, when we sing and speak we breathe much less often.[12] Steven Pinker put it neatly: 'Syntax overrides carbon dioxide: we suppress the delicately tuned feed-back loop that controls our breathing rate to regulate oxygen intake, and instead we time our exhalations to the length of the phrase or sentence we intend to utter'.[13] It's as if the voice's needs are so pressing, so irresistible, we're even willing to tolerate a touch of respiratory discomfort to meet them.

To compensate for not being able to breathe so often, we take in more air when we breathe to speak than is needed purely for life support.[14] The air goes through the pharynx (throat), down the trachea (windpipe) and into our lungs. In order to produce speech, that air is then expelled from the lungs by the contraction of the diaphragm. Now under pressure, it jets back up the trachea until it reaches the larynx. At the larynx the air meets the obstacle of the vocal folds.

Normally, when we exhale the air can't be heard[15] because the vocal folds (two flaps of retractable muscle tissue) are pulled open, like a pair of curtains.[16] When we speak, though, the vocal folds vibrate. After each opening to release the air, the elasticity of the vocal folds ensures that they return to their closed position, until the air pressure builds up once more and the cycle begins again. The sound of the human voice is produced by this rapid opening and closing of the vocal folds, hundreds of times per second, producing a succession of bursts of air that we hear as a buzz or hiss.[17]

The thickness, tension, and length of the vocal folds help determine their frequency. The longer, thinner, and more taut they are, the more frequently they vibrate and the higher the frequency or pitch of sound they produce. The shorter, thicker, and more lax they are, the less frequently they vibrate, and the lower the pitch produced.[18] Frequencies are measured in cycles per second, known as hertz or Hz, after the nineteenth-century physicist Heinrich Hertz. The rate of puffs is very fast – from 32 cycles per second or hertz (Hz) for a very low bass to over 1,000 Hz for a soprano. Normal speaking is between 98–262 Hz, and this rate is known as fundamental frequency.

## RESOUNDING SOUNDS

If the voice emerged straight out of the larynx, it would make a poor, thin sound.[19] It needs to resonate, and the characteristics of the buzz or frequency are shaped by what happens to the air once it has passed through the vocal folds – through its central space or slit, called the glottis – into the rest of the vocal tract (i.e., everything above the larynx). The vocal tract forms a long tube, closed at the bottom end, and opening out into the mouth at the other. Through the movement of the jaw, tongue, lips (the articulators) and pharyngeal cavities (the resonators) it can contort itself into many different shapes and sizes. These alter the resonance of the voice to produce the wide range of human sounds, just as blowing into a half-filled bottle makes a different sound from blowing into a quarter-filled one.[20] Or in the same way as, when we fill a jug of water from the tap, we know by sound alone when to turn it off: the water, gushing into a column of air, sets up a vibration, and as the water rises and the column of air shortens, the resonance changes and the sound bubbles up the scale[21] like a practising contralto.

But no part of the vocal apparatus is as supple and mobile as the mouth. Its most flexible constituent is the tongue, whose tip, edges and centre can move independently, and whose complex musculature can propel it rapidly backwards, forwards, up and down – virtually everything but pirouette. The lips, too, can veer from closed to rounded to spread, while the palate (in the upper gums, the roof of the mouth, and the muscular soft palate at the back) has a dance of its own, being raised in speech and lowered in silence.

Any alteration to the mouth can change the voice, even smiling. Because it widens the mouth and shortens the vocal tract, smiling can actually be heard. So, in order to sound upbeat and positive, smiling while you talk into the mouthpiece of a phone isn't as absurd as it might seem.[22]

The act of speaking requires remarkably complex and skilful collaboration between the different muscle systems (seven of which are involved in speech), whose actions must be synchronised.[23] The larynx not only has to partition words and separate phonemes (small sounds within words), but also coordinate with the movements of the lips and tongue to a millisecond's accuracy.[24] No moon-shot could demand greater precision. Its very complexity makes one marvel at how we ever manage to utter a sound. Mercifully, we're entirely unconscious of the action of our vocal folds, which are guided not directly by our will, only indirectly by our mental perception of the sound they produce.[25]

## SOUNDS AND SWEET AIRS
## THAT GIVE DELIGHT

And what an array that is: the average larynx can stretch to over two octaves in range, sometimes even three.[26] This adds up to twenty-five notes (if we include semitones). In addition, the mouth can form thirteen vowel sounds (creating thirteen different resonances). Thirteen vowels on twenty-five different musical notes produce a full complement of 325 different sounds. We make far more sounds in speaking, the linguist Edward Sapir once suggested, than we ever realise.[27]

At the same time few of us employ more than one octave when we speak, mostly staying in the lower part of our total voice range.[28] So the total number of possible sounds is far greater, as Sapir also points out, than those in actual use – for psychological and cultural reasons as well as for mechanical ones. When we're very young, the muscles of our speech organs get used to the positions required to produce the traditional sounds of our mother tongue, and at the same time learn to inhibit other settings. As a result, most of us develop what Sapir calls a 'strange rigidity' over the voluntary control of the speech organs, leading to difficulties

learning the new sounds of foreign languages where the relationship between lips and tongue is different.[29]

What we manage to do with sound is nonetheless astonishing.
With the exception of the muscles around the eyes, those of the
human larynx have more nerves than any other muscles in the
human body, including the hands and the face, even though we
only use around one-third of their capacity in speaking.[30] Each can
produce a different balance of forces in the larynx, generating a
different pulse wave and sound quality.[31] They're our vocal pal-
ette: through them we colour our voices with affection, bitterness,
pleasure, disgust, etc. With such a formidable range, we are in
effect Leonardos of the larynx.

## A GOOD EAR

Ear and voice complement each other: both are activated by the
movement of air. Just as air makes the larynx vibrate, so it's the air
in the form of sound waves that causes the eardrum to vibrate – the
basis of hearing. To listen to someone's voice is therefore 'a
partnership of vibration'.[32] The ear has even been called a 'sonic
mouth', since some of the same organs in the body form them
both.[33] By ensuring that the ear canal resonates at the same
frequencies as the vocal tract, nature has thoughtfully matched
the reception organ with the production one,[34] and developed a
human ear with the precise properties best needed to hear the
human voice.

Hearing is one of the first senses to develop in a foetus – the ear
has already begun to be formed in an eight-week-old foetus, and
three months later is structurally complete. When we go to sleep,
our perception of sound is the last sense to close off, and when we
awake the first to start up again.[35] And hearing, it's claimed, is the
last sense to die at the end of life.

Yet some of the business of hearing is still frankly unfathomable.
Although we know, for instance, that the brain processes acoustic
vibrations into neural signs, some of our phenomenal auditory

talents, like translating the physical properties of sound into the abstract realm of meaning, remain a profound mystery,[36] all the more so since this occurs within 150 milliseconds after the beginning of the sound.

Sound travels from the outer ear, via the middle ear, and into the inner ear. In the course of this journey it's converted from sound waves into mechanical vibrations (in the eardrum), from mechanical vibrations into pressure waves (in the cochlea), and thence into electrical signals firing towards the brain.

The outer ear (the cup-shaped part we can see, known as the pinna) collects sound and amplifies it, by as much as three decibels.[37] From here the voice travels down the external auditory canal – a tube that acts as a resonator of its own, with its own frequencies. It's this characteristic of the external auditory canal that makes us sensitive to certain frequencies of the voice. Most adults are sensitive to frequencies roughly between 16 and 16,000 Hz (higher than a whale's moan and lower than a bat's squeak). Healthy young people, if they haven't spent too long clubbing, can hear up to 20,000 Hz, while older people tend to lose the ability to hear higher frequencies, at least in our culture – one reason for their common complaint, 'Everyone mumbles these days'.[38]

We can hear a wider range of frequencies than we can speak, but the frequencies to which the external auditory canal are most sensitive lie roughly in the middle of our hearing range, around 1000–3000 Hz.[39] In other words, at normal volume, we detect the middle frequencies more easily than the high ones. In this way human beings have developed the most favourable capacities for speaking to and hearing each other.

From the external auditory canal sound travels to the eardrum – a thin, taut membrane protecting the three smallest bones in the human body, the ossicles in the middle ear. These hinged bones, known poetically as the hammer, anvil, and stirrup, send the vibrations of air on to the fluid-filled cochlea or inner ear. Here lies the organ of Corti,[40] the so-called microphone of hearing. A spiral structure, it contains some of the bundles of delicate, microscopic hair cells or cilia that are ranged in rows from the top of the cochlea to the bottom like tufts on a rug. A young person with normal hearing has around 15,000 such hair cells in each ear, each sensitive to a particular frequency at a particular loudness, with the hair cells at one end reacting to high-pitched sounds, and those at the other to low-pitched ones.[41] The vibrations of air move the fluid in the cochlea, which bends the hair cells, stimulating nerve signals carried by the auditory nerve to the brain.[42]

No piece of technology or Swiss precision-measuring instrument has ever come near the extraordinary sensitivity of the ear in its abilities to detect nano-changes in loudness and frequency or pitch. (Frequency is an acoustic measurement of the voice's vibrations; pitch is a perceptual term – how those frequencies sound to us). If you play a pure tone (where the pattern of vibration keeps repeating itself, like a tuning fork) at a single level of loudness, the ear can perceive 1,400 different pitches. If, on the other hand, you keep to one frequency but change the volume or intensity, the ear is capable of identifying 280 different levels of loudness. That means that, if both the frequency and intensity are changed, the ear has a repertoire of between 300,000 and 400,000 distinguishable tones.[43] Does the planet contain a more discriminating organ?

Our auditory skills certainly demand just as many superlatives as our vocal abilities. We hear because the pressure of the air is great enough to move the eardrum, and yet the extraordinary responsiveness of the ear to tiny variations in pressure almost defies belief. The force brought to bear on the ear is measured in dynes – very small units of force. To support the weight of one ounce against the force of gravity, for instance, we have to exert an upward force of 28,000 dynes. Yet the ear can detect a force of 0.0002 dynes. In other words, the force acting on the eardrum is around 140 million times smaller than the force needed to lift a weight of one ounce. At the threshold of hearing, the eardrum moves roughly one-tenth of the diameter of a hydrogen molecule.[44] It's even been argued (astonishingly) that, if there were nothing between you and an airport ten miles away, with no competing sounds or intervening objects for the sound to reflect off, in theory you could hear a piece of chalk drop at the airport.[45]

## THE BRAINY VOICE

Perhaps the most critical area for our vocal skills lies above our vocal apparatus, in the brain. It's the central grey area that synchronises the activities of our respiratory tract and oral and facial muscles, like a conductor bringing the various sections of the orchestra into simultaneous play. It's also in our cerebral cortex that the nerve impulses conveyed from the ear are interpreted and turned into meaningful messages.

Although the alchemy wrought by the human auditory system still remains relatively mysterious, it's becoming increasingly clear that certain parts of our brain are particularly sensitive to the human voice. A Canadian team scanned the brains of adults listening to vocal sounds produced by male and female speakers of different ages, including non-speech (laughs, sighs, and coughs), as well as sounds made by animals or mechanical objects. The vocal sounds elicited significantly greater neuronal activity than the non-vocal sounds. Another experiment, comparing voices with bells, handclaps, white noise, and scrambled voices, produced similar results, as did a third filtering out high and low frequencies. Together this trio of experiments provides strong evidence that certain regions of the human brain – the superior temporal sulcus (STS) or groove running along the whole temporal lobe – are not only sensitive to the human voice, but also picked out and activated by it.[46]

Yet this pleasing study raises as many questions as it answers. Though some will say that we're hard-wired to respond to human voices ('hard-wired' being a favourite term of popular science, bestowing genetic certainty on even the most speculative notion), the greater activity in the STS could be the result, and not cause, of the psycho-social fact that human voices are important to us. Who knows whether different voices don't produce different degrees of neuronal activity? Or whether a lonely widow's cat's miaow might not generate more neuronal activity than the voice of the television weather forecaster?

## GENETICALLY SPEAKING

Over the past twenty years it's become clear that the voice is also shaped by genetic factors. This isn't surprising if we remember that our voices are the product of the length, shape, and structure of our vocal folds and vocal tract, themselves determined at least partly by our genes. The example of identical twins is usually given here. When presented with recordings of their own voice and that of their identical twin in random order, twins have difficulty identifying themselves.[47]

On the other hand the interaction between genotype and the environment is complex, and most twins – like most siblings – share the same environment.[48] I once managed to fool my future brother-in-law for an entire fifteen-minute phone conversation into thinking that I was my own, non-identical older sister, so similar did we (apparently) sound over the telephone. Yet the emerging field of the genetics of voice offers the hope that a genetic origin might be found for some vocal problems like nodules, opening up possible new treatments.[49]

## HEARING VOICES

We're both the originators and hearers of our voice. Most people, when presented for the first time with a recording of their own voice, find the experience not only excruciating but also incredible. This, it's usually claimed, is because it doesn't correspond with the sound that we think we make, since we hear our own voices not just through air conduction but also through bone conduction, i.e., via our own head. (Alfred Tomatis taught his patients to listen to their voice through air conduction. I've occasionally heard my own voice this way – it's not a pleasant experience.) Chapter 8 proposes some other reasons why most of us are so shocked by the sound of our voices.

The very act of speaking is a powerful experience. Tomatis believed that 'it is through [a person] hearing his own voice that the notion of life penetrates him'.[50] This suggests that it's not only the (f)act of vocalising but the audio-feedback too which, together, play an important part in creating our sense of self, our idea of ourselves as a person in the world. But it's also clear that talking is a physiologically demanding activity, with major effects on the cardiovascular system. It increases blood pressure, heart rate, and arrhythmia. When the cardiovascular effects of talking were compared with listening, it was found that talking, even about a non-stressful subject, increased the heart rate and blood pressure more than listening did, especially in the case of women. Slowing speech and taking time to breathe, by contrast, lowered blood pressure and heart rate.[51]

In fact a constant self-monitoring takes place when we speak. We repeatedly compare the quality of sounds we make with those we intended to and adjust them so that they match.[52] Treating opera singers who were losing their voices, Tomatis guessed that their own loud voices (at full tilt over 110 decibels, and sounding louder within their own skull) had damaged their ears, so that they could no longer hear the range they were physiologically capable of singing – victims of an occupational deafness every bit as extensive as assembly-line machine operators'.

The voice, he claimed, can only reproduce what the ear can hear, and the larynx emits only those harmonics that a person themselves can discern. So if you change your ear's ability to listen, you'll inevitably also change your voice – truly we speak and sing with our ears.[53]

A hundred years after Garcia, Tomatis, the archaeologist of phonation, set about liberating the ears. The voice was beginning to divulge its origins.

# 3

## How We Colour Our Voices with Pitch, Volume, and Tempo

THE WRITER DOROTHY PARKER was alone and bored at a party one evening. Whenever anyone she vaguely knew came up to her and asked, 'How are you? What have you been doing?' she replied, 'I've just killed my husband with an axe and I feel fine.' Because she said it with the intonation usually used for party small talk, every one of them simply smiled, nodded and, unastonished, drifted on.[1]

Parker demonstrated how pitch, loudness, and tempo – some of the features that make up paralanguage – can be more important than language itself. For paralanguage doesn't just support words but gives them life. Without it, they'd be inert, flat, lacking in emotion. Paralanguage is what separates a real human voice from a synthesised one.

So important are intonation and stress – sometimes known as prosody – in shaping our reactions that they can make us back off from someone or warm to them without knowing anything about them. Prosody is the audio version of our personality, our sonic self.

Yet though, 'It's not what you say, it's the way that you say it,' now belongs in the realm of cliché, paralanguage is relatively young as a recognised field of study. Until the 1940s linguists dwelt almost exclusively on the verbal aspects of speech. Then a pair of major studies of American and British intonation,[2] followed by a pioneering 1958 paper on paralanguage itself, showed that prosody could be studied and analysed as much as language.[3] Almost fifty years on the field has grown vastly, although it's still mined with disputes. The researchers can't even agree on terminology – 107 different names have been used to identify register alone.[4]

## LET ME STRESS

What counts as language and what as paralanguage isn't fixed – it's subject to argument and change, for intonation has a linguistic function as well as a non-linguistic one.[5] It allows us, for example, to distinguish between 'conduct' as a noun and as a verb, or between a question and a statement (even though teen 'uptalk' has been busily destroying this distinction). Merely by the way it's said, the meaning of a word can be either reversed or reinforced. And, depending where you place the emphasis, you can say the sentence, 'I didn't want to go home,' with five different meanings.

Consider the case of the 19-year-old illiterate Londoner Derek Bentley, hung in 1953 for the murder of a policeman. Bentley was convicted primarily on the basis of the testimony of another policeman, who claimed to have heard him say to his fellow defendant, 'Let him have it, Chris.' The prosecution maintained that those words proved that Bentley had encouraged his fellow criminal to shoot the policeman, but his defence lawyer argued (unsuccessfully) at his trial that the same words might have been spoken (with the emphasis presumably on the 'let' rather than the 'have') to mean, 'Give him the gun, Chris.'[6]

So a person's fate can depend on the stress of a word,[7] and in tone languages like Chinese where a pitch change indicates an entirely different word, the potential for mistakes is prodigious. Someone learning Beijing Mandarin Chinese may think, for example, that by using the word '*ma*' (level tone) they're saying 'mother', but if they employ a falling-rising tone they're actually saying 'horse'.[8]

Paralanguage as a concept includes not only vocal features but also non-vocal ones, like facial expression and body movement. In fact it's now something of a dustbin category, into which eyebrow movements, pitch, gesture, and any other vocal attribute that can't be put anywhere else is tossed.[9] Yet despite its imprecision, para-language is still a useful concept, if only because it helps us appreciate the sheer opulence of the human voice, and the number and variety of its components.[10]

## HOW YOU PITCH IT

Pitch is the falling or rising tone heard in the voice: it creates our voice's melody. It can alter by the syllable, by the word, or over a stretch of speech (where it's known as the pitch-range). Patterns of pitch are called intonation.[11] Pitch is an auditory sensation: it's what we hear, or think we hear, and is affected by judgement, volume, and other factors. It's distinct from frequency, an acoustic feature that can be measured by a spectrograph: you can change pitch without changing frequency, and vice versa.[12] In other words pitch is subjective while frequency is objective. Pitch is also a musical term,[13] but music is of limited use in helping us to understand the pitch of speech – after all, we can't speak out of tune,[14] and our voices have far more gradations than the notes of the musical scale.[15]

Intonation draws attention to new information, for example, 'I just heard a *great* singer.' Falling pitch also signals that one speaker has finished speaking and another can begin – it oils the turn-taking through which conversation proceeds. Without the infor-mation supplied by intonation, conversation would be an infinitely greater cacophony of overlapping, competing voices, and yet we usually change pitch without thinking. Our brain circuits are charged with the extraordinary task of continuously adjusting the mass, length, and tension of the vocal folds, so as to produce the variations in intonation patterns we want to convey.[16]

Average pitch levels vary according to age and gender. Although the least tiring one is three or four tones above the lowest we can clearly produce[17] – i.e., near, but not at, the lower end of our physiological vocal range[18] – our speaking pitch ranges between two or three octaves, and none of us has an habitual or stable speaking pitch.[19] Over the course of a year, our pitch can cover as wide a range as that between different speakers,[20] and can vary by as much as 18 per cent from one day to another.[21] Indeed male (but not, interestingly, female) voices fluctuate significantly within the course of a single day, between morning, early and late afternoon.[22]

Pitch plays a huge role in shaping how we appear to others, as well as to ourselves. Some have even gone so far as to claim that future academic success can be predicted by voice pitch and range: in two studies teachers judged slow-speaking and low pitch as a sign of academic failure, and found higher pitch and lower volume among the academically more successful.[23] Whether the pitch was cause or effect of low achievement (or unconsciously affected teachers' judgements), what's disturbing about these findings is the implication that only one kind of voice allows you to flourish academically. If it's true, inequality of opportunity is audible.

Certain pitches become identified with particular qualities. The British broadcaster Jon Snow is sure that 'it's bass registers that give authority to a voice'.[24] Satirist Rory Bremner thinks that politicians like Michael Portillo and TV performers like Ian Hislop talk with a deeper than natural register in order to sound authoritative.[25] To play Othello at the Old Vic in 1964 Laurence Olivier added a whole octave at the bass end of his voice.[26]

When American university students were asked to write down words that best described the pitch of well-known people, they judged the low inflections of conservative commentator William Buckley as 'authoritative, aggressive, in command, decisive, controlling'. Writer Truman Capote's rising inflections, on the other hand, were considered 'effeminate, ambivalent, indecisive, questioning, unsure', while former President Lyndon Johnson's all too stable pitch suggested 'loss of affect, boredom, no emotionality, fatigued'.[27]

Intonation plays an important symbolic role. It can convey volatility, or act as a safety-net. The level, almost monotonous, pitch of veteran American news anchor, Dan Rather, emits gravitas and decency: its restrained, almost presidential cadences evoke old values of trust and authority, casting back to the acoustic of a pre-cable and satellite TV world before upbeat intimacy became the norm. In Jon Snow's words, his voice 'brought anchorage and security to viewers in a very insecure age'.[28]

Compared with Rather, most American news-broadcasting voices today arch and buck, changing pitch for what seems like capricious reasons. Pitch now seems dissociated from meaning, and has become part of corporate style instead. Says Rather, 'I don't believe in having extremes of pitch and register. When I hear this in other people, it's off-putting to me. There's a belief today that to keep people from being bored you need to move your voice up and down the scale pretty regularly. Everything in me shouts against that.'[29]

The voices of American broadcasters are now hardly distinguishable from those of waitresses, air stewards, or receptionists.[30] I've sat in a seafood restaurant in Connecticut and heard a waitress deliver her spiel about the specials with such rhythmic rapture, her voice all a-swooping-and-a-leaping, that to derail her by interrupting in a downbeat British way would have been an act of heartlessness.

American intonation seems relentlessly 'up' compared with British. When I go to the United States I'm always struck by the mismatch of melodies: I hear my own intonation anew when I hear it through American ears. After a few days I invariably shift my intonation 'up' a few notches, Americanising not my accent but my inflections. Modifying vocal melody like this makes communication easier; not to would sound like a pompous display of Britishness, or as if I were auditioning for a part in Masterpiece Theatre. George Bernard Shaw is alleged (but has never been proved) to have said that Britain and America were two countries divided by the same language. As much as anything else, it's our different intonation that does the dividing.

Not that British broadcasting is without its mannerisms. Listen to a Radio 4 announcer or a BBC Radio 3 morning presenter and you hear a voice that also gambols in an unnatural rhythm and with abnormal cheerfulness – partly to disguise the fact that a script is being read. 'And now' (normal, low pitch) 'we're' (voice has soared like a frenzied finch) 'going to hear' (it plunges rapidly) 'a Mozart' (stays low) 'sonata' (rapid ascent again, followed by a moderate fall) 'no. 15 in F' (low) 'written' (starts travelling up again) 'in' (in the clouds) '1788' (and back down again). The bizarre pattern of pauses sounds like the unfortunate consequence of an incurable disease.

Most of us grow into expert decoders of each other's cadences. We also quickly learn how to read those of politicians and other public figures. *Newsweek* reported that Robert J. McCloskey, a spokesman for the State Department in the Nixon administration, had three distinct ways of saying, 'I would not speculate.' 'Spoken without accent, it means the department doesn't know for sure; emphasis on the "I" means "I wouldn't, but you may – and with some assurance"; accent on "speculate" indicates that the questioner's premise is probably wrong.'[31]

Pitch varies between countries. As the linguist Edward Sapir put it:

> Society tells us to limit ourselves to a certain range of intonation and to certain characteristic cadences . . . If we were to compare the speech of an English country gentleman with that of a Kentucky farmer, we would find the intonational habits of the two to be notably different, though there are certain resemblances due to the fact that the language they speak is essentially the same.[32]

We can't read the intonation of someone from another culture properly, argues Sapir, unless we're familiar with their country of origin – otherwise we're likely to hear an Italian as temperamental, rather than simply sharing the normal melodic curves of all Italian speakers. For languages create their own rhythms. 'If a Frenchman accented his words in our English fashion, we might be justified in making certain inferences as to his nervous condition.'[33]

Does the persistence of British anti-German and anti-Japanese prejudice so long after the end of the Second World War have anything to do with how harsh German sounds to British ears, or with the Japanese use of pitch contours and glottal attack that in English would be a sign of aggression?

In a globalised world where labour is increasingly mobile, sensitivity to intonation is becoming more and more important. Consider the experience of a group of Indian women working in the staff cafeteria at a major British airport, who were seen as surly and uncooperative by both their supervisor and the cargo handlers they served. Even though they said little, it was soon clear that their intonation was creating friction. When British canteen staff asked a cargo handler who'd chosen meat if he wanted gravy, they'd say, 'Gravy?' using rising intonation. The Indian assistants, on the other hand, said the word with a falling intonation, until the supervisor pointed out that it was as if they were saying, 'This is gravy.' Since that was obvious, it was interpreted as rude and insulting.[34]

Certain forms of expression are almost entirely a property of the voice. Sarcasm, for example, allows you to undermine and ridicule through tone rather than explicitly through language – it can express insubordination without leaving any incriminating words in its wake. Sarcasm forces the person at whom it is aimed either to ignore it or to draw attention to it ('Don't you use that tone of voice with me, young man'). Either way the target is left as victim rather than victor of the interchange. Nonverbal channels of communication, by being inherently ambiguous, 'Allow us to express our feelings . . . without taking full responsibility for them. It is undeniably easier to retract what has been expressed nonverbally than what has been expressed verbally.'[35]

So powerful a weapon is sarcasm (an essential component of the

teenage arsenal) that in American 'brat camps' there's a No
Sarcasm rule. A recent Anti-Social Behaviour Order served on a
British man also stipulated that he wasn't allowed to use sarcasm.
In 'The Wall' Pink Floyd sang of 'dark sarcasm in the classroom'.
Generations of teachers used it as a teaching aid.


## LOUDSPEAKER

Volume, like pitch, is both physiological and psychological. The
level of intensity or loudness that we hear doesn't necessarily match
the amplitude with which the vocal folds vibrate.[36] (And we can be
peculiarly insensitive to our own volume – it's always other people
who are loud, never ourselves.) Volume also plays an important
role in managing the interplay of conversation: we often speak
more softly as we approach the end of a sentence, for instance.[37]
And it has a strong cultural dimension. As a general rule, Asians are
softly spoken (in Asian countries a soft voice signals deference) and
find Americans loud. (So do Britons.) Arabs, on the other hand,
sound loud to Americans, while to Arabs the American voice is too
quiet and sounds insincere.[38] But volume is changing. Americans
used to speak louder in public than the British, who tended to treat
public space as though it were a library where you had to whisper.
The mobile phone, though, has produced a whole generation of
Britons uninhibited about blaring private matters in public.

Of course we're not born loud- or soft-speakers – we learn to use
the volume level that prevails in our culture, and then turn it up or
lower it depending on our subculture and peer group. Indeed
regulating the volume of our voices according to the situation is
an important social skill. Nor are we mono-volume creatures – we
can be soft at home and loud outside it, or vice versa. Emerging
from the enforced quietude of the classroom, children erupt into
noisy exuberance: in the clamour of the playground lungs and
larynx, as well as limbs, are exercised. One of a swimming-pool's
most important functions is surely to allow children to scream
without fear of reprimand.

We use our voices differently in a group. A 14-year-old girl says of outings with her schoolfriends:

> I always try to make them be quiet – lots of them go really loud and really obnoxious to everyone else – it's a bit harsh and we have quite posh accents, and people give us looks. People who go on the bus together make a lot of noise in a small space and it's embarrassing. [Being loud with a group of people in public can be liberating] but it can also be quite overwhelming if everyone's speaking to get their point across and you're not.[39]

Loudness can be an important part of the collective public identity of a group. By marking out a shared acoustic space, group members assert their right to be heard. Among those whose voices have been silenced historically – groups of young girls or teenage Asian boys, for example – volume can be a sign of defiance. Though the buildings might not belong to them, they're able to impose their voices on the street and lay claim to the air.

But volume can also be inversely related to status. To compel total attention some powerful people speak so softly that their listeners are obliged to lean forward to hear. (An American politician told me that he speaks softly because a politician talking loudly jars with an audience, and makes them tune out.[40]) Speaking quietly can be a display of arrogance, but then again it can also mean that someone believes that they've nothing interesting to say. Reversing the norm can be powerful, too. Since loudness is traditionally associated with rage, and softness with intimacy and confidentiality, quietly expressed anger can be devastating.

How loud is a loud voice? The average conversational voice, with speaker and listener some three feet apart, is 60 dB. Quiet speech hovers around 35–40 dB, while shouting rises to 75 dB. Rustling leaves measure 10 dB, and loud radio music 80 dB. Around 120 decibels creates a sensation akin to touch, and above this we hit the pain threshold.[41]

In normal human conversation the voice can't be raised to a

point where it might endanger the ear (over 90 decibels), yet somehow a loud voice seems to provoke more anger than other equally loud sounds. In 1914 the city of Bern in Switzerland enacted a by-law against 'carpet-beating and noisy children'. In 1971 a Hare Krishna sect was arrested in downtown Vancouver and convicted under the Noise Abatement Act for their public chanting. They were standing next to a construction site where demolition equipment produced noise levels of 90 decibels. Needless to say the construction firm wasn't fined.[42]

Politicians sometimes try to make a more powerful impression by increasing the volume and attack of their voice, almost invariably resulting in a croak.[43] If they understood that consonants gave meaning to the words and vowels supplied the emotion, they could achieve the desired effect without strain. The feelings behind a Shakespeare speech (though not its meaning) can be communicated by pronouncing only the vowels (try it with 'To be or not to be'). When luvvies go, 'Daaarling,' they're extending the vowels to extract the maximum emotion from them. At the same time, if the final consonants in a speech aren't sounded clearly, the meaning isn't conveyed. By taking care with final consonants and extending vowels, public speakers can sound forceful without having to rely on volume or shout.[44]

## THE TEMPO OF TALK

A person's pace seems so characteristic of the way they speak yet well-known politicians and entertainers can still be recognised even when the rate is altered.[45] The average tempo of an adult American or British adult is 120 to 150 words per minute, or about six syllables per second.[46] 90–100 wpm suggests incapacity, dignity, or vanity. Roosevelt spoke far slower in his *Fireside Chats* than any other political orator on the air in the 1930s, but after the bombing of Pearl Harbour in 1941 his rate dipped to 88 words per minute – the normal radio tempo of the time was 175–200 wpm.[47]

Speed can be used to prevent interruption, or to put distance between oneself and a difficult subject. We also accelerate as we become more excited, like horse-race commentators.[48] The belief that some languages are spoken more rapidly than others is an illusion caused by different styles of speech and types of language.[49]

Our speech rate reveals profound differences in outlook, confidence, self-image and style, but our reaction to other people's tempo is stuffed with prejudice. Slow speakers' abilities are consistently denigrated. They're judged less truthful, less persuasive, but also colder and weaker.[50] Fast speakers, on the others hand, are considered cleverer.[51] You might think that, because speaking quickly makes speech less clear, it also makes it less persuasive. In fact the opposite is true: fast speakers are seen as more knowledgeable and trustworthy.[52] But our judgements are also influenced by how we sound ourselves. We think people who speak in a similar tempo to our own are more competent and attractive than those who go slower or faster.[53] People whose speech rates are mismatched can have difficulties communicating or even dislike each other.[54]

Tempo, like volume, works in subtle and sometimes contradictory ways. The American politician I interviewed told me, 'I often speak more slowly than I'm comfortable with. I think faster than I speak.' To which his wife retorted, 'That's such a power ploy, so that you've got everyone waiting for your next word.'[55]

It's only in drama that speech is always fluent – the rest of us fluff, repeat, prolong, or stress weirdly some of the time. The difference between normal hesitancy and stuttering can be a matter of just a few per cent, yet we're still unforgiving as a culture in our attitudes to stutterers and those who aren't fluent.

But there are cultural differences in tempo too. In Zulu society, for example, slow speech is a sign of respect and sincerity.[56] Americans, by contrast, tend to speak quickly, and American radio is strikingly faster than British: American broadcasters treat breath like a foreign body. Pace and tempo, argues Dan Rather, are important factors in establishing clarity and intelligibility. 'I am

conscious of speaking slower than other broadcasters but it would
be a mistake to try and change it. I speak at that pace where I can
best be understood and hold interest. I have done sports broadcasting and play-by-play [ball-by-ball commentary] – you have to
speak faster in order to cover the action, especially with basketball.
I do remember feeling that the pace and tempo of play-by-play was
not commensurate with clarity and having people understand the
news. I've never had any pressure to speak faster, though I do know
that the trend is to speak faster.'[57]

American television now routinely speeds up sitcoms and com-
presses speech in order to fit in more ads. If 'Frasier' seems to talk
faster on one affiliate station than on another, that's probably
because he is talking faster – the station has accelerated the epi-
sode.[58] Rather thinks that there's been a spill-over from taped to live
broadcasting, putting pressure on broadcasters to speak more
rapidly. 'I was listening to a colleague on a cable TV station speaking
so fast and I thought, why does he do that? It was all pudding to me.
There is a theory that audiences, particularly young audiences, the
most coveted these days, expect people to speak faster.'[59]

Extreme speed or languor can be comic: John Cleese, in his *I'm
Sorry I'll Read That Again* radio monologues, raced up to 200
wpm.[60] The Guinness World Records reported a man speaking at
637.4 words per minute on a British television programme in
1990.[61] But starting slow and speeding up can also be a powerful
oratorical device: Martin Luther-King's 'I have a dream' speech
began at 92 words per minute and ended at 145. In a thinly
disguised anti-Semitic speech on American radio in 1938, Father
Coughlin, broadcaster of weekly right-wing sermons, veered from
100 to 275.[62] His accelerando alone must have been enough to
raise listeners' blood pressure and instil fear.

## OVERTALK

Voice specialists call uncontrollable or excessive talking 'logor-
rhoea', a word that carries unwelcome echoes of diarrhoea. It's

also known as manic speech – 'loud, rapid, and difficult to interrupt. Individuals may talk non-stop . . . without regard to others' wish to communicate'.[63]

Unusual quantities of speech have been linked with anxiety. On the other hand, when professors and student counsellors were asked to choose the most mentally healthy college students they chose those who talked more.[64] Those who speak more are often perceived as possessing higher leadership ability too,[65] yet most of us apply different measures in public and private life to judge whether a person is masterful or just over-voluble.

Many of the people I interviewed brought up the question of balance in conversation, in work and in relationships. 'My partner is generally much more voluble than me,' said a 43-year-old man, 'but when we come to be actually talking to each other, it's very even – that's one of the things I like about it.'[66] A 51-year-old academic observed, 'I try and force myself to listen to other people's voices – to give a space for other people to talk. I think one of the dangers, because I have a strong voice, is to be dominant.'[67] And a 49-year-old woman admitted, 'I have a real horror of situations where one person talks too much and seems oblivious of the fact. I can't stand it in discussions when someone hogs the talk time and you notice that others haven't spoke once, perhaps because I could easily talk non-stop but I monitor myself.'[68]

A 53-year-old man recalled with horror a long car drive with his father-in-law who, in response to a question, talked for half an hour without pausing to allow anyone else to contribute. The son-in-law was shocked to discover that not everyone possessed an internal gauge that told them how much they were speaking.[69] A 22-year-old teacher says that she can tell if she's talking too much by the amount of fidgeting done by her 7-year-old pupils.[70]

Verbosity, of course, can serve all sorts of defensive purposes. The characters in the stories of the great Yiddish writer Sholem Aleichem pour out a torrent of words to defend themselves against chaos. They seem to be saying, 'I talk, therefore I exist.'[71] When one of his story-tellers breaks off his unfinished tale because it's time for him to change trains, his fellow passengers can't believe that a Jew like themselves would rather stop talking in the middle of a sentence than miss his connection.[72]

If over-talk can disturb, so also can over-fluency. A voice produced too easily casts doubt on its own authenticity. We may all be actors, but the over-fluent voice draws uncomfortable attention to the degree of performance. Without doubts, nuance or slips its arguments sound pre-assembled: hearing it is like being trapped in honey.

## I'D KNOW THAT VOICE ANYWHERE

The Roman rhetorician Quintilian believed, 'Every human being possesses a distinctive voice of his own, which is as easily distinguished by the ear as are facial characteristics by the eye.'[73] Though chapter 16 challenges him and the idea of a 'voiceprint', timbre nevertheless enables us to identify someone we haven't seen for years after only a few words.[74] Also known as voice quality or voice set, timbre is idiosyncratic, biologically controlled,[75] and present no matter whether a speaker is yelling or whispering. It's why actors almost always remain recognisable: whatever role they assume,[76] strains of Judi Dench or Tom Hanks invariably peep through. How dexterous the voice has to be – to make the same words spoken by different people sound distinctive enough to allow the speaker to be swiftly identified, but at the same time to maintain enough that's similar about the words so that we can recognise and understand them even when they're spoken at a different pitch and volume.[77]

What makes this all the more phenomenal is that we can recognise another person's voice in milliseconds, and we hear a person's timbre as a consistent sound even when they're walking through different-sized rooms whose different reflecting conditions make the real sound of their voice fluctuate. In processing other people's voices we constantly, selectively add and filter out certain characteristics in order to compensate for local and temporal changes: it's as if our brains are committed to transforming uneven bits into a smooth whole.[78]

Why do no two people sound alike? The voice is shaped not only by the length and thickness of the vocal folds, and the muscles of our lips, mouth, and tongue (all of which differ between individuals), but also by our different bone structures that produce differently shaped resonating chambers in the throat and nasal cavities. The potential combinations of these factors create the infinite profusion of human voices.

# 4

# *What Makes the Voice Distinctly Human*

W E'RE NOT VOCALLY unique creatures, just rare ones. By looking at how the human voice differs from the vocalisations of other species, and of early humans, we can begin to understand which of our vocal talents are truly distinct, and which we share with other creatures and have only modified or elaborated. Are our vocal skills the product perhaps of our keener sense of hearing, or have they been forged by the specialisation of the brain into different hemispheres processing language and intonation? Research into animal vocalising, human fossils, and the brain together has helped create a kind of archaeology of the voice, and unearthed some extraordinary shared origins – we may have more genetically in common with hummingbirds than we realise.

## VOCAL SPECIES

Though complex communication sounds are emitted by a whole range of animals, only three groups of mammals: humans, cetacea (whales and dolphins) and bats, and three groups of birds: parrots, hummingbirds, and songbirds have to learn them. In all the rest they're innate.[1] Cats' miaows and dogs' barks, for example, are instinctive and not learned. Vervets use a repertoire of thirty-six different calls to warn of the degree of danger from predators,[2] but they don't go through a period of gradual vocal development. Humans are the only primates to do this, and even though whales and dolphins are vocal mimics and some bats can learn to vocalise, none of the few non-human mammal vocal learners is as consummately gifted at it as we are.[3]

The group that produces sounds most like human sounds is the songbird. Birdsong (as distinct from bird calls, which are brief and not usually learned) can last from a few seconds to tens of seconds: like speech it consists of ordered strings of sounds separated by a few silent intervals. Birdsong is used to communicate information between parents and young, between siblings, and even between different kinds of birds.[4] It can tell a bird whether they're friend or foe, be used to attract a mate and signal ownership of territory.

There are all kinds of parallels between the ways in which songbirds sing and we speak. Just as air is filtered through our larynx to produce the human voice, so birds sing when air flows to their upper vocal tract through an organ called the syrinx. The width of their beak-opening acts rather like our mouth in determining sound frequency, and like us they need extraordinary neural control.[5] In fact birds' vocalisations are vastly more complex and precisely modulated than those of any other animal apart from humans.[6] They vary not just in rate, rhythm, and loudness, but also in timbre or tone quality and vowel quality. These factors can be combined into variables of staggering complexity[7] – no wonder chaffinches are thought to be able to identify a rival by the individuality of its voice, and gannets their mates by the same means.

Most remarkably, songbirds too learn from experience. They come to use their voices only if they're exposed to the communicative signals of adults of their species. They even have their own form of babble when they're learning. Just as human infants who don't hear the human voice or speech don't develop normal vocal capacities of their own, so songbirds raised in isolation produce abnormal 'isolate' songs. Contrast this with, say, chickens, who still turn into perfectly competent cluckers even when raised in acoustic isolation. For humans and birds, hearing others (as well as hearing ourselves) vocalise is critical to our own vocal learning.

The resemblance doesn't end there either: there are striking cultural similarities in the way humans and birds use their voices. Human culture is so routinely contrasted with animal nature that it's remarkable to discover that some songbirds have 'dialects' – defined, localised, particular acoustic features that are culturally transmitted.[8] Some birds, like the white-crowned sparrow, have even been dubbed 'bilingual', in that they defend their territory by learning the songs of a local neighbour or a migrating white-crowned sparrow passing through.[9]

## BIRD BRAIN

In a nature reserve in a Brazilian tropical forest Erich Jarvis, a neurobiologist from Duke University in North Carolina, sets a sugar-water bottle in a cage on a branch near a tree where hummingbirds frequently sing. The hummingbirds captured in the trap are immediately killed and their forebrain nuclei processed for gene expression.

What Jarvis discovered should for ever put paid to the pejorative term 'bird brain'. In the avian brain he found glutamate receptors, a class of gene that's responsible for building connections between nerve cells. When Joseph Conrad wrote that 'man appears a mere talking animal not much more wonderful than a parrot',[10] he may have been closer to the truth than he realised, for Jarvis went on to map equivalent vocal-communication areas in the brains of songbirds and parrots. Only vocal learners, argued Jarvis, have this kind of forebrain region dedicated to vocal learning – indeed birds may have similar brain structures for generating song as humans have for learning vocalisation.[11]

The research raises intriguing questions. Did vocal learning emerge independently in the three groups of mammals and birds within the past 65 million years? Did they all share a common ancestry (are we genetically related to hummingbirds?) and then diverge? Even more crucially, is the similar pattern in these brain receptors a cause or result of vocal learning?[12]

Certainly, despite the uniqueness of human communication skills, the evolutionary origins of our voices are undeniable: we share many of our repertoire of screams and groans, growls and whimpers with apes, monkeys, and other creatures,[13] although our culture proscribes such uninhibited vocal sounds and they only survive in rituals or reappear when we're *in extremis* – for example, in pain or grief.[14] So in this sense human beings are not only vocal learners but also vocal unlearners: an important part of acquiring social skills is learning to suppress emotional expression in the voice – expression that resembles that of macaque from which our evolution diverged some 20 million years ago.[15]

## PITCH AND POWER

Is there an evolutionary basis to intonation, too? Across every language and culture a question has a rising intonation, and a statement has a falling one,[16] causing some to believe that this must be either a genetic inheritance or an intrinsic human trait.[17] 'Sound symbolism', according to one school of thought, guides our use of pitch. Research across fifty-six avian and mammalian species has found that confident, aggressive, and dominant individuals use low-pitched, often harsh sounds, while submissive or subordinate individuals use high-pitched tones. This so-called 'frequency code' extends from the frog to the rhinoceros. It evolved, say its exponents, because animals know that in fights the larger creature wins. By signalling size through sound, disputes can be settled without the need for bloodshed. Bluff also comes into play, with some animals trying to suggest that they're larger than they are, and others who want to capitulate, trying to make as small, non-threatening, and therefore high-pitched a tone as possible.

Is this why questions, which depend on others for answers, use a rising pitch, while statements, which express certainty, use a low one? Some say that by using a high pitch to ask a question or show deference we're not much different from a whining dog that is trying to elicit goodwill and cooperation.[18] To its advocates, sound symbolism may even help explain the origin of the smile, since this shapes the mouth in such a way as to alter resonance, and push the pitch up.[19] (Smiling and saying 'Hello' in a deep growl feels not only counter-intuitive but also quite forced.) Assertiveness-training classes are also enlisted to support the frequency code, since these teach people to use as low a pitch as they're comfortable with in order to create an image of self-confidence. And the theatre, too, supposedly proves the theory: imagine how ludicrous a Hamlet or a Richard III would sound played by an actor with a high-pitched voice.

To move from a whining dog to Hamlet would seem to require more than just one intervening step. Yet, although the frequency code is reductive, it still offers some intriguing pointers to understanding similarities between animal and human use of intonation.[20] Alternatively, as one neuro-psychologist has pointed out, sounds are merely changing air pressure on a tiny hypersensitive membrane in the ear. Through the process of interpretation,[21] we humans turn them and tones of voice into auditory metaphors. So instead of turning to whining dogs for an explanation of our intonational differences, we might perhaps do better to look to human history.

## WHEN WE FIRST SPOKE

Researchers are poring excitedly over fossilised skulls, using archaeology to try to date the period when human beings first became capable of speech. Hominid fossils provide all sorts of anatomical clues.

The trail starts from a peculiar physiological fact: that the vocal tract of a newborn is incapable of producing the full range of speech sounds[22] – indeed it's more like that of an adult chimpanzee than an adult human.[23] In newborn babies, as in amphibians, reptiles, birds, and mammals, the tongue is situated entirely in the mouth, and the larynx is much closer to the pharynx (or throat), giving the air a short journey from larynx through pharynx and straight into the nose. The child's growing mastery of the sounds of speech is partly the result of the gradual descent of its larynx and the root of its tongue during the first two to six years of its life.

In this respect humans are unique among air-breathing vertebrates: because our tongue and larynx lie so low in our neck and our pharynx is so long, we can turn our tongue and vocal tract into a huge variety of different, sound-altering lengths and shapes – the prerequisite of speech. Apes, dogs, and monkeys, on the other hand, have mouths and tongues that allow them to swallow food without risk of blocking their larynx, and eat and drink while they breathe, but not talk.[24] So this is the trade-off: talking makes us breathe and eat less efficiently, and splutter when drunk. Though we can speak, we can also easily choke on our food. Apes do neither, which is why they don't need to learn the Heimlich manoeuvre.

It's been proposed, extraordinarily, that the descent of the human larynx in a human child mirrors the evolution of the larynx in our human ancestors. Australopithecine hominids (the famous 'Lucy' was one) and *Homo habilis* (the so-called handy man) had skull bases very much like chimpanzees and orang-utans today, and a newborn baby's vocal tract seems to be similar to this stage of human evolution. By *Homo erectus*, 700–500,000 years ago, the vocal tract was beginning to resemble ours: the tongue and larynx had begun their descent (the partially descended larynx of a 3-year-old child is roughly equivalent). The early archaic Homo sapiens of 300–100,000 years ago was the first ancestor with a vocal tract able to articulate a large repertoire of speech sounds, while the early modern *Homo sapiens* fossil skull 30–20,000 years old found at Cro-Magnon in France has a skull identical to that of present-day humans, its vocal tract equivalent to our mature human one.[25]

There's something shocking about characterising the newborn

human as some kind of unevolved creature, a throwback to pre-history, rather than in the more usual way as a perfectly formed, bonzai adult. The idea has also had a controversial afterlife, generating a heated debate about whether we're descended from the Neanderthals, and whether they could speak.

According to one theory, the Neanderthals, who lived in Europe some 130,000 years ago, simply couldn't have been able to produce the range of sounds that we do because of the position of their larynx – they could only have had the vocal capacities of a 2-year-old human child, and we must have descended from a separate branch of the species.[26]

But when archaeologists excavated a Neanderthal burial site in the Kebara Cave on Mount Carmel in Israel in 1987, they found a human skeleton some 60,000 years old that possessed a complete and well-preserved hyoid bone, the first ever discovered in a fossil hominid.[27] This small, U-shaped bone, located between the lower jaw and the spinal column, is supposedly a good indicator of the anatomical position of the vocal tract, because the tract and larynx always sit beneath it. On the evidence of this bone, the Nean-derthals had a low-lying larynx and so were anatomically capable of speaking.[28]

It's not the larynx or hyoid bone that's vital for speech, accord-ing to a third set of anthropologists, but the hypoglossal canal, which transmits the nerve that supplies the muscles of the tongue in mammals. This is larger in humans than in African apes, presum-ably because the human tongue is richer in motor nerves. Fossils show that the hypoglossal canals of Australopithecus and *Homo habilis* were also smaller than ours (and roughly on a par with those of African apes), but Neanderthals' are the same as ours today. By this reckoning, hominids may already have had the capacity to speak 400,000 years ago.[29]

Then again hypoglossal canals relatively similar in range to those of modern humans have been found, by a fourth team of research-ers, in non-human primates and even Australopithecus. If the existence of this feature is proof of the ability to speak then hominids 3.2 million years ago, whom we know couldn't even make stone tools, would have had the capacity for speech. The speech capabilities of the Neanderthal remain open.[30]

These theories are based on organs merely centimetres – anatomically speaking – apart. Trying to guess at function from anatomy is a fraught business, not so very different from trying to deduce a family's behaviour from the layout of their rooms: it can supply some valuable pointers, but also lead one horribly astray, especially because the voice isn't simply a product of anatomy.

Yet in among these competing accounts of our vocal origins a remarkable common point remains: although some aspects of modern spoken language and consonant articulation (the vowels came later) may have appeared as early as two million years ago,[31] truly modern speech – both words and the sounds that our voices make to say them – is probably less than 100,000 years old.

## THE SEARCH FOR GENES

Another way of trying to identify when and how humans diverged from apes is by comparing the human genome with that of chimpanzees. Since human and chimp DNA is nearly 99 per cent identical, it's the differences that are critical. When more than 7,500 human, chimpanzee, and mouse gene sequences were compared, twenty-one genes specific to humans that are connected with hearing were found.[32] This might suggest that finely tuned hearing acuity is essential for understanding human speech, but the researchers sensibly went on to caution, 'Although it is tempting to conclude that this will constitute a list of genes that "make us human", one has to take a step back to see the gulf that exists between understanding at this narrowly focused molecular level and at the organismal level'[33] – a caveat most science writers were deaf to.

Instead, many of them immediately concluded, with the *New York Times* reporter, that since 'among the genes that show accelerated evolution in humans are those involved in hearing

. . . it seems that just a handful of genes might define the essence of humanity'.[34] Of course popular culture demands that 'the essence of humanity' – that chimera – be something discrete, dramatically discoverable, and preferably pithy enough to fit into a headline: in this respect, genes never let us down. But fascinating though current research into birds and primates is, humanity can't be reduced to a membrane in the ear, and the more that we study them, the more human vocal skills tantalise with their infinite complexity. Nevertheless we can be sure that, even now, the hearing abilities of apes are being probed with indecent enthusiasm by researchers around the world.

## THE LEFT SIDE

On 12 April 1861 a 51-year-old patient called Monsieur Leborgne was brought before the French surgeon and neurologist Pierre Paul Broca at the Hospital Bicetre in Paris. Leborgne had lost the capacity to speak twenty-one years earlier. Although he could still understand most of what was said to him, from the age of 30 he'd responded to every question with a single syllable – 'tan' – hence his hospital nickname of Tan. Five days after Broca first saw him, Tan died of gangrene.

Conducting an immediate autopsy, Broca found that a small part of Tan's brain had been destroyed by a neurosyphilitic lesion, leaving a cavity 'capable of holding a chicken egg'.[35] In a presentation to the Anthropological Society of Paris about his findings the following day, Broca generated enormous excitement by his assertion that the human faculty of spoken language resided in the second or third frontal convolution of the brain.[36] The speech centre in the frontal lobe of the left hemisphere was named after him and is now known as Broca's area, and Tan's kind of speech disorder is called Broca's aphasia.[37] Broca is credited with launching the scientific theory of the localisation of functions in the brain, in particular the concept of left-hemisphere dominance for speech.[38]

Broca was lucky to have made his claims at a moment when the scientific community was prepared to take them seriously,[39] especially since they'd been badly burned by the whole phrenology saga – that nineteenth-century craze for divining character from bumps on the head and the shape of the brain that had temporarily found scientific favour in Europe and America.[40]

Broca's discovery was soon followed by another of equal importance. Carl Wernicke, a young German clinical neuro-psychiatrist, also described function in the healthy brain by studying those of unhealthy people who later died.[41] One of his patients, despite having suffered a stroke, spoke fluently and had unimpaired hearing, yet could barely understand anything said to him.

After the man's death in 1873, Wernicke found a lesion in the rear temporal region of his brain in the left hemisphere. This region, lying close to the auditory part of the brain and now known as Wernicke's area, is involved in speech comprehension, he concluded. It's here, he and his successors believed, that sound-images of words used to represent objects and concepts are lodged.

In fact what Wernicke's area actually does is today subject to considerable dispute. It's unlikely, it now seems, that Wernicke's area represents the sounds of words – indeed, there's probably nowhere in the brain where the sounds of words are represented: they're probably constructed on the fly, and exactly how the brain interprets sounds as meaning remains one of the major unanswered questions.[42] Yet at a time when most scientists still thought that the brain functioned as a whole, Broca and Wernicke helped to show that the left hemisphere was dominant for language. Their discoveries have profound implications for the human voice.

## LATERAL THINKING

One of the main ways in which we differ from most other mammals is through lateral specialisation of the brain, yet sometimes it's hard to remember that hemisphere asymmetry is a serious business, so degraded has it become through popular psychology. In the 1970s and '80s left hemisphere and right hemisphere became not so much a description of the specialisation in brain function as freighted adjectives ('Are you left-hemisphere or right-hemisphere?'), conveying instant judgements about how intuitive, non-verbal, visual, spatial (RH) or driven, verbal, analytical, logical (LH) a person was. The hemispheres were treated like personality or even lifestyle attributes, the twentieth century's own variant of phrenology.

Apart from the absurd over-simplification, characterising the brain like this polarises it into two almost warring halves rather than the cooperating, complementary, interconnecting networks they really are. Because speech comes from organs in the middle of the body, it makes sense for the two halves of the brain to share out some of the functions in order to minimise 'neurological indecision' or conflict between them.[43] Yet more than a century after Broca and Wernicke, we now know that, though the left hemisphere is dominant for language (and auditory processing[44]), the right hemisphere has important language-related functions too.[45] Indeed, from the point of view of the human voice, the right hemisphere is paramount.

This is where most of our intonation-decoding abilities lie. It's our right hemisphere that lets us distinguish the voice of a happy person from a sad one, and intuit the metaphorical meaning of a phrase rather than its literal one. Labelled our 'emotion processor', the right hemisphere allows us to assign social and emotional significance to speech melody.[46] Some studies even claim that, when a person listens primarily with their left ear (controlled by the right hemisphere), they're more likely to make judgements based on a speaker's tone of voice rather than meaning.[47]

The right hemisphere not only helps us understand other people's voices but also affects the way we use our own: it supplies the left hemisphere with the most expressive element of the human voice, its prosody. Prosody, as we have seen, isn't some postscript added on to the verbal message, but an intrinsic part of the whole communicative process – sometimes the most important part. One neurologist has gone so far as to argue that prosody constitutes a third element of speech, sitting beside grammar and semantics.[48]

People with right-hemisphere damage often have difficulty understanding context and connotation, and seem particularly unable to recognise fear in the voice.[49] They also have problems detecting inconsistency between words and tone, and find it hard to distinguish between lies and jokes,[50] or between 'Do you want to go outside?' spoken as an invitation and as a threat.[51] Sometimes they confuse words whose meaning changes depending on the stress, for example, green house v. greenhouse, dark room v. darkroom, and can have difficulties recognising familiar voices.[52]

Right-hemisphere damage also causes people to lose control over perhaps the most important non-linguistic aspect of their own voice – the ability to vary pitch. They use intonation that's often inappropriate, and metaphor and irony pass them by. Right-hemisphere damage can produce a monotone known as 'aprosodia' (a term meaning disorders in emotional speech). A 47-year-old man was admitted to hospital with paralysis of the left side of his body: he was fully alert but his voice was flat and monotonous. Whether he was discussing his sadness about his illness, or his pleasure at his imminent weekend pass, there was no difference to his intonation and rhythm. Asked to say a sentence angrily, tearfully, or happily, all he could do was slightly quieten or raise his voice at the end of it.[53] He was like a painter suddenly unable to use colour.

Yet although (in right-handed people) the left hemisphere is dominant for language and the right for emotions, identifying and understanding emotional inflections in the voice is decidedly a joint project, mediated by the corpus callosum, the thick bundle of nerves lying between the two hemispheres. It used to be thought that the corpus callosum kept the hemispheres apart. Now we understand that it connects them, helping information move between the two halves. Emotion is detected in the right hemisphere and conveyed through the corpus callosum into the left hemisphere, from where both words and intonation are generated.[54]

Reading seems to be much more right-lateralised than speech, but almost all imaging studies of spoken-word and language processing show strong bilateral activation; the right hemisphere loves speech, just as the left does. The question is, what's each hemisphere doing with the speech? How different is the processing carried out in the two hemispheres?

In a recent Belgian study thirty-six students and hospital workers were hooked up to ultrasound monitors and played tapes of people saying sentences, first neutrally, and then in discordant ways, for example, 'He really enjoys that funny cartoon' was spoken sadly, and 'The little girl lost both her parents' happily. When they were asked to concentrate on the meaning of the sentence, blood rushed to the left hemisphere of their brain. When asked to identify the emotion expressed by the prosody, there was a marked increase in blood-flow velocity in the right hemisphere, but no decrease in the left hemisphere. Perhaps this was because the participants were being asked to label the emotion, i.e., put it into words, but both sides of the brain were engaged in the task of prosody-reading.[55]

Other research also warns against too neat an identification between the left hemisphere and speech. One study found an enlarged left planum temporale, the part of the temporal cortex in Wernicke's area believed to control language in the left hemisphere, in chimpanzees too.[56] Problems with prosody can also arise following damage to other parts of the brain, not just the cerebral hemispheres.[57] And some kinds of prosody – like understanding the difference between 'If you need me when you get there, call me' and 'If you need me, when you get there call me' – seem to be accomplished equally by both ears and hemispheres.[58]

But still the popular culture searches for an organ of empathy. The best-seller *Emotional Intelligence* made a star out of the amygdale,[59] which came to be seen as the seat of emotional understanding and passion that bypasses the neo-cortex or thinking part of the brain. Various cases have been reported of people who, after suffering lesions in the amygdala, can't understand the intonation patterns connected with feelings. One woman in this condition, despite having normal hearing, could barely recognise fear and anger in other people's tone of voice.[60]

Yet could evolution have left us so dependent on a single organ for the detection of fear? Though it creates a compelling image, this is an Adam Smith model of the human organism, as if the human brain were like the market, and mental function operated with a specialised division of labour.[61] But the amygdala's role in prosody is unlikely to be so central.[62] Increasingly, those who championed it as chief executive of Human Brain Inc. are beginning to recognise that a socialist model – the cooperative – may be a more apt guide to how the brain processes vocal emotion. Neural networks across different regions of the brain work together to transmit feelings acoustically and produce an understanding of the emotional aspects of the voice.[63] The mid-brain periaqueductal gray, for instance, is now thought to play a major role in shaping vocal cues, creating primitive cries of pain and distress, including a newborn baby's cry.[64]

New information about the contribution of different cerebral regions to the creation of the voice is emerging all the time. Trying to pin our ability to produce and detect vocal emotion on to one part of the human brain or allocating prosody to the right hemisphere and language to the left limits rather than enhances our understanding. And while the evolutionary development of the voice is astounding, so too is the dynamic interplay between mother and baby out of which, within just a couple of years, the child's voice evolves.

# 5

# *The Impact of the Mother's Voice (even in the Womb)*

FROM PSYCHOANALYSIS TO neuroscience, remarkable claims have been made for the importance of the mother's voice. It acts as a kind of umbilical cord.[1] It's a 'sonorous envelope' that 'surrounds, sustains, and cherishes the child'.[2] Or it's a 'sound bath', an audio-phonic skin, the first psychic space, in which the mother expresses to the infant something about itself that forms the basis of its own developing sense of self and ego.[3] Because the aural field is 360 degrees, we can't shut out sounds as we can sight[4] – only smell can suffuse the child in the same way.[5] At its best, the maternal voice serves as a container for the child, a sonic version of amniotic fluid.

It's certainly an exquisitely sensitive instrument. In a forty-second videotape of a mother playing with her 18-week-old daughter, sixteen different voice-quality settings were identified. Analysed, these were found to give a far more accurate guide to changes in the mother's feelings about the child than the mother's facial expression.[6] The auditory sphere is even believed to play a key role in the creation of the superego[7] (that 'conscience' or 'ought' part of the self where parental values and social rules are enshrined). The actual acoustic sound of the parent heard by the child is surely intimately related to the eventual tenor of its own internal voice.

The maternal voice may also have an important role to play in making separation bearable. When South American rodents were separated from their parents, there were changes in the limbic system in their brain – changes which maternal vocalisations had the power to suppress.[8]

The ability of the mother's voice to comfort a child – even in her absence – is now being used to ease the pain of unavoidable separation. Mothers about to be hospitalised for cancer treatment are advised by an American paediatrician to make a tape recording of themselves reading or singing to their children that can be played to them while she's away.[9] In Britain, Holloway Prison's Story Book Mums scheme encourages incarcerated mothers to record stories on tape to send to their children.

One woman noticed that her son, 10 months old when she began her sentence, was confused when he visited her in prison, even though he'd always recognised her in photographs:

> My husband would say, 'Give Mummy a kiss,' and Stephen would automatically look around for the photograph because he couldn't make the connection between the person in the photograph and me. It was upsetting, and I was worried that he was forgetting me. But sending him the tape helped a lot. My husband says that Stephen points and smiles when he hears my voice and he can't keep away from it. When he comes in to see me now, he gives me a kiss and a cuddle – and I'm sure it's partly because he recognises my voice.[10]

An American mother, in the throes of a painful divorce that brought regular absences from her children, developed a similar, if more saccharine, idea – a 'Mommy doll' containing a device to record the mother's voice singing or saying, 'I love you.'[11] Hospitalised children needing mechanical ventilation have been found to be calmer if a recording of their mother's voice accompanies the soothing music. Recordings like this, some believe, might eventually reduce the need for medication.[12]

## THE DISCRIMINATING FOETUS

> Now at this time Mary . . . entered the house of Zacharias and
> greeted Elizabeth. And it came about that when Elizabeth heard
> Mary's greeting, the baby leaped in her womb . . . And she cried
> out, 'When the sound of your greeting reached my ears, the baby
> leaped in my womb for joy.'
>
> Luke 1: 39–44

Anecdotal beliefs that foetuses can hear and react to voices date
back to biblical times, and yet it wasn't until the late nineteenth
century that they began to appear in the scientific literature. Even
then the prevailing view was that the human foetus was effectively
deaf.[13] Only with the development of new ultrasound techniques at
the end of the twentieth century did it become clear that foetuses
begin to react to some sounds as early as 14 weeks, and from about
28 weeks' gestation respond to auditory stimulation.[14] Remarkably, although the foetus can't speak or understand speech, it's
already able to recognise voices.

The mother's voice affects the foetus's heart rate, slowing it
down significantly, according to some studies,[15] which suggests
that the soothing capacity is present even before the baby is born.
In other studies foetuses got more and not less excited when they
heard their mother speaking.[16] Either way, they're able not only to
discriminate between their mother's voice and other people's, even
of the same sex, but also seem to remember the maternal voice and
respond to its familiarity. Even foetuses learn from experience.[17]

They're also, it seems, able to distinguish prenatally between
male and female voices *within a few syllables*, and shortly before
they're born already demonstrate a precocious ability to discriminate pitch.[18] The mother's voice *in utero* may also be helping to
shape the foetus's brain.[19]

But what exactly does the foetus hear through the abdominal
wall of amniotic fluids? Low-frequency sounds pass through more
easily than high, and vowels are heard more clearly than con-

sonants,[20] so that to a baby in the womb the mother's voice sounds 'like Lauren Bacall speaking from behind a heavy curtain'.[21] The mother's prosody also appears to be preserved in the womb: remarkably, adults have no difficulty recognising their intonation in playbacks of recordings of speech made in a woman's uterus – the melody as well as the words was intact.[22]

Of course, to the foetus, the maternal voice is louder and more audible than all other voices because it's conducted not just through the air but also through the mother's body, passing from abdominal tissues into amniotic fluid, via her spine and pelvic arch.[23] One psychoanalyst has suggested, intriguingly, that the foetus experiences sound as contact, possibly because sound waves from a person's voice create tiny yet distinct impressions on the eardrum and skin, so that vibrations are felt as well as heard.[24] When mothers speak, therefore, they really are engaged in a body-to-body experience with the baby in their womb, and aren't just producing some localised vocal or auditory experience.

Because sound to a foetus is solid, almost tangible, a really loud and threatening noise may cause it to fear 'auditory extinction'.[25] The relationship between hearing and touch are closer, perhaps, than we realise. This has huge psychological implications, and might go some way to explaining the intimate connection between maternal voice and secure attachment.

## LISTENING FOR MOTHER

The prenatal studies may be fascinating, but the research on newborns reveals even more compellingly that the human baby arrives in the world with a preference for the human voice over other sounds,[26] and fully equipped with the tools to distinguish between voices – blowing a raspberry at the Aristotelian idea of the neonate as a tabula rasa.

The newborn, it seems, knows what it wants to listen to – and mostly it's Mother. Infants less than two hours old react and orient more to their mother's voice than to those of other women.[27]

Newborns younger than three days old also seem to prefer her voice to that of another female – even if they've spent most of their short lives up till then in a nursery and so have been barely exposed to it.[28]

Findings of this kind are based on the intensity of a baby's sucking on a teat,[29] but watching and listening to babies is also revealing. During the first week of life the sound of a rattle or bell stops a baby crying as successfully as the human voice, but by the second week a human voice has become the most effective way of calming them, and by the third week the female voice is more effective than the male,[30] any other sound or even the sight of her face.

Hearing the female voice not only has a quietening effect on a baby but also makes them smile. After birth babies smile when they hear their mother's voice but not when they see her face.[31] By the fifth week the mother's voice has become so successful at eliciting smiles (much more so than the father's voice or anyone else's) that when in one study a baby feeding from a bottle – even in the first, hungry minute – heard a woman's voice, the child interrupted its sucking and gave a broad smile before returning to its food.[32]

The infant isn't a passive recipient of the maternal voice, but responds by producing comfort sounds of its own, which in turn make the mother talk more. In this way the mother's voice initiates an audio-phonic feedback loop, the baby develops a sense of its ability to summon the mother, and mother and child participate in a kind of mutual cooing.[33]

The past twenty-five years have seen a cascade of other remarkable findings about babies' reactions to voices. This is what we know:

Newborns prefer to hear their mother's voice filtered in the way that it would have sounded to them in the womb.[34] (The maternal voice provides the only continuity between their prenatal and postnatal life[35] – no wonder it can make them feel safe after their turbulent arrival in a strange new world.)

They can also discriminate between two unfamiliar voices on the basis of the different prosody used to say *a four-syllable sentence*.[36]

Babies as young as one month can detect changes in average pitch and timbre.[37]

So attuned are small babies to their mother's intonation that they suck faster when they hear their mother if, and only if, *the mother has recorded a message specially addressed to them*.[38]

In other words, babies are supremely gifted voice-readers. In the early months – before they're subject to the imperialism of language and the despotism of the eye – they rely much more on voices than faces for identification[39]: they can pick out their mother's voice from other voices well before they're able to distinguish her face from other faces.[40]

Sadly, although it continues well into the preschool/nursery years, infants' preference for voice reverses as they age: by the time they turn 4, children are already switching between auditory and visual modes. And once they've grown up, the contest is over: adults rely much more on vision than hearing. When it comes to the nuances of vocal melody, growing up in some sense means growing deaf.

All sorts of reasons for this switch have been touted. Babies begin to hear (at around 28 weeks of gestation) before they start to see (at birth), so perhaps it takes sight a while to catch up.[41] Or maybe the privileged status of hearing is due to its vital role in helping us learn language, and once that's achieved – at least according to this theory – it's done its central job.[42] Or could it be that, as they become part of adult culture, they begin to recognise and share our disdain for hearing, and our society's apparent indifference to the voice?

Whatever the explanation, there's no denying that the glorious sensitivity to the human voice exhibited by infants and young children atrophies steadily on the route to adulthood. Yet the piece of research I find most exciting is much more suggestive and upbeat. It regards our stunning pre- and postnatal auditory abilities not as talents that we eventually lose but as harbingers of skills to come. When they heard happy vocal expressions, babies ranging from 12 to 72 hours old either opened their eyes or kept them open, and when they heard sad vocal expressions, they sucked rhythmically or stuck their tongue out, clearly able to discriminate between the contrasting sounds of the two emotions.[43]

This early aptitude seems to result from their exposure in the womb to the distinctive prosody (pitch, volume, rhythm) of their mother as she expresses different feelings. When she talks angrily her breathing, heart rate, and degree of muscular tension, as well as the movement of her diaphragm, change, so that the baby not only *hears* its mother's angry voice prenatally but *feels* it too. And, astonishingly, this capacity to discriminate between the vocal sounds of different emotions and respond differently to them is evident within hours of birth.[44]

Darwin suggested in 1877 that the infant at a very early age understands the feelings of those who tend it in part from their intonation.[45] Many studies since have shown that infants from birth onwards are fabulously sensitive to other people's feelings, especially those expressed through the voice, and in particular positive emotions.[46] This ability of newborns to decode the acoustic communications of adults is the beginning, according to one psychoanalyst, and may be the prototype, of all future discriminatory learning.[47] In other words, our remarkably precocious vocal understanding of whether our parents are happy, angry or sad might be launching us on to the path of not only emotional intelligence, but also other equally vital kinds of intelligence.

There's more. Women, across different cultures and down through history (just look at paintings, statues, and sculptures), tend to cradle their babies on their left side. Girls playing with dolls do it too. The right side of the brain, as we've seen, is where the interpretation of emotions and vocal intonation is centred, so left-cradling allows information about the baby's psychological state to flow swiftly from the infant via its mother's left ear and eye to the centre for emotional decoding – her right hemisphere.[48] By cradling her baby on her left side a mother can better understand the feelings it's expressing through its face and voice.

Equally importantly, although the right ear of a baby cradled on the left side is buried in the crook of its mother's arm, its left ear remains free – to gather up the sounds of the mother's voice *for processing in the baby's own right hemisphere*. Decoding its mother's prosody and interpreting her inflections is no mere idle pastime for a baby, but an issue of survival: from the very beginning, it needs to be able to secure for itself care and nurturance, and voice-reading is a powerful aid. No wonder babies are so gifted at it.

## IGNORING DAD

The few experiments on paternal voices that have been conducted have uncovered an almost shocking indifference in babies. While they can distinguish between their father's voice and that of an unfamiliar man, they don't prefer their father's.[49] Even at 4 months old, infants express no more interest in the voice of their father than in that of a male stranger.[50]

Perhaps this isn't surprising since, as we've seen, newborns prefer their mother's voice even if they've hardly heard it since birth. When it comes to the importance of a voice, prenatal experience, it seems, counts for more than postnatal. Babies have done some important auditory perceptual learning *in utero*, which no amount of billing and cooing received after birth can rival.[51]

## THE HUMAN METRONOME

Plato defined rhythm as 'the order in movement'.[52] Rhythm, by promising repetition over time, allows us to anticipate what's going to happen, and so helps us process and structure our emotional dealings with other human beings. In our early lives, rhythm also enables us to experience absence and discontinuity in a safe, manageable way – sucking itself has a rhythm, a pulling and letting go.[53] And vocal rhythm is a kind of non-verbal language: through it we communicate how much proximity and reciprocity we want from another human being.

Babies can detect rhythm, and from early on they do things rhythmically. Prenatal life is played out to the accompaniment of the mother's heartbeat, which serves to structure the baby's world[54] but which is lost suddenly at the moment of birth. Newborns exposed repeatedly to the sound of the beat of a human heart cry less and gain more weight.[55] And when the cooing sounds made by a 2-month-old baby were analysed, they had a rhythmical structure corresponding to the heartbeat of an adult.[56]

There's a tempo to almost all basic human activities. Babbling babies make syllable-like sounds that last roughly 350 milliseconds, no matter what culture they're from. In early 'proto-conversation', when the infant is 6 months old, turn-taking takes the form of a slow adagio – 1 beat in 900 milliseconds, or 70 per minute. Within a month or two, shared vocal play accelerates to andante (1 beat in 700 milliseconds, or 90 per minute), or moderato (1 beat in 500 milliseconds, or 120 per minute).[57] The heart beats on average 72 times per minute, but we don't know by what processes the body manages to maintain all these rhythms – perhaps by means of biological 'oscillators' or neural clocks.[58]

Since a baby's sucking and the cycle of attention between a baby and its carer last between 3 and 6 seconds, and an average musical phrase, a line of spoken poetry, a spoken phrase and a breath cycle all last between 2 and 7 seconds, the cognitive psychologist Paul Fraisse has gone so far as to suggest that the 3- to 6-second temporal period might be fundamental to human motor function and perception.[59]

One researcher discovered that there was even a regular rhythm to oral tests in primary school: because the teacher expected the child to answer on the downbeat they didn't hear and therefore ignored correct answers that weren't produced in this regular, metrical interval (about 1 second apart). When child and teacher don't achieve rhythmic synchrony misunderstandings can result.[60]

Infants have a precocious understanding of time. From the first few weeks and months of life, they can distinguish between different lengths and intervals of time, as well as between simple rhythms.[61] Their delight in rhythm is often expressed through music. Nursery songs across different cultures and languages use the same rhythms and melodious forms, as if babies were teaching adults some universal principle of rhythm.[62] The repetition of sound and gesture intrinsic to rhymes, games, rocking, chants, bouncing, clapping, and song soon begins to emerge in words and speech.[63] Listening to infants, you realise just how arbitrary it is to separate music from speech, when both depend so crucially on rhythm. Babies, as we'll see, use rhythm to help them learn language. Indeed the point at which they begin to babble coincides with the peak period of rhythmic activity, the moment when they shake rattles most.[64]

## A SINGLE PULSE

A mother is talking to her baby in short bursts. The baby 'replies', matching not just her pitch but also her rhythm. Although it obviously can't understand the meaning of her words, through pitch and tempo it's already an equal participant in the communication process, responding to the metrical and melodic qualities in her voice. For at least some of the time mothers and babies exist in a vocal harmony[65] that not only expresses their emotional connection but also helps to create it. For the preverbal infant, the key aspect of being in communication with a parent may be sharing the same rhythm, which in turn provides the basis for the development of empathy in the child.[66]

Over thirty years ago a landmark piece of research found that, *as early as the first day of life*, the body movements of the human neonate are synchronised with adult speech. When William Condon and Louis Sander analysed, frame by frame, film of babies moving and adults talking, they found an extraordinarily close relationship between them, undetectable at normal speed. Nor did it only extend over the course of a few words, but through an entire sequence of 125 words.

Condon and Sander called this 'interactional synchrony', and believed it was unconscious and not deliberate. The same effect was observed whether the language was English or Chinese.[67] They claimed they'd found evidence of interactional synchrony occurring within twenty minutes of birth! Although there have been several unsuccessful attempts since 1974 to replicate their findings,[68] even the possibility of such sustained coordination at this micro-level raises profound questions about the role of auditory ability and rhythm in human development. Those elements in the baby's psycho-neural organisation that control the timing and form of body movements seem to react keenly to another person's changing rhythms and accents.[69] The human voice penetrates deep into our sinews, and long before we're able to say a single word, our bodies have bobbed millions of times in the precise, repeated, shared rhythm of our mother tongue and native culture. A baby begins to speak through its limbs before it's able to do so with its mouth.

Again we see how sound can create a powerful sense of closeness. As Condon put it, 'We're almost in auditory touch. When I speak to you, my thoughts are translated into muscle movements and then into airways that hit your ear, and your eardrum starts to oscillate in absolute synchrony with my voice . . . it takes only a few milliseconds for a sound to register in the brain stem, 14 milliseconds for it to reach the left hemisphere.'[70]

Attunement extends in both directions. The gazes of 6-week-old babies are coordinated with the rhythm of adults' voices and vice versa.[71] Babies can start vocalising in response to the mother's voice as early as three days after birth.[72] For the first two years of life, according to one study, babies' vocalisations and pauses last for almost exactly the same length of time as their mother's.[73]

Mothers, too, possess formidable imitative capacities and astonishing perceptual sensitivity: they can match their baby's pitch, pitch contour, loudness and almost every other aspect of their baby's voice.[74] (In this respect we're all Rory Bremners.) Most mothers use vocal rhythm intuitively to help regulate their baby's moods – speaking faster to stimulate them, and slower to calm them.

Synchronicity can even be used deliberately to make a connection. A 12-month-old girl spent most of the time throwing toys on the floor. But when a female researcher imitated the child's 'spit' sound, 'Suddenly, I had her attention. She offered another spit sound and this time I roughly matched it, and then slightly elaborated it. She then offered her own elaboration, and then we slowly moved into a vocal dialogue, which gradually took on increasingly rich and evocative variations.'[75]

The timing of the preverbal, vocal exchanges that babies engage in with their mothers is remarkably similar to the timing of verbal dialogues between adults[76] – not for nothing are they known as 'proto-conversations'.[77] Babies learn to speak with rhythm far earlier than with words. This interpersonal sophistication isn't necessarily 'innate' or 'hard-wired' into the biology of the neonate – it might be the result of the major developmental achievements that occur in the first few weeks of postnatal life.[78] For the baby, 'the music comes before the lyrics'.[79]

It's obviously a thrilling business for babies, this ability to share a common pulse, a correspondence of rhythm, intensity, and tempo. And yet irregularities of rhythm are essential too. A major study found that, purely through the pattern of vocal rhythm between mothers and their 4-month-old babies, it was possible to predict which babies would be the most securely attached at 12 months. But the most secure infants weren't to be found, as one might have supposed, in the most rhythmically coordinated pairs. On the contrary, these were the ones who seemed to be least securely connected to their mother – over-monitoring her, wary, vigilant, and rigid in interactions with her, already perhaps trying to offset some disturbance in the relationship. In the face of too much uncertainty, these infants had had to find ways of making communication with their mother predictable.

The least vocally coordinated pairs also seemed to lack enough coherence and predictability for the babies to attach securely, leaving them trying to avoid attachment. The most secure infants occupied a middle position, with space for flexibility and variation. Their mothers provided enough regularity of vocal rhythm and tempo to allow them to predict the pace and shape of their interaction, but also enough variability and novelty to be exciting. As a result, the mid-range pairs could be playful with each other.[80]

Timing gives partners feedback about the other. It can help control the flow of information from mother to baby, regulating the speed of her response and ensuring this doesn't exceed what the baby can reasonably handle without feeling overwhelmed or under-stimulated. Patterns of vocal rhythms organise the infant's experience of being with another person.[81] Again and again, in this pioneering research area, studies have found mothers and babies sharing cycles of attention and inattention, of looking and turning away, of being 'on' or 'off', in which they adjust to and match each other's rhythms. Through this they learn reciprocity.[82] Vocal-rhythm coordination may even in itself be an important means of forming attachments.[83] 'We suggest that the capacity for interpersonal timing is a necessary (though not sufficient) condition for two human beings to enter into an effective dialogue with one another, be they 6 weeks or 60 years old.'[84] The corollary may also be true: that when each communicating partner insists on their own rhythm, producing a vocal 'arrhythmia', they're liable to be in 'serious interactional trouble'.[85]

Although mothers use facial expressions and bodily gestures as well as their voice to attune to their infants, the voice seems to be the critical component.[86] It certainly seems to set in earlier, causing some researchers to argue that vocal exchanges play a special role in early human experience.[87] Vocal matching also seems to be specific to humans.[88]

## THE VITALITY OF SPEECH

Synchrony isn't only about timing and duration but also, crucially, intensity. This was first recognised by psychiatrist Daniel Stern, who coined the phrase 'vitality affects' to express qualities of feeling like 'surging', 'fading away', 'fleeting', 'explosive', 'crescendo', 'decrescendo', etc.[89] What's important to babies about parental behaviour, he argued, is the level of excitement it produces. Vitality affects take the form of a 'rush' of anger or joy, a wave of feeling – the texture of experience, for which we have almost no vocabulary.

Stern's examples conjure states that only exist in time, and have an auditory dimension: they track the development of a feeling, small shifts in arousal, the ebb and flow of excitement. (They also beautifully describe the human voice.) When we hear someone speak, it's this trajectory of energy that we monitor, often more than the fixed meaning of any individual words. Stern believes that infants experience their world in patterned flows of feeling that can last for seconds or fractions of seconds. 'Mother and infant . . . interact in a split-second world.'[90]

For example:

A 9-month-old boy bangs his hand on a soft toy, at first in anger, but gradually with pleasure, exuberance, and in a steady rhythm. His mother falls into his rhythm and says, 'Kaaaa-bam, kaaaa-bam,' the 'bam' falling on the stroke, and the 'kaaaa' on the upswing.

An 8½ -month-old boy tenses his body to stretch for a toy just beyond his reach. His mother says, 'Uuuuuh . . . uuuuuh!' with a crescendo of vocal effort matching the child's accelerating physical effort.

A 10-month-old girl finally places a piece in a jigsaw puzzle, and with an energetic punch of her arm partly raises herself off the ground in triumph. Her mother says, 'YES, thatta girl,' the 'YES' with an explosive rise that matches the girl's own flinging gesture.

These examples are a kind of interactional synchrony in reverse, this time with the mother's voice and volume matching the baby's movements rather than the other way round. Attunement like this, argues Stern, plays an important role in helping a baby realise that its internal feelings can be shared with others: it's how parents contour and contain the psychic experiences of the child. 'The experience becomes a "we" experience, not only a "me" experi-ence.'[91]

Of course parents and carers aren't always able to tune into a child's mood or emotions, sometimes over-valuing one aspect or under-matching another. Infants, Stern contends, recognise these mismatches and mixed messages because they can decipher tone of voice. Although gradations of feeling are also expressed through facial expression and gesture, voice is a prime conduit through which a parent conveys to their child just how engaged they are with the infant's state, how much they've noticed and are acknowl-edging it.


## CULTIVATING RHYTHM

Although rhythmic patterns seem to be partly innate, there are also cultural differences. When an Italian psychoanalyst, during a long stay in South Africa, observed a newborn baby and his mother for the first three months of his life, she noticed how different the rhythm of their interactions was from that found in Western countries. The baby, Bambata, was fed very often, spoken to only occasionally by its mother but in a rhythmic, swinging tone of voice, and patted to sleep to a rhythm that matched the call of one of the most common birds in South Africa. 'Both mother and child were immersed in . . . [a] shared rhythmical and musical conso-nance.'[92] Up to the age of 3 months, Bambata was never heard to vocalise: he didn't need to, since his mother was always to hand. Compare this rhythmic continuity with the separation present between most Western mothers and babies, to some extent even from birth.

If cultures pattern themselves round different temporal rhythms, what happens to mothers who move from one culture to another? Indian mothers, according to one study, tend to have more vocal overlap with their babies than French and American mothers, pausing less, and producing more non-verbal vocal sounds. But the communicative style of Indian mothers who'd recently emigrated to the United States had begun to change. Positioned halfway between the Indian and Euro-American approaches, they used less overlap but more pauses and verbal sounds than Indian mothers use in India (although they still differed from the French and American mothers). In the process of becoming vocally acculturated, they experienced a conflict between the practices of their native and host cultures. Struggling to juggle two different ways of being with their children, these women had lost their supportive social networks and were now losing confidence in their mothering abilities. As a result they were getting depressed. Immigration might disturb the temporal rhythm and synchrony of mother–baby interaction, leaving mothers less attuned to their infants.[93]

## DEPRESSING SOUNDS

Postnatal depression is audible. The voices of depressed mothers are more likely to be flat, and they usually talk less than non-depressed mothers, using shorter utterances as well as longer pauses.

A depressed woman's voice, some have suggested, might even impact upon her foetus, since the unborn child is receiving clues to its mother's state of mind through the sound of her voice. Whereas the maternal voice normally stimulates or soothes the foetus, the depressed mother's slower rhythm, weaker tone and lower pitch might already, prenatally, be depressing it.[94] (So here's another thing for a pregnant woman to anguish over: that when her baby is still *in utero*, her feeling low – a mood her pregnancy may itself have helped create – might already be causing psychic damage.)

Depression continues to affect the voice after birth. Non-depressed mothers have been found to respond nearly twice as fast as depressed mothers to their 3-month-old babies' vocalisations.[95] Depressed mothers' timing in response to their infants is less predictable.[96] Depression, by affecting the mother's ability to co-ordinate her vocal behaviour with that of her infant, might reduce the synchrony between them.

If we're persuaded of the benefits that follow from attunement, then clearly its absence must also produce consequences, yet the conclusions of some researchers are more than a little disturbing. 'If a mother does not react with precisely attuned rhythms of speech and facial expressions, a 2-month-old becomes withdrawn and distressed. When a mother has postnatal depression she finds it very hard to satisfy this critical attention of her baby.'[97] Given the extent of at least mild postnatal depression and mothers' differing abilities to attune to their babies, you might wonder how any small baby ever gets stimulated or avoids becoming depressed itself.

Pitch too is affected by depression, with an unmodulated voice supposedly less effective at stimulating a child's attention and promoting learning.[98] But mothers with postnatal depression seem to be damned if they use unmodulated voices and equally damned if they use over-modulated ones, because another study found that depressed mothers used significantly *higher* pitch and *more* modulation when talking to their babies than undepressed ones.[99]

Why could this be? Research into her mothering skills is anxiety-inducing for any woman, let alone a postnatally depressed one. A woman already doubting her ability to mother may try to make herself sound more competent than she really feels by raising her pitch and over-modulating her voice.

Maternal depression undoubtedly has an impact on children, and the maternal voice may well be one route through which it gets communicated, but findings like this probably cause further depression. Couldn't the researchers have built into their research design some major support and confidence-building for the mothers, as well as reassurance about the fundamental resilience of children, and seen what effects this too had on maternal pitch?

## AFTER BABYHOOD

Long beyond infancy, the parental voice remains a major force in children's lives. A pioneering American study of how ordinary families talk to their very young children observed some parents using 'a loose social bond of talk in which words were used less to organize conversation than to create and sustain social closeness'.[100] These kinds of parents talk for sociability and to signal to the child that, whatever else they're doing at the time – washing up or tying the child's shoelace – they're still intimately involved with them. Their voice acts as a connective tissue.[101]

The tenor of the voice – affirming and encouraging, or consistently negative – also has powerful consequences. 'Parents . . . instruct, enthuse, yell, or cajole. Children capture equally the emotional tone and the sound pattern of the words they hear. Long before they begin using words, they begin learning about how families interact in the culture, what people are like, and who and how valued the children themselves are.'[102]

Reading bedtime stories can develop intimacy:

The adult's voice isn't just a carrier wave, bringing to the child's ear the same fixed batch of words they could obtain if they read the book to themselves. The adult voice creates the book for the child. It brings it to life between the speaker and the hearer, as a shared possession both can enjoy . . . When the adult voice performs the story, it is doing some of the work of deciding what the world of the story is like.[103]

Increasingly, as children get older, the parental voice takes on a directive function. One 46-year-old woman found it easy to control her first child through physical means, until the second arrived. 'And then I realised that I could only control my daughter now with my voice. My son was a babe in arms, so I couldn't use them with my daughter, so I had to rely on my voice.'[104] For a parent, the voice may be all that stands between domestic order and chaos.[105]

As we age, our voices tend to start resembling those of our same-sex parent, as many of the people I interviewed (often ruefully) testified. A 44-year-old woman admitted, 'My voice is very similar to my mother's in intonation. It's strange – it's like a physical memory: I have a sense of what it must be like to be in her body saying those same things because it's so exactly the same.'[106]

Cadences are carried into successive generations, too. A 42-year-old woman finds it 'spooky when I sometimes hear my intonation coming from the mouth of my 7-year-old daughter when she speaks to her younger brother. She's internalised my voice but I've also to some extent internalised my mother's voice – all the stuff I don't like in it. Sometimes I wonder, whose voice are you speaking in?'[107]

## MOTHER SUPERIOR?

Given the importance of mothers' voices, perhaps it's not surprising that they're so often either idealised or demonised – sometimes by mothers themselves. I never had such a long, willing queue of interviewees as when I was writing an article for a national newspaper (anonymity guaranteed) on shouting at one's children, after a piece of research purported to show that the practice could permanently and significantly alter the structure of children's brains (and not for the better).[108] In a kind of negative competition mothers tried to outbid each other in convincing me that they shouted more persistently and louder than the others.[109] (In a village in the South Pacific, they apparently practise a unique form of logging. When a tree is too big to be felled by an axe, a villager with special powers comes each morning and screams loudly at it. After thirty days of screaming, the tree dies and falls over. The villagers claim that it works because screaming at living things kills their spirit.[110])

Mothers' self-criticism isn't surprising, given the number of newspaper articles recommending that parents should go back to school 'to relearn how to talk to their offspring',[111] and child-care manuals instructing parents on what kind of cadences to use. 'When telling a child off, parents should use low-key voices, a monotone that states calmly but firmly what the child has done wrong and what the punishment will be.'[112]

So the maternal voice is no longer regarded as an all-purpose comforter: in the popular imagination it's been reformulated as a problem, one which can only be solved by learning a whole new set of vocal skills.

Sometimes the mother's voice is depicted as actively malign. A monotonic, metallic voice, it's been said, can have a powerful effect on a child. 'Such a voice disturbs the constitution of the self: the sound-bath no longer envelops the subject . . . It contains holes as well as producing them.'[113] This French analyst went so far as to claim that the mother of a schizophrenic could be often recognised by the discomfort her voice produced in the doctor or psychologist consulted. Much of this work is highly speculative, articulating prejudice as much as demonstrable fact. More anxious and angry maternal voices might well produce more irritable, insecure children – as one ambitious and provocative American study found.[114] But, as its authors themselves acknowledged, who's to say in what direction the influence flows? A child's individual characteristics also affect its mother's emotional reaction: some babies are more intrinsically lovable than others, a fact probably echoed in the mother's voice, which may betray far more conflict, inconsistency, and ambivalence than our idealised image of the maternal voice is able to tolerate.[115]

Nevertheless, the smothering rather than mothering maternal voice has become a modern villain – less of a nurturing umbilical cord than an 'umbilical web' that allows 'no chance of autonomy to the subject trapped' in it.[116] In Hitchcock's *Psycho* the mother's voice is so powerful that it survives even her death. By the end of the film Norman Bates (Anthony Perkins) lies pathetic in a police cell, reduced to a ventriloquist's dummy, completely controlled by his mother's voice.

The triumphant maternal voice emerging from the son's body in

*Psycho* is a grotesque reversal of the usual order. Instead of helping the child to produce sounds of its own, the mother annihilates them. Psycho is mother and not child.[117] Here, all that can't be tolerated in the male voice is projected on to the mother's. Her voice becomes identified with the pre-linguistic, subjective state that must be repudiated (it's commonly believed) if a man is to be able to develop a masculine or paternal voice.[118]

Although it's the omnipresence of the mother's voice that's indicted here, could it be the exact opposite – its absence – that is really so enraging? The ebb and flow of the mother's voice – a kind of 'sound-object' – may give the foetus its first experience of separation. 'At times the voice speaks, and at times it is silent. It *is* an external object, as unpredictable and uncontrollable as the breast will be after birth . . . and can therefore be a source of both well-being and frustration and anxiety . . . the disappearance of the enlivening and stimulating voice might give the child a proto-experience of absence and loss.'[119]

The mother's voice starts and stops – this may be its true crime.

And yet our fantasies about the maternal voice are irresistible. Almost everyone I interviewed, no matter how difficult their relationship with their mother when she was alive, seemed to preserve an idealised memory of her voice after her death. One 43-year-old woman put it neatly: 'I hear my mother's voice all the time, and it's so sweet, and it comforts me. I think of it being tender much more than it probably was. I don't have many flashbacks of the neurosis that would come zapping with the voice, oy, oy.'[120] Both before and after we come face-to-face with our mothers, her voice leaves an ineradicable trace.

# 6

# Mothertalk: the Melody of Intimacy

WHEN WE SPEAK to babies and toddlers we alter our voices dramatically: listen to an adult talking to a child and you hear a voice forever swooping and soaring.[1] But although baby talk – sometimes called 'motherese', 'parentese', or 'infant-directed speech' – may seem like an intuitive response to the challenge of communicating with a preverbal child, or even an example of how easily adults regress, it's actually a highly sophisticated register operating on many different levels. Through motherese our voices swathe babies in warmth while helping them understand language and learn the rules of conversation. It's hard to imagine another human faculty that could achieve so much developmentally in such an apparently effortless way.

First identified by linguists in 1964,[2] baby talk has now been recorded and analysed all around the world. It has seven main features. Addressing a 2-year-old, we tend to use a much higher pitch – around an octave higher – than when we're talking to an adult, or even a 5-year-old. Our pitch range is also wider than usual,[3] and we use a 'rising final pitch terminal', raising our voices at the end of a sentence even when we're not asking a question. We're more inclined to whisper, too, as well as speak more slowly in shorter sentences, stressing two different syllables in a word that usually has only one.[4] Remarkably, even American Sign Language has its motherese, involving signing more slowly and using exaggerated movements when addressing infants rather than adults.[5]

You don't need to be a mother, a parent, or even an adult to use it.[6] Children as young as 3 modify their speech when talking to younger children,[7] and one child used its extra-high pitch in talking to her dolls when she was only 2 years and 3 months old herself.[8] Baby talk is so sensitive to the characteristics of individual children that a mother was heard to alter her voice according to which 3-month-old twin son she was talking to, using a higher pitch and rising intonation contour when speaking to the less responsive baby.[9]

## WHY THE FUNNY VOICE?

Reams of research points to the same fact: that, given a choice, young infants clearly prefer baby talk to adult-directed speech.[10] There are many different possible reasons. One is simply novelty – babies are stimulated by auditory change, so the exaggerated sweeps of the mother's voice and her wide pitch range probably excite them.[11] (As we've seen, by the age of 3 to 5 days, infants can discriminate pitch and volume, and are more sensitive to high pitch.[12]) When everything has been stripped away from baby talk but its pitch characteristics, infants still prefer it to speech designed for adults.

Baby talk's features are all geared to get the child's attention so as to engage in conversation – its distinctive melodies make it more easily picked out against other background sounds.[13] By going up at the end of a sentence, mothers are also introducing their babies to another skill essential for successful conversation – turn-taking. Their final rise announces, 'Now it's your turn.'

Babies, as we'll see, mimic adult voices – it's how they learn to speak. Motherese is the reverse: it gets adults to imitate babies' voices. So 'the baby talk of adults seems to have originated in the talk of babies to adults'.[14]

But babies already talk like babies, so what's the point of parents doing the same?[15] In motherese the difference between what the baby is capable of producing and what it hears from its mother is reduced.[16] By borrowing the child's own emerging sounds, adults when they baby talk are acknowledging and legitimating them.

## REPEAT AFTER ME

Motherese is babies' first dictionary, because they respond to intonation before they understand other aspects of language. A German researcher in the 1920s reported a child of 9 months who not only looked at the clock after hearing, 'Where is the clock?', but also did the same when the intonation pattern stayed the same only different words were used.[17]

Adult-directed speech, being a somewhat sloppy affair, is a poor way to learn language.[18] When trying to articulate vowels and consonants we often overshoot our targets: in the rush of daily conversation it's often hard to detect where one word ends and another begins. Speaking to babies slowly, though, gives them more time to process what they're hearing,[19] and over-enunciating and emphasising a couple of words in a sentence helps them divide up the speech stream into comprehensible units.[20]

Parents in the United States, Russia, and Sweden (whose languages contain quite different amounts of vowel complexity) all stretch their vowels when speaking to their infants, particularly 'ah', 'ee', and 'oo' – the three vowel sounds common to all the world's languages.[21] Exaggerating vowels in this way increases the acoustic distance between them, making each one more distinct from the next. 'Beads', when stretched, is much less likely to be confused with 'bed' or 'bid', 'coooold' is more easily distinguished from 'kind', 'caught', and 'can't'. And this also helps babies ignore meaningless variations in the way one individual or another says the word.

Mothers instinctively change their cadences and intonation as their child ages, to accommodate its growing capacity to comprehend.[22] When we talk to newborns, for instance, we use our voice mainly to soothe and calm. By 3 months babies are more socially responsive, so the mother uses baby talk to encourage their social and emotional development.

Her pitch is at its highest when the baby is 6 months old.[23] At 9 months both pitch and warmth decline because by now babies are listening to the sounds of their mother tongue more selectively, more like adults.[24] As the baby reaches its first birthday, its mother's voice becomes less emotional and more informational and directional.[25] It changes again when the child is between the ages of 2 and 5, when her pitch and pitch range decrease once more. By the time a child turns 5, their attention span is so much better that subtler ways of communicating can be used.[26]

According to this argument, mothers, by instinctively using and modifying baby talk, are giving their infants a customised tutorial in language, a free grammar lesson. (With its talk of the 'mommy linguist', some of the literature on motherese is seriously patronising – as if mommies could never be qualified academic linguists, only intuitive ones.)

Noam Chomsky claimed that the ability to learn language must be innate because the child suffers from such 'poverty of stimulus' that it couldn't possibly acquire such a complex system purely from exposure to the limited examples of speech in its environment. On the contrary, said others, the baby is exposed to a wealth of linguistic stimulus merely through motherese, and the modifications in a mother's voice are as important as innate universal grammar in helping a child learn the rules of its language and therefore to speak.[27]

To the psychologist Steven Pinker this line of thinking is just 'folklore' – another example of American parents' ceaseless exertions to prevent their 'helpless infant from falling behind in the great race of life'.[28] But Pinker tries to have it both ways: as a devout Chomskyist he subscribes to the 'poverty of input' theory, and then goes and blasts motherese for being terribly complex. Baby talk isn't designed to guide infants through the language curriculum, he insists, while at the same time acknowledging that it does mark out sentence boundaries and highlight new words. Pinker is an evolutionary psychologist, and therefore much happier comparing motherese to the vocalisations 'that other animals direct to their young'.[29] Note the 'other'.

It's paradoxical that motherese became a hammer with which to bash the Chomskyist universalists on the head because motherese itself is now held by many to be innate and universal – and has been challenged in turn on precisely those grounds.

## GREETINGS FROM ON HIGH

The list of explanations for baby talk never stops growing. Its acoustic patterns have been compared to those of poetry, with our ability to respond, at even a few weeks old, to the poetic features of language in motherese supposedly laying the foundation for our later ability to produce and appreciate literature.[30]

To others, baby talk is a kind of living, vocal fossil, with its origins in evolution. The increasingly large brains of early hominins (human ancestors), if they'd been allowed to mature in the womb, would have made childbirth more difficult. Instead human infants were born earlier, less mature, and so exceptionally helpless – without even the ability (which chimps develop at 2 months) to cling to their mothers.

Since human mothers also lacked body hair that their babies could easily grasp, and stood up on two feet, making any vestigial clinging ability that their babies possessed at birth less effective, they had to put their infants down when they foraged for food. Was prosody used by bipedal hominin mothers to compensate for this increase in distance and keep in touch with their baby? Did motherese act as a kind of 'vocal rocking', a way of reassuring babies that mother was near – an extension of the mother's cradling arms?

Motherese, according to this theory, is genetically driven, the product of past selective pressures, and retains the melody and prosodic features that humans used before speech.[31] Yet if motherese had an evolutionary function, wouldn't it have died out in cultures where babies are carried around by mothers in slings or in those where, on the contrary, very small infants spend most of their time in nurseries, pre-school, or looked after by nannies and child-minders?

## LOVER TALK

There's also an emotional aspect to baby talk. As we've seen, long before the infant can process words, the melodies of its mother's speech give it access to her feelings and intentions. One feature of motherese that especially enthrals babies is the amount of emotion it expresses. Compared with speech addressed to other adults, where politeness and restraint are usually de rigueur, motherese is a much better guide to the feelings of the speaker.[32]

Most research into motherese misleadingly compares speech addressed to a baby with that spoken to an adult acquaintance (rather than, say, to a lover). But when various similar emotions – love, comfort, fear, and surprise – expressed to a baby and to an adult were compared, there were astonishingly few differences between them. Baby talk turned out to be remarkably like lover talk: an adult saying lovingly, 'Honey, come over here,' to a baby sounds acoustically almost identical to them saying it to another adult.

So perhaps infant-directed speech isn't a special register or tone at all, but just reflects our greater expressiveness and vocal freedom when we communicate with babies. Powerful social taboos prevent us from sounding loving or enormously comforting to adults with whom we're not on kissing terms, but when we do express emotion to other adults, we use almost exactly the same acoustic features as when we speak to babies.[33] Motherese is truly the melody of intimacy.

Interestingly, although motherese is cross-cultural, those few cultures where it doesn't exist are also societies that frown upon uninhibited emotional speech: their adults aren't only exercising acoustic restraint in communicating with children but also with other adults.

Baby talk's main emotional purpose may be even more specific: to express happiness. When we speak to adults in a tone as positive as that used in baby talk, 6-month-old babies respond equally well. Even more telling, when babies have to choose between happy adult-directed speech and neutral baby talk, *they prefer the adult-directed speech*. By this reckoning, motherese is fundamentally a response to babies' delight in happy voices.[34]

## UNIVERSAL SING-SONG

Is baby talk universal or shaped by culture? As well as English it's been found in French, Italian, German, and Spanish-speaking societies.[35] Arabic has it, and so too do Cocopa, Gilyak and Comanche, as well as Marathi,[36] Greek, Hidatsa, Kannada, Kipsigis, Latvian, Luo, Maltese, Marathi, Romanian,[37] and Berber.

In tonal languages like Mandarin Chinese, as we've seen, a change in pitch can also change the meaning of words. Yet when Mandarin mothers address their babies, they use a much higher frequency than normal, slow down, and use longer pauses, just like other baby talkers.[38] In fact they risk transgressing linguistic rules in order to use motherese.[39]

But then there's the experience of the half a million people living in the western highland region of Guatemala who speak the Mayan language of Quiché. Quiché mothers, who are mainly responsible for the care of babies, keep their infants close to them at all times, either strapped to their back, in a cradle of rags near by, or beside them when they sleep. Since they interpret any sound from their baby as a sign that it needs feeding, vocal interaction between infants and parents is minimal. The idea of talking to babies to stimulate their linguistic development is quite alien here: infants are ignored in conversation until the arrival of their first two words – their mothers don't even recite nursery rhymes or lullabies to them. Quiché parents don't use baby talk, or any other simplifying register, and yet their babies grow into fully competent speakers of their native language (advantage Chomsky).

The Quiché experience suggests that culture does have a role in shaping motherese. It may be a worldwide, instinctive, unconscious practice with many near-universal features, but these are inevitably modified by a country's social conventions about how children should be treated – even by differing cultural ideas of what babies are,[40] as well as by a society's rules governing the public expression of emotion. Hence the fact that middle-class Americans, who show the most extreme prosodic changes when they speak to children, are citizens of a country where emotional expressiveness isn't only tolerated but actually expected, as compared with Asian cultures where there are strong social constraints against exaggerated vocal expression.[41]

## PETS AND THE PATRONISED

Baby talk isn't just for babies: we also speak in a high pitch to our dogs[42] and rabbits,[43] using pet talk or so-called doggerel.[44] There are all sorts of factors that affect how and when pet talk is used. Women, for instance, pet talk to their dogs more than men do, but both men and women pet talk more to unfamiliar dogs.[45]

Despite the apparent absurdity of studying how pet owners speak to their pets, comparisons between baby talk and pet talk are actually quite useful: they can help elucidate what's distinctive about talking to babies, and whether motherese really does have a linguistic function. An American linguist who claimed never to use baby talk suddenly realised that she used pet talk almost constantly to her cat. Clearly she wasn't trying to teach it to speak English, only to express affection.[46] On the other hand one study found that mothers exaggerated their vowels to their infants but not to their pets, confirming the theory that stretching the vowel or 'hyper-articulating' makes the distinctions between the different vowels sound larger so that babies can tell them apart.

Though women speak more affectionately to their children than to their pets, they use a high pitch to address both.[47] On some level, we don't differentiate between our child and our chihuahua.

Baby talk also has something in common with a vicar's delivery of a sermon (slow, with clear enunciation),[48] and with foreigner talk – the slow, distinct speech with frequent pauses and exaggerated stress that we think, perhaps misguidedly, makes us sound more like the person to whom we're talking.[49] Speech addressed to people with learning difficulties, hospital patients, old people, and even plants shares some of these features too and has been called 'secondary baby talk'.[50] In all these cases (except, perhaps, the plants) there's an element of coaxing and wheedling to soften the directness of commands, providing a kind of smile in the voice for people we care about.[51] Or think of as being less competent (babies, people with learning difficulties, old people, foreigners): secondary baby talk might also be an expression of condescension,[52] a powerful way of infantilising people.[53]

In fact, in terms of their high pitch and pitch range, *baby talk and secondary baby talk to elderly adults are paralinguistically the same.* Secondary baby talk may not only be conveying to the elderly their childlike status but also reinforcing it.[54] Again, if the evidence of our voices is to be believed, we regard the very old and the very young as indistinguishable.

## FATHERESE

Finally, the most charged question of them all: do fathers also speak motherese? Or is there something in the way that mothers address their infants – either intrinsic or learned – that's different?

Some studies have found only minimal differences.[55] In others, mothers talking to their babies increase their average pitch significantly more than fathers, who use the same intonation range for talking to babies as to adults.[56]

Indeed, when recordings of parents talking to their infants were fed into BabyEars, a machine designed to detect acoustic differences, it was able more accurately to classify which emotion was being expressed when it came to female speech.[57] Was this because mothers are more skilful and practised in baby talk? Were the fathers more self-conscious engaging in baby talk in a laboratory? Or did they use different prosodic features to convey feelings?

'Daddyspeak' has begun to appear in newspapers and magazines. There are anecdotal accounts of macho men with normally bass booms protesting that they never baby talk, only to turn to their child and say, in a voice a full octave higher, 'Do I, sweetie? No, I dooon't talk baby talk to YOOOOU.'[58] Of new fathers using diminutive words and exaggerated intonation, sometimes in inappropriate settings (i.e., to checkout assistants in supermarkets).

Do fondly self-mocking articles like this mark a new stage in male self-representation? Some men have begun to appropriate the stereotypes of dizziness previously pinned on to women, but are wearing them proudly as an expression of strength. We fathers, they seem to be saying, are so confident in our masculinity that we even use mushy voices.

'The descent into Daddyspeak is inevitable', claimed one. 'Adjectives mutate into gibberish. Pronunciation dies a slow death', to be replaced by rhyming sillyisms. This father claimed – in an argument with impeccable academic support – that, 'Daddyspeak strengthens the tentative bond between father and child. For one, it helps level the playing field. My son can't speak yet and so, to be fair, I act as if I can't either . . . Also, while the adult male voice is often demanding and even scary, Daddyspeak is soft and inviting.'[59]

But it's more complicated than this, for when white, middle-class mothers in suburban Atlanta speaking to their 2-year-olds were compared with fathers, both men and women were found to raise their pitch – *the men even more than the women*. Yet when it came to the older children, *the fathers didn't differentiate between 5-year-olds and adult listeners in terms of pitch*. The mothers, on the other hand, were still raising their pitch significantly when they spoke to the older children.[60]

If the sole purpose of exaggerated intonation was to get the child's attention, then why wasn't everyone addressing all children of the same age in roughly the same way? Perhaps it's a question of fathers' cultural expectations: maybe men revert to the more limited prosody in order to avoid being tarnished by more stereotypically female speech patterns.[61] Taken together, these studies suggest that, despite the exceptions, one of the major factors inhibiting men from using baby talk is still *the overriding male imperative not to sound like a woman.*

So is baby talk a way of getting infants' attention, teaching them social and grammatical skills, expressing affection or condescension, or providing security while foraging for food? However many of these functions it manages to accomplish at the same time, there's no doubt that motherese plays an important developmental function: our voices are shepherding children through their early years. Yet ultimately it remains an enigma – to adults and researchers, if not to babies.

# 7

# *The Emergence of the Baby's Voice*

IN AMONGST THE marvel and mucus of birth, one astounding fact is taken for granted: babies are born with voices that they immediately know how to use. Though crying is actually a complex endeavour,[1] newborns don't need to be taught how to do it. Right from the moment of birth they're in possession of their most powerful weapon: their ability to raise the roof. Exercising it is probably their first truly independent act. And a highly symbolic one: the newborn baby's use of its voice is the rite of passage through which it is conferred membership of the human species. With that first cry, it is welcomed into the community of humanity.

## TIMELINE: THE BABY'S VOICE

Remarkably, cries have been reported from human foetuses – usually following a medical procedure where the foetus was inadvertently touched. Known as *vagitus uterinus* ('squalling in the womb'), there are examples stretching back to ancient Egypt, Greece, and Rome. In 1923, an American physician, George Ryder, heard the sound of a baby crying after he had applied traction with forceps. Listening via a stethoscope, his assistant and nurses said the sounds were 'high and squealing, much like the mew of a kitten'.[2] The human foetus may be practising the use of its voice already *in utero*.[3]

## THE FIRST CRY

Humans, according to the philosopher Kant, are the only species to emit a sound at birth – one reason, perhaps, that the newborn baby's cries have been invested with such purity, innocence and almost mystical power. Hegel, on the other hand, thought the first cry reflected the 'horror of the spirit at its subjection to nature'.[4] The reality is more prosaic. The first audible cry clears the respiratory tract of mucus for the first breath. It's also an instinctive reaction to a shocking change of temperature, light, and air.[5]

Immediately after being born, healthy newborn babies everywhere produce an almost identical reflex cry – one which, intriguingly, matches the frequency of the international standard tone for tuning musical instruments.[6] So essential is that first cry as a sign of independent life that, if it is absent, a baby is often – still today – slapped into producing it. The pitch, tone, and dynamic strength of the cry are assessed within one minute of the infant entering the world, and then again at five minutes. The APGAR scale judges health, allocating two points for a good strong cry, but only one for a weak (whimpering or grunting) one.

The human cry is a remarkably efficient signal.[7] In 1969, after studying infants over their first 6 months of life, a Boston paediatrician claimed that the cause of a baby's cry could be identified by its acoustic characteristics. A hunger cry, for instance, is rhythmic and lasts about 0.65 seconds, while a pain cry is typically 3 to 4 seconds at first, followed by 7 seconds of breath-holding before the next explosion.[8] When they're less than 3 weeks old, healthy babies already have four acoustically distinct cries – that of hunger, anger, pain, and frustration.[9]

But the cries of babies with diseases affecting the central nervous system, with brain damage, cleft palate, hydrocephalus, hypoglycaemia, meningitis, chromosomal abnormalities, are different from normal crying – more high-pitched.[10] There are also significant differences between the cries of healthy full-term babies and small premature ones, and even between healthy premature babies and sick ones.[11] And there's a distinctive sound in the cry of a baby with a diabetic mother,[12] or one exposed in the womb to cocaine.[13] Extraordinarily, a relationship has also been found between the characteristics of a baby's cry and the amount of marijuana its mother used when she was pregnant.[14] So the infant cry can also be an effective, non-invasive diagnostic tool.[15]

From the very beginning, the baby is linked to its parents in an audio-phonic loop.[16] A baby's cry is tuned to the 3,000 hertz band where its mother's ear is at its most sensitive.[17] The cry is one of the few acts under the infant's control: its most powerful, and often only, means of communication.

Newborns can discriminate between their own cries and those of other babies. They get upset when they hear other babies cry,[18] which is probably why, when one baby cries in a maternity ward, the others inevitably follow. It's always assumed that they're simply copying each other, but they're probably also pained by the sounds of distress. Human empathy develops early, and it's expressed vocally.

Studies into the reactions of newborns to cries also cast fascinating light on their developing sense of self. When 1-day-old babies were played audio tapes of another neonate crying, as well as recordings of the wails of an 11-month-old, and a tape of their own cries, they cried most to howls of the neonate, *but didn't respond to the playback of their own cries.*[19] Already at birth, it seems, babies can discriminate vocally between me and not-me, and are most sensitive to the group that most resembles them.

Babies' cries can also be a guide to their psychological state. Entering a ward full of battered babies, voice teacher Patsy Rodenburg heard strangulated cries – 'their experience of violence had already pierced their voices'.[20]

Yet the common belief that mothers can intuitively tell why their baby is crying is a fallacy. In a major study, mothers couldn't identify with any accuracy whether their child – or indeed anyone else's – was crying through hunger, pain, or being startled. All of them tended to over-identify the cries as hunger![21]

Perhaps this isn't surprising, since mothers vary in their abilities to read their baby's cues,[22] and those abilities change as they become more experienced. Some babies, too, send clearer signals than others.[23] In any case, parents rarely rely on cry characteristics alone to help them work out what their baby is trying to tell them, but use other cues as well, like the time of day, when the baby was last fed, etc.[24] Cries may reveal how distressed a baby is, but not necessarily the cause of the distress.[25]

The crying baby has been held up as a model of how all of us should use our voices because, from the moment they're born, babies exploit the full vocal range available to them,[26] using their whole body to make a sound: the shoulders and neck are free, the mouth open, and the breath travels freely from the lower abdomen.[27]

Babies may be small and need to sleep a lot but their voices don't tire easily.[28] And unlike adults, they aren't self-conscious about their voices. They don't try to make them sound beautiful,[29] but just use them instinctively.

But according to this argument, as life and its pressures infiltrate the child's breath, the natural voice begins to slip away, to be rediscovered only in heightened moments.[30] The extraordinarily high pitch that even a healthy, inconsolable screaming infant can attain inevitably gets lost over time.[31]

Because we make sounds and music before words, the voice has come to be associated with our preverbal, primitive selves,[32] and the baby's cry is often seen as somehow lying beyond culture. Yet although the cries of healthy babies everywhere may be remarkably similar in pitch and melody,[33] there are cultural differences to baby's crying patterns. The Japanese, for instance, hold crying-baby competitions, based on the belief that crying is good for an infant's health. Some contests even pit amateur sumo wrestlers against each other to see who can elicit the loudest crying by gently shaking the babies or raising them up high.[34] In Britain they'd probably risk prosecution for causing shaken baby syndrome, but in Japan the babies that cry the loudest are crowned the winners.

Both the tolerance of infant crying and the meaning attributed to it varies from culture to culture – depending, for example, on how much physical contact a mother has with her baby. The Third World baby carried by its mother in a sling is able to use body movements to communicate to her: she can respond before it gets vexed enough to cry. In cultures like these the cry functions as a real distress signal. Babies in Western societies, on the other hand, develop a wider repertoire of cries, conveying gradations of discontent.

Voice therapists who rue the loss of rawness and purity that comes with the development of the acculturated speaking voice tend to ignore these factors, contrasting instead the primitive or baby voice, which, they claim, is the 'real' voice, with the social voice that they regard as disconnected from its natural richness.[35]

But the idea that an infant voice could exist beyond the reach of human society belongs in fables. Even Darwin acknowledged:

The wants of an infant are at first made intelligible by instinctive cries, which after a time are modified in part unconsciously, and in part, as I believe, voluntarily as a means of communication, by the unconscious expression of the features, by gestures and in a marked manner by different intonations, lastly by words of a general nature invented by himself, then of a more precise nature imitated from those which he hears.[36]

Simply by living in human society our voices change.

Perhaps it's more productive to see the idealised commentaries on the baby's cry as a lament for the adult voice, and the constricted bodies and conflicted selves that produce it. Of course voices can be freer or tenser, but the nostalgia for an entirely uninhibited, untrammelled voice – one that we may only have possessed in our first few days or weeks of life – is a fantasy of life beyond other people, whose own voices inevitably help shape ours.

In fact, the voice's most important role is as connective tissue, binding together a baby and its carers. The infant learns to 'speak' through its cry. Sometime during the second and third month it discovers that the sound it produces by itself has the power to conjure up its mother's presence – its voice can summon people, pleasure, and comfort. From then on the infant's cry is purposeful, and acts as a kind of speech.[37]

Infants' cries are higher-pitched when they're in pain, and those caring for babies respond much more urgently to higher-pitched cries by picking the baby up, holding it, rocking it, and sometimes also stroking it. The greater distress that the baby communicates in its cry, the more immediate the response it elicits in the carer.[38]

Is the baby's cry an evolutionary feature, to ensure that it gets from care-givers the nurturance it needs? Perhaps, but it's also clear that babies learn from experience. Newborns' cries are desperate because they haven't yet realised that they'll produce a response. But already, six to eight weeks later, they sometimes go quiet after a bout of crying, to listen for their parent's footsteps. If they don't hear them, they then resume crying.[39] The cry is no longer simply a reflex – now it's begun to be one side of a conversation. Within a couple of months of being born, babies leave a space for a response to their cry: they've learned the art of turn-taking.

Yet even here, personal and social factors – a mother's own early experience as a baby, fashions in child-rearing – modify the circuit of cry and response. (As does the amount of crying. Parents exposed to long bouts of crying often feel hostile towards their baby and their distress can end up being greater than the baby's.[40]) So babies' cries, as well as helping to elicit the care they need, also reveal something of the social world into which they've been born.

## THE FIRST 3 MONTHS

To speak, you need to control the movement of the larynx, glottis, soft palate, jaw, lips, and tongue, as well as be able to synchronise the respiratory cycle with the activity of the vocal cords. Saying,

'Hello, how are you?' alone requires the coordinated use of more than 100 muscles. It's scarcely surprising, then, that newborns can't do it,[41] especially since their vocal tract is very different from adults' – more like that of a non-human primate. Babies, as chapter 2 described, re-enact in their vocal development the journey that their own species has taken, lending something epic to vocal development.

Yet despite their anatomical limitations, newborn babies arrive in the world with astonishing auditory, vocal, and intellectual capacities. Thirty years ago neonates were regarded by developmental psychologists as an almost alien species, lacking the vocal skills and abilities possessed by their parents and older siblings. But it's now clear that newborns come not only with a set of innate abilities that allows them to communicate from the very beginning, but also with a ferocious desire to connect vocally with other human beings, which, if it's reciprocated, turns them rapidly into quasi-speakers.

Despite their immature vocal tract, even the tiniest premature infant, born at 22–24 weeks' gestation, can vocalise.[42] In one experiment, a 42-minute-old baby was able to copy mouthing movements – truly an innate ability.[43]

And these talents also develop at dizzying speed: when they're still under 3 months of age, babies can take part intelligently in 'proto-conversation' with an attentive, loving parent.[44] At about 4 weeks, whether they're born in San Francisco or Shanghai, Oslo or Pretoria (or even deaf[45]), babies first gurgle and coo, mainly in response to a voice. Cooing allows them to discover the particular auditory consequences of certain mouth and lip movements:[46] they're developing a map that connects the mouth shapes they make with the sounds that they hear – one which, as we'll see below, will allow them to develop a fabulously precocious understanding of vowels and consonants.

From around 6 weeks, a 'parent's inviting vocalisations'[47] evoke matching sounds from the baby, and together they can produce 10–15 exchanges.[48] Research using intonagrams, which measure the pitch, duration, and volume of vocal cues, found that by 2 months infants produce intonation patterns similar to those of adults.

Astonishingly, this can take place even earlier. When the falling intonation contour of a mother's 'Happy boy' to her *2-day-old* son, Peter, was compared with his, the change in his tone was remarkably similar.[49] Before they can express themselves through language, babies communicate their desires and intentions through intonation and gesture. Via pitch and intonation, they're 'learning how to mean'.[50]

But the learning process is crucially dependent on their environment – as we've seen, humans can't learn to vocalise in acoustic isolation,[51] suggesting that, despite the evidence that our ability to vocalise is innate and 'hard-wired', human infants need to hear voices from the people around them to develop their own.[52]

## 3–6 MONTHS

Crying now subsides, and is replaced by non-crying sounds (except in deaf infants, who can't hear their own voice). Within a few months of birth, spontaneously and incessantly, human infants begin experimenting with speech production, sampling most of the range of sounds that they'll later need in order to speak. No other mammal produces even a fraction of the kind of original vocal play that human babies do.[53]

At 3 months, the larynx begins its slow descent to the (lower) adult position. It finally arrives there between the ages of 3 and 7. (A second descent takes place at puberty). Other dramatic changes to the infant's vocal tract take place in its first 6 months:[54] the palate is lowered and moves forward, so that it can close off the passage of air into the nose. The tongue lengthens and its muscles get stronger, while the opening of the pharynx allows it to move from front to back. The result is that, at 5 months, babies gain control of their breathing, and can use their larynxes more or less like adults.[55] Since they also have greater neuro-motor control, they're able to modulate the pitch, duration, and volume of their voices.[56]

## 6–8 MONTHS

Requests now begin to appear in babies' repertoire of intonation, followed by calling and commands. All these patterns are still being conveyed entirely vocally, rather than through language.[57] Babies of this age can also squeal, growl, and blow raspberries, all in preparation for babbling. With vocal folds just 3 millimetres long they can achieve an extraordinary range of volume and harmonising or distorting sounds.[58]

From 6 months babies start to become skilled imitators of the intonations of the people around them.[59] When a group of 7-month-old babies heard a woman speaking in high and then low pitch, the average pitch and duration of their own sounds changed.[60] Similarly, babies have been heard adjusting their pitch to that of the parent talking to them.[61]

Between the 7th and 10th month – whatever language the child's parents or carers speak – 'canonical babbling' sets in. Although you can get babies to babble more by responding to their vocalisations with your own sounds,[62] babies everywhere (no matter how vigorously they're encouraged or discouraged) begin babbling at roughly the same age.[63]

Babbling used to be considered a random collection of sounds, bearing no relation to words.[64] Now, with its repetition of strings of vowels, consonants and syllables, babbling has been identified as the precursor of speech.[65] Intriguingly, babies babble more out of the right side of their mouths (although we don't normally notice this because our brains correct it), which suggests that babbling engages the language-processing centre in the left hemisphere of the brain.[66]

So babbling seems to be a kind of rehearsal for speech, a playing with its components. The babbling child happens upon combinations of syllables that figure in actual speech, and over time these accidental syllables become transformed into intentional ones. Remarkably, although canonical babbling begins later in deaf children, and is different in duration and timing to that of hearing children,[67] deaf children produce 'manual babbling' at the same time as hearing children's vocal babbling.[68] Babbling is uniquely human.[69]

Whatever the language spoken by their parents, all babbling babies tend to make the sounds 'b','p','m','d', and 'n' first. The reason, according to an ingenious explanation, is that, when their lips and mouth are pressed up against the mother's breast or feeding bottle, the only sound that babies can make is a slight nasal murmur. As a result this sound becomes associated with food, and soon the baby begins to make it whenever food or the breast hasn't arrived, or whenever any desire is ungratified, so that gradually it becomes a term for the object of the baby's greatest longings – mother.[70] In this way what starts out as simply a by-product of anatomy quickly acquires (Pavlovian) associations, then becomes transformed into a phonetic representation, and ends up a near-universal word for the most important person in the infant's world.

Babbles may be quasi-words, but they're phonetically similar in English, French, Thai, Chinese, and Dutch – at this stage babies still have their own Esperanto.[71] But by 8 to 10 months, babbling has become much more language-specific.


## 12 MONTHS

During their first year of life, infants gain control of their pitch levels.[72] By now babbling increasingly resembles the baby's own native language, and even imitates the length of its words. Single-syllable words are also starting to appear. At the end of 'infancy', toddlers are beginning to sound like native speakers of their language.[73]

## VOWEL MOVEMENTS

Voice is often contrasted with words, yet the human voice plays a tremendously important role in helping us become verbally proficient. From a remarkably early age, babies can imitate spoken vowels (as well as sung pitch). In one study, babies at 12 weeks old tried to mimic a woman making three vowel-like sounds, different in volume, length of time, and pitch, even though they were anatomically incapable of producing the full range of sounds themselves.[74] In fact it only needs *five minutes a day of exposure to a specific vowel for three days* to get infants under 20 weeks of age responding with matching sounds.

Where vowels lead, consonants aren't far behind. Even younger babies, ranging from 24 hours to 7 days old, can make the right mouth openings for consonants after seeing them made by an adult,[75] while 1–4-month-olds can discriminate between a 'b' and a 'p' sound. Despite their limited experience, babies are able not only to make fine discriminations between speech sounds but also, it seems, to hear them in a very similar way to adults.[76]

And they learn quickly to connect sounds and shapes. At 4 months babies already seem to know that an 'a' sound goes with a face mouthing an 'a', and an 'i' vowel goes with a face mouthing an 'i'.[77] And they can spot the similarities across a wide range of speakers – high-pitched, low-pitched, male or female, even those with a heavy cold. Not even computers or algorithms can do yet what 2-year-olds or even newborns can – disregard the acoustic differences between speakers.

And yet when they need to, they can do the opposite – remember individual voices and use the acoustic differences between them actively to help them learn to talk.[78] When they hear words again from stories they've heard a fortnight before, they listen longer if they're spoken by the person who'd told them the story. This is probably because they've retained information about the characteristics of that same speaker's voice.[79] Far from filtering out the voice as some bothersome additional information, babies are able to use it to help them recall the words.

## LEARNING THROUGH LOSING

In the course of their first 6 months, linguistic experience has a profound effect on babies' ability to distinguish sounds, but not in the direction one might expect. At birth, no matter what language community they're born into, all babies perceive phonetics in the same way. As time passes, however, the range of sounds they hear *diminishes* until, at 6 months, they can only hear properly those that are salient in what will become their mother tongue. So while 6- and 8-month-old Japanese babies can hear 'r' and 'l' in the same way as English babies and adults, from around 10–12 months they no longer differentiate between the 'r' and 'l' because in Japanese those sounds belong to the same category.[80]

Similarly, 6–18-month-old babies reared in English-speaking households can hear syllables that are distinct in Hindi but not in English – syllables that adult English speakers can no longer differentiate[81]: their phonetic perception has been altered by linguistic experience. In some real phonetic sense, therefore, growing up entails loss. We become deaf to certain sounds: in order to master one language, we have to lose our sensitivity to all of them.[82]

Something parallel seems to happen with pitch. Babies begin life with absolute or perfect pitch, and between 3 and 6 months can imitate pitched tones.[83] At 8 months they can detect slight changes in a musical sequence, a capacity that they lose over time (adults tend not to notice such small changes, and where they do, use relative pitch to do it instead).

Losing absolute pitch brings advantages as well as disadvantages. Pitch helps provide us in those early months with a map of words. And yet, as we develop, the information it provides is too fine-grained to be of use in daily life, except to speakers of the world's tonal languages (like Mandarin, Cantonese, Thai, and

Vietnamese) where absolute pitch helps you hear the subtle differences in similar-sounding words.[84]

As infant-pitch researchers have pointed out, 'It may be that
pitch-matching is *usual* in babies and that, for most of them, the
ability is lost with the onset of language or *with the failure of many
environments to support its continuation* (my italics).'[85] By this
reckoning, vocal sensitivity isn't something that needs to be
learned, rather not lost. Though the panoply of vocal and auditory
talents with which we're born diminishes as we age to enable us to
learn our mother tongue, if their importance in human communication were recognised, we might be able – with proper encouragement – to hold on to at least some of them.

## THE MELODIC LESSON

Babies are born, it seems, with a capacity to focus on the rhythm of
speech, especially rising and falling intonation, and stressed and
unstressed syllables. These help them to divide up the stream of
speech into units – sentences, words, and syllables – long before
they understand what those components mean. (Motherese, as
we've seen, also eases the task.)

By between 7 and 9 months babies appear to be using some kind
of 'statistical' learning to help them break up speech into individual
words. Having worked out that the majority of words in English
begin with the stress on the first syllable, they then make use of this
acoustic cue to help them identify where one word ends and
another begins.[86] Similarly, by registering that falling pitch, length-
ened syllables, and a pause signal the end of a clause, they're able to
slice up a sentence into clauses. So babies exploit the melody of the
language as a grammatical tool. Prosody helps them begin to
understand the linguistic components submerged beneath the
rapids of speech.

Of course their prenatal life has primed them to this task. In
some real sense they've already been tutored by cadence, by the
prosody that crosses the abdominal wall. The rhythm of speech

might be the bridge connecting their prenatal with their postnatal lives.[87]

## GOING NATIVE

Babies' ability to discriminate between different languages is awesome. Newborns open their eyes most widely at happy expressions *in the language to which they've been most exposed prenatally*, whatever it is. They can only do this because they've detected the distinctive prosodic features of different emotions expressed in their native language.[88] They've learned *in utero* not just what happiness and anger sound like, but how happiness or anger sound like in the language that their mother speaks.

More remarkably, *babies can detect differences even between languages that they've never heard*. If you filter sentences in different foreign languages, newborns can differentiate between those languages purely on the basis of prosodic information. So newborns from French-speaking families can tell the difference between English and Japanese (but not between English and Dutch, which have similar rhythmic properties).[89]

Two-day-olds, it seems, already have distinct linguistic tastes, preferring their native language to any other. Through sucking they show that they want to hear this rather than a foreign language, perhaps (once again) because they've got used prenatally to the intonation patterns characteristic of their native language.[90]

## BORN TO TALK?

Are these skills innate, or the product of a formidable capacity to learn? The behavioural psychologist B.F. Skinner believed that infants are a tabula rasa, and learn to speak in the same way as rats learn to push a bar. Noam Chomsky, on the other hand, proposed that they arrive with innate grammar. Increasingly researchers into babies and speech argue for something more subtle – that language is innate but modified by experience, a form of 'innately guided learning'. What's innate isn't a universal grammar or phonetics but 'inherent perceptual biases' that place constraints on perception and learning. Babies arrive as 'citizens of the world', 'with abilities highly conducive to the development of language',[91] but the 'neural commitment' to one language that they make very early on shapes the way in which they process linguistic information. So simply by listening to language – to the voice – infants acquire sophisticated (even statistical) knowledge about its properties, and detect patterns.[92] What's given by nature interweaves with what's gained by experience.[93]

(This is why learning a second language is hard. 'Henry Kissinger was not born with a German accent, nor Chomsky born with a Philadelphian one. These are not innate characteristics; once acquired, however, they have persisted over decades.'[94])

One pair of researchers has come up with a particularly elegant formulation. By the 6th month of gestation, they argue, a baby's peripheral auditory system is fully functioning, but connections between different areas of its brain haven't yet been formed. We know that the human voice penetrates into the womb, so perhaps at this formative stage the neural substrate is becoming organised to respond preferentially to sounds that could be produced by the human vocal tract. Hearing the voice and speech from such an early stage, according to this exciting view, modifies the brain, remodelling it both before and after birth in such a way as to make it particularly sensitive to the human voice, and to allow the child itself to vocalise with increasing control.[95] So babies' brains might actually be altered by hearing adults' voices, with the brain and experience *working together* to produce both the remarkable discriminatory skills that we've seen as well as the child's own emerging voice.

## PLAYING WITH WORDS

The voice helps us in the serious business of acquiring language, but it gives babies and children something else that's vital –

pleasure. Speaking is a sensual business. Vocalising and speech develop from sucking; they're another aspect of orality.[96]

The early years are permeated by vocal fun. A kind of labial play occurs between parents and infants, long before more obviously recognisable play begins. It takes the form of growls, giggles, and acoustic jokes, of games of imitation and response.[97] When we play with our babies, we adults are – briefly – able to recover some of the labial pleasure that we experienced as infants. Indulging in vocal silliness, we open and stretch our mouths beyond the usual adult limits. In vocalising with our infants, we can become playful babbling babies once again.

PART TWO

# 8

# *Do I Really Sound Like That?*

IN THE REAMS of academic research on the human voice, remarkably little has been devoted to people's feelings about their own. With a few exceptions the comments of a voice specialist in 1955 – 'The psychology of the voice has been, for the most part, treated from the standpoint of the listener'[1] – remain true today. What makes this all the more strange is that an individual's speaking tune is not only a crucial expression of their personal and social identity, but also helps to create it. We *are* our jokey cadences, our odd stresses, our fluting tones. Bogart's 'Play it, Sam', Diane Keaton's 'Annie Hall' 'Lah di dah', are as famous for their inflections as for their words. So potentially sensitive an instrument is the voice that a psychotherapist has described how her singing teacher 'could determine my general personality and various mood changes, plus intrapsychic conflicts that unknown to her I was then working on in my own analysis, all according to the way I was singing that day'.[2]

Our voice gives birth to our thoughts: we use it to think with. Vocalising makes words and concepts concrete. 'Isn't it extraordinary how ideas occur to me when I'm talking to you, and not when I'm thinking by myself?' a patient remarked to her therapist. Another commented, 'Talking helps. It gives my thoughts air. They were smothering inside.'[3]

The very act of creating audible sounds and hearing them is a powerful experience that helps to distinguish 'me' from 'not-me'. And just as our voice moves physically from the interior of our body to the exterior, so it's also an important way through which our internal life is externalised. The voice bridges the inner and the outer, the self with the other, and (through public speaking and broadcasting) the individual with the wider world. For some people the very act of using their voice, its kinetic attack, reminds them that they're alive.

We're both speakers and hearers of our own voice and most of us react with horror when we hear recordings of ourselves. As a 47-year-old man – a professional opera singer – put it, 'I hear a much darker, richer sound in my speaking voice and then when I hear it back it sounds tinny, like I'm speaking two octaves higher.'[4] At a workshop on the voice where everyone was asked to say something about themselves, the greatest mirth was provoked by a man who announced that he loved the sound of his own voice. He came after dozens of people had confessed to hating theirs. This is partly because we hear our voice not through the air like other listeners, but conducted through the bones of our skull and the distorting vibrations conveyed by the Eustachian tube.[5] Cheap recording equipment, and even some of the newer digital answering-machines, make the voice sound thinner and less resonant than it really is.[6]

But almost forty years ago a pair of researchers advanced another, more persuasive theory. The most disconcerting element of listening to one's voice, they claimed, is that one 'hears not only what is unfamiliar as his voice . . . but also what is quite familiar in his voice . . . part of the disturbance that people experience when they hear their own voices is accounted for by the unsuccessful or incomplete editing of aspects of themselves that they did not consciously intend to express and which they now hear in the recording'.[7] Freud talked of 'the return of the repressed' – buried desires, impulses, and traumas that find a way of re-emerging through dreams, symptoms, or parapraxes (verbal slips). Another avenue through which the repressed makes its return is the voice.

It was only when I began interviewing that I realised how intimate a thing it is to quiz people about their feelings about their own voice and that of those close to them. I was asking them, in effect, to divulge very personal information: what they dislike (as well as like) in themselves, and how they manage their relationships through the voice.

## A WHINE OF INJUSTICE

Consider this 48-year-old woman, married with three children, and with a successful career:

> I hate my voice because it's harsh and strident and squeaky; that associates with horrid, whiny children. Now with my husband it's my voice telling him – ordering him, in his eyes – to go to the shop, be organised. So the same aspect has turned from the little girl whingeing to an adult, a nagging parent – the same fault somehow magnified. It makes me feel very bad because it's very hard to control.
>
> My kids tell me that when I speak on the phone to clients I have a deep client voice, so clearly I can control it. Yet even when I'm trying to control it with my husband it's still perceived as this high, demanding, strident tone. It's hurtful because I don't want to be a demanding, ordering person.

Where did she think this voice comes from?

> I was a whiny child. I used to come up with, 'It's not fair, why should she always get the better deal?' about my sister. I felt a huge need to get more of my mother's attention than I got. Mum perceived me as whining and demanding from the word go. There's something painful and very deeply omnipresent I hear in my voice that's somehow shameful, something my fault and yet

not my fault . . . It's a constant reminder that whatever you think you've achieved, it's an illusion, I'm just where I ever was – my voice shows me up.[8]

Enshrined in this woman's voice is the dissatisfied girl who was labelled 'difficult' early on. Though there's an adult side to her that's expressed in her work ('the deep client voice'), she can't activate it in her intimate relationships: the little girl insists on making herself heard, the child voice is too powerful.

Above all, this woman recognises that some arrested part of her, some aspect of her psychological development for ever frozen in childhood, is expressed in her voice. It therefore becomes a source of shame and self-blame but also, as she's fleetingly able to acknowledge ('and yet not my fault'), a marker of some ancient unsatisfied need.

An American analyst suggested, in 1943, that 'since the tone of voice is such an important medium of expression in infancy and childhood, it can later readily express unconscious emotional conflicts originating at that time'.[9] Though this woman has been able to bring the conflicts into consciousness, she hasn't yet managed to integrate the different aspects of herself into a cohesive whole.

Our voices often embody a split-off part of ourselves that we haven't metabolised. In the course of another interview, a 57-year-old woman tries several different ways of disowning her voice:

I hate my voice – I sound like Princess Anne with a cold. It's not a voice that I identify with at all, not in any way associated with my personality or who I feel I am – it's just grafted on. I don't know where it came from. I know that my cheekbones come from my heritage but my voice doesn't seem to, it seems entirely artificial. It's a very upper-class voice and there's nothing in my background that should suggest that . . . I think it must have been formed by schoolfriends who had posh voices . . . It also feels false to change it, and I don't have control over it.[10]

Unlike the woman before, who recognised herself all too clearly in her voice, this one feels thoroughly alienated from hers, and believes any other voice she might develop would be equally inauthentic. The phenomenon of 'false self' is well known: perhaps 'false voice' is its acoustic counterpart. A number of the people I interviewed felt this way about their voice. One, a 38-year-old woman, said:

> I used to have a very false self inside a lot of the time and one's voice echoes that . . . My mother and I were as false as each other with our voices; we put all our energy into squashing our anger and power. My father and my sister were much more real in the expressiveness of their voices – very angry and powerful. Perhaps because of that I'm very attuned to whether there's a discrepancy between how people present themselves and how they really are and I think a large part of that is how they use their voice.[11]

Another woman, aged 42, remembered:

> I noticed after my first baby was born that I started talking in this peculiar way, with a slightly funny accent. It felt like when I opened my mouth someone else's voice came out, though nobody else seemed to notice. I realised that this artificial voice was connected with my struggles to see myself as a mother: I felt I had to take on a different persona, become Mother with a capital 'm', and my voice was reflecting that, like I was impersonating someone.
>
> When my baby was around 3 months and I started mothering my own way, realising that I could be me and be a mother, that peculiar voice vanished. Now, very occasionally, I hear that same odd quality – like I'm giving a performance – and I immediately start to think about what's going on, why has my own voice been displaced?[12]

## I LIKE WHAT I HEAR

Vocal self-dislike isn't universal – some of those I interviewed were happy with their voice, less for its acoustic qualities than for what it enabled them to do. A 7-year-old girl likes her voice 'because it can go high and low and make different sounds . . . I like my voice because it can express itself, it can tell people how I'm feeling'.[13]

Almost everyone who confessed to liking their voice did so a little sheepishly, as if to apologise for contravening the social norm. The 7-year-old was sometimes frightened by her own vocal power. 'When I shout, it feels like I'm a big giant and I get to smash all the pieces up in the world . . . I get scared that it might boom me . . . I want to run away from it but I can't, it's part of my body . . . I feel as if it might do some harm to me or others.'[14]

A vital voice can galvanise an entire hall. A 38-year-old woman spoke at a public meeting to save her village school:

> The man who spoke before me knew his stuff but had such a boring voice I could see him lose the audience almost completely. I knew that my most important task was to revive them, and I did it through my voice. It was partly because I was talking about something I felt so passionate about . . . concerning my children's future, and I just allowed all that passion to come into my voice. I could see the audience come back to life – it was a wonderful feeling, and of course my words helped, but it was the attack and energy in my voice that mostly did it.[15]

## THE VOICE AT WORK

We manage our personal and social relationships through our voices. We may not have a shared, public language in which to talk about the voice, but my interviews revealed how much individual private awareness there is about its importance, especially in the world of work. From scenic artist to politician, physiotherapist to judge, people described how they use their voice to establish authority, make themselves approachable, and make it more likely that what they say is heard, absorbed and remembered. We tend to think of certain obvious occupations as relying on the voice, but these interviews convinced me that this category is almost infinitely expandable. Today we're all professional voice-users.

A 51-year-old physiotherapist recognises that he uses his voice differently at work and at home:

> With my patients I have to be able to produce something that is often not consistent with how I'm feeling. I can't come in and just be low-key and tired and depressed . . . the voice has to be imbued with some kind of energy . . . you might have to crank it up a bit. The people I deal with are often deaf or they're at various states of deterioration in their cognitive or mental states such that tone probably conveys more than content.[16]

One commentator pointed out in the 1950s that, 'Certain occupations seem to develop an indigenous speech melody. That is why it is so easy to parody a politician, a teacher, a minister. It is as if the individual wished to hide his particular shortcomings behind the mask of a socially approved stereotyped score.'[17]

A judge, after listening to advocates in court down the years, feels that he now knows what works vocally in the law and what doesn't:

> In describing to the jury an incident of rape you find advocates using almost the same timbre of speech, the same pace, the same modulation at the beginning of the story as they do at the end. They haven't set aside the appalling moment it actually happened, which they could have done easily by altering the pace and gravity of their tone . . . You don't get any stirring or shuffling of papers in the court at a moment like that . . . I think the voice is the most important tool in the courtroom, and the great advocates are those who know how to use it. The ham

actors fail, and those unable to create any atmosphere with their voice fail.[18]

He himself, when sending repeat drug offenders to rehabilitation programmes, uses a warm and soothing conversational tone, but is deliberately cold and flat if, after breaking an order, they return to court again for sentencing.[19]

A British Member of Parliament worries that her public voice doesn't convey the passion that she feels. 'I tend to have a very controlled and rational voice. I'd like the passion to come through but I also don't want to bare my soul with the outside world. I suppose your voice functions as the guardian of that private space.[20]

## THE PLACATING VOICE

Most of the people I interviewed were conscious of the role played by their voice in that private space. A father described what happens when his 11-year-old son 'loses it':

> He gets himself slightly out of control, and you'll see him in the middle of it looking for some way to get out of it, and he doesn't know how to so you have to offer him lifelines – just reduce the temperature, give him the time and space to calm down, and you do that partly through the voice. I'll often really lower it and sound quite sympathetic or conciliatory, just so he doesn't feel he's being confronted.[21]

Many of us try to regulate problematic relationships at least partly through the voice. A 38-year-old woman observed that, in talking to her younger sister, 'I've got to control my voice – there's a sort of strangulation in the throat where I'm somehow trying to curb what I'm thinking.'[22]

In conflict the voice can be an incendiary device, setting off explosions between warring couples or parent and child, ratcheting up the levels of hostility, or it can soothe and defuse. The voice of the UN General Secretary, Kofi Annan, seems to embody the institution's mediating, tension-lowering aspirations in its levelness and solidity. A 36-year-old woman feels:

> fairly certain that somehow my voice developed as a way of controlling the level of expressed emotion and tension in the family. When I began working with groups of people who were violent and out of control, as an adult, I heard my calming voice again – saying what I wanted to say, but in a way that didn't inflame anyone's emotions – and realised that this was how I used to speak in my family, trying to control explosiveness in the others, and also of course in myself. I'm good at calming down groups where the tensions are getting high, but I sometimes squash the potential for argument between people that they need to have – so it's a skill, but it's also a deficit.[23]

## OUR SPEAKING SELVES

We don't just have one voice but many – a 15-year-old girl immediately identified five of hers:

1. Talking to people I don't know on the phone – here I use a formal, carefully enunciated voice.
2. When I'm babysitting, I'm trying to relate to the children, and to sound as much like them without being patronising. But when I'm angry with them, I try and put on an authoritative tone.
3. Talking to people I don't really like and am trying to ignore, I'm, yeah – monotonous, without much energy in my voice. My parents phoned me on my mobile recently when I was at a friend's house, and they were trying to get me to come home earlier, and I was talking in such a monotone that all my friends were falling about laughing.

4. When I'm having a deep conversation with someone on a
   difficult subject to broach, I might drop back to my low voice,
   but with a lot of energy.
5. When I'm shouting at my mother – well, I shout. But then so
   does she. She sounds like a posh hawk when she shouts, and
   she says I sound like a battering ram.[24]

Some kinds of voices are shamefully pleasurable to use, if hard to
admit to – like the 'hobby-horse' voice or rant, where you talk at
others rather than to them. A 36-year-old said, 'Sometimes the
slightly manic voice in me takes over. It loses the capacity to be in
tune with other voices. I feel embarrassed about it . . . but actually
when I'm in it I quite often enjoy being in it because it feels
powerful and energising.'[25]

The sound of our own voices, as one study found, raises our
blood pressure, even if no one else is present.[26] Talking is quite
literally an exciting business. A 49-year-old man admitted that:

> sometimes when I speak I bore myself, but other times when I
> start speaking I really fascinate myself and then you can't shut me
> up. It can be like a runaway train, my voice, and you know, one
> of the things I have trouble with is volume control, at least
> according to some people around me. My wife and daughter are
> embarrassed by it . . . but there's nothing more pleasurable than
> letting my voice go . . . it makes me feel not small, not depressed,
> not withdrawn . . . I have this explosive way about me in a lot of
> situations that will become vocally expressed.[27]

This man's explosive voice, he believes, once saved his life. Having
just returned from work, he was lying down in his New York
apartment at five in the afternoon with his shirt off:

> Our apartment had very noisy floorboards and I heard what was
> the unmistakable sound of a human footstep in the house . . . I
> made a very quick decision that the best policy was an aggressive

one. So I decided to charge out of the room and face this person but I realised there would still be a distance between me and him and, rather than actually mount a physical charge, I would use my voice as my first line of assault. And I came out with the loudest sound I could make, saying, 'What are you doing here?' Even as I turned the corner . . . I saw that he flinched and stepped back. And in that moment, I realised that the effect of my voice meant that I had the upper hand. He tried to speak himself but I kept overwhelming him with this loud voice as I crossed toward him . . . and I grabbed him by the collar . . . he was fairly large but I was able to push him back out of the apartment.[28]

Non-Western cultures acknowledge the impact on a speaker of speaking. The Ojebway Indians won't even say their own name, believing that to do so would stunt their growth. They don't mind other people saying it – it's simply the owner who can't. According to the anthropologist James Frazer, 'When a man lets his own name pass his lips, he is parting with a living piece of himself.'[29]

## TALKING TO ONESELF

Spending a lot of time alone can lead one to forget the actual sound of one's own voice, and hear only the inner one. Going for a long time without speaking, one loses touch with the self-in-the-world carried by the voice. A voice needs a listener. Although you can talk to yourself, you then have to be both speaker and listener.

The answering-machine finesses these torments. The comic writer Michael Frayn has described recording an answering-machine message:

'Hello', you begin, certainly – but as soon as you've said it you realise you haven't said it in the usual way. There was no upward inflection, no note of query. Your voice fell instead of rising. You have no sense of an audience. You know you are talking to yourself, and you have begun to feel rather foolish . . . You are not speaking politely or impatiently, confidently or cautiously. You are speaking slowly and carefully . . . you are speaking portentously . . . you seem to believe you are making a statement which may be used in evidence in some future court case. You are broadcasting a last message to the world from the besieged city. You are speaking to posterity.[30]

The caller, meanwhile, is in a lather of embarrassment of their own, unsure whether the beep has already sounded, whether to speak with all the normal conversational cadences used when someone's actually there (a pointless performance – an eavesdropper can almost always tell when someone is speaking into an answer-phone rather than to a living person), or whether to adopt a special answering-machine register. Despite their ubiquity, *there's no good way to leave a message on an answer-phone*. However much we try and coax our voice to sound natural, the knowledge that the machine's owner can replay the message with all its slips and pretensions induces in most of us a fatal degree of self-conscious-ness.

It's a similar experience talking to someone who's had a stroke or who's in a coma: you hear your own voice prattling on normally in the hope – as the evidence suggests – that the other person can hear, but without getting feedback. You speak as if there's been a response – a duologue for one.

Voices need regular exercise. Unused, they rust up and creak. (People who live alone in isolated places or speak little develop shorter expiration times.[31]) Those who don't speak a language for a long time lose the oral sensations accompanying it – the moues and trills particular to each language. Returning to it is like kissing an old friend.

One of the greatest boons of having a pet, surely, is that it allows one to talk to oneself. Since we're the first hearers of our own voice, the vocal endearments we lavish on our pet grace our own ears too. Having a pet and petting it activates a loving voice that cossets not just the pet but also its owner.

## DEVOICED

In 1900 the 18-year-old Ida Bauer, who suffered from recurrent aphonia or loss of voice, reluctantly agreed to be treated by Sigmund Freud in Vienna; as Dora she became the subject of probably his most famous case history.

Dora had been the object of sexual advances by Herr K., a family friend. When she was 14, Herr K. had 'suddenly clasped the girl to him and pressed a kiss upon her lips',[32] creating a 'violent feeling of disgust' in the girl.

Dora's hysteria, Freud concluded, was a 'displacement of sensation': when Herr K. pressed himself against her, instead of feeling the pleasure that Freud thought a healthy girl would, 'it was dismissed from her memory, repressed, and replaced by the innocent sensation of pressure upon her thorax, which in turn derived an excessive intensity from its repressed source'.[33] This was 'a displacement from the lower part of the body to the upper',[34] a way of transforming her own covert sexual feelings towards Herr K. (along with her fantasies about what her father was getting up to with Frau K., with whom he was having a sexual relationship) into disgust, the conversion of a psychic experience into a physical one, and an oral experience – the attempted kiss – into an oral absence.

Though 'Dora' became a honey-pot for feminist debate about sexual difference and desire,[35] it wasn't the first time that the mouth and throat were identified as a site for the expression of trauma and hysteria. In 1880 Josef Breuer, Freud's collaborator and co-founder of psychoanalysis, had treated 21-year-old Bertha Pappenheim (known in Breuer's case history as Anna O.) for recurring loss of speech and disordered speech due to anxiety.[36] As a result of 'deep hypnosis' Breuer elicited streams of material from her unconscious, which, he argued, had produced the symptoms: 'Finally her disturbances of speech were "talked away".'[37]

Together, the cases of Anna O. and Dora represent the beginnings of psychoanalysis, and it's fascinating that both involve the voice. But what's curious is how little interest Freud and Breuer showed in the voice per se. Freud seemed concerned exclusively with what Dora converted *from* rather than *into*. For him the symptom remained a symptom. His interest lay in how the unconscious spoke through the body, rather than how the body itself spoke; in the voice's absence more than its presence.

What makes this all the more bizarre is that, together with Breuer, he was the inventor of the 'talking cure' (Anna O.'s term for the therapeutic process).[38] Remarkably, he developed a way of accessing the unconscious ultimately through the medium of the voice. (Ironically he was himself to die of cancer of the jaw.) And yet Freud paradoxically seemed to be much more interested in visual hysterical symptoms than vocal ones. The index to the standard edition of his complete works lists one single entry for 'Voices', in contrast to over 100 for disturbances connected with vision. On the subject of vocalising, it seems that Freud himself in some sense suffered from loss of voice.

## SILENCED

And yet Freud and Breuer established an important association between vocal problems and psychological states. Today vocal problems like aphonia (loss of voice) and dysphonia (voice disorder) are routinely traced back to states of emotional tension that put pressure on the body, and 'psychogenic' loss of voice, caused by underlying psychological factors, has attracted a considerable literature.

Much of it is frighteningly crude – moral judgements parading as science. Indeed there are pejorative attitudes enshrined in the very language of loss of voice. It's commonly linked with 'the hysterical personality',[39] 'the neurotic personality', and other 'personality disorders'. Non-organic symptoms like hoarseness and pressure on the throat are often known as 'globus hystericus',[40] and children suffering from vocal strain or (hyperkinetic dysphonia) have been labelled aggressive and immature.[41] In these circumstances it's amazing that they have any voice at all.

In reality the relationship between psyche and voice is far more complex. A more sensitive approach sees voice loss as something precipitated by conflicts associated with anxiety, fear, and aggressive feelings.[42] A 38-year-old woman described how, within two months of becoming head of department (a promotion she'd been striving for), she lost her voice. 'I was absolutely terrified about being that powerful and having that degree of responsibility in such a grown-up job. Losing my voice seriously interfered with my status, power and position – effectively I used my voice to demote myself.'[43]

The voice both reflects and mediates our relationship with the outside world, and can be used to express attitudes and feelings that would be derided or dangerous if articulated through words. Losing one's voice can be a way of going on strike, a withdrawal from the social world, sometimes deliberately. Analysts distinguish between elective mutism, where a person is able to speak but chooses not to, and the traumatic mutism that follows shock or injury. A psychiatrist who treated an elective mute boy observed that he 'found not speaking the only way he could find a voice in his family. He spoke to me first, saying that if he spoke to his family they would stop listening. He found a way of speaking to them through me and then totally regained his speech.'[44] The singer Shirley Bassey experienced traumatic mutism, temporarily losing her voice after the tragic death of her 21-year-old daughter. 'It was the combination of the guilt and the grief and the nervous strain. I was grieving through my vocal cords.'[45]

Trauma can rupture the circuit that makes up the vocal process, disturbing the boundary between inside and outside. Making sounds is an act of trust: to allow the intake and expulsion of air you must open up the body. A traumatised person finds such openings too risky. Psychotherapists and voice teachers often work with the 'crashed voice' to resolve deeper psychological issues or psychic trauma.[46]

## LOSING IT

A person who loses their voice permanently is excluded from normal social life just as surely as the immobile are by a flight of steps. After surgery to remove his cancer left him effectively voiceless, the British journalist and broadcaster John Diamond wrote eloquently about its social consequences: having to point at things in shops like a tourist,[47] relishing the Internet because there 'I am just as articulate as I ever was',[48] and dreaming that he hears himself using his old voice.[49]

Even more disorientating was the ontological effect of losing his voice. If we are what we sound like, then the loss of that sound diminishes some core aspect of the self. 'In not being able to talk, I am not me. Or, at least, not the me I think of when I think of me.'[50] Diamond lost not just the capacity to speak but also his former ability to speak with great fluency. 'Thus I am forced to entertain the unwonted thought . . . would the people I love love me, know me, have taken trouble with me, if this is how I was when they first met me?'[51]

## MUTING ONESELF

But we devoice ourselves daily in far less dramatic fashion. Since our emotions leak so easily into our voice, speaking is an intrinsically revealing and potentially dangerous experience. Most of us use a variety of strategies to protect ourselves from exposure. Are Prince Charles's tortured circumlocutions and near-stutter connected with decades of paternal criticism? The actor Mel Gibson said, 'I . . . tend to use the bottom register of the voice a lot . . . It's a security thing – you don't want to express yourself vocally too much to other people.'[52]

An uninflected voice often serves as a psychological defence, the dirge and drone of uniform pitch protecting a speaker from the risks of emotional display. A 42-year-old man complained that his father-in-law's voice is 'so monotonous that just five minutes of listening to it makes me lose the will to live. I know that he had a tough childhood and probably learned to protect himself with this very level, unmodulated voice. But hearing him makes me as depressed as he must be, to be able to produce such a lifeless sound.'[53]

Vowel movements are as much of a discharge as bowel movements: a person can be vocally retentive just as much as anally retentive. The fear of letting the voice out has been called 'phonophobia'.[54] 'For many people the fear of being too loud or emotionally committed creates a common habit of pulling the vowel back in moments when volume is required. The vowel starts on its natural pattern of release but then is denied and trapped by either swallowing the sound or clenching the jaw . . . All the energy stays bottled in the throat.'[55]

This form of swallowing the words sounds as if the speaker is attempting to retrieve what they've just said, or can't fully commit themselves to it.[56] Of course there are powerful social pressures that often inhibit such commitment, as an American voice teacher has suggested. 'How many times people have said, well, I can't talk to you when you're feeling so upset – just calm down and we'll talk about this,' as if it's wrong to speak with feelings.'[57] A 36-year-old woman remarks:

> I'm a pretty volatile person myself but when I hear my husband shout at the children or me it drives me mad because he doesn't actually let his anger out: it rises up but then he shouts it back down into himself so that he sounds like a bad imitation of Hitler. And that's what I find so infuriating: he seems to be letting it all out by shouting, but the actual sound of the shout locks it back in again, which makes it much more frightening. Perhaps that's why his shouting doesn't clear the air.[58]

This is the Basil Fawlty school of rage – simultaneously released and repressed. The mismatch in harmonics between this husband

and this wife's voices might be contributing an additional, hidden difficulty to the way they resolve conflicts.

Creating small, often inaudible and unresonant sounds can be a method of grovelling, or might be the result of having had your voice suppressed in childhood. This reduces not just the volume of the voice but also its energy: the vocal folds quiver rather than vibrate. 'Devoicing disconnects the speaker in the throat, making it hard to express levels of emotion and sound truthful . . . Imagine being defended by a lawyer who devoiced!'[59]

An American voice teacher describes it like this:

> There's this 2-year-old who comes into the kitchen with a life or death need for a chocolate-chip cookie and says, with all the passion in her little soul, 'I want a chocolate-chip cookie, I want it right now, I have to have it.' And the mother/father/carer says, 'Well, that's not a very nice way to ask for a chocolate-chip cookie. I don't give chocolate-chip cookies to little boys or girls who don't know how to ask for it nicely. You go away and come back when you know how to ask for it nicely.' So later that day, the child comes in. Their need for a chocolate-chip cookie is even deeper because it wasn't satisfied before and they're about to speak with all the passion in their heart, and in taking that breath, they remember, you know, nice little girl, and they go, 'May I have a chocolate-chip cookie, please?' And their voices get stuffed into this tiny high little place, which probably is a lot to do with the tongue muscles bunching up. And so what's happened is that their initial passionate impulse, their primary impulse, has been redirected into a secondary, maybe even a tertiary expression.[60]

If the mouth is a gatekeeper, then sometimes the lips are recruited to help prevent emotions from exiting. You can hear it among speakers with tears-in-the-voice, who sound as if sobs are trapped in their vocal tract, a characteristic often detectable in frequent smilers. The woman who thinks she sounds like Princess Anne with a cold may have painful feelings that are literally stuck in her throat. With little prompting she volunteered that, 'I speak from the base of my throat – I can feel that's where the strongest vibrations are.'[61]

Of course we all do it, this repudiating with our voice. An aunt gives me a brooch for my birthday similar to half-a-dozen other brooches that she's forgotten she's given me down the years. 'Fantastic,' I respond, the 'Fan' exploding with excessive enthusiasm, a sharp drop of energy on the 'tas' as I hear my enthusiasm, note that it sounds false, and try to reduce it. But my voice plummeted so dramatically that she will have detected it too, and I must decide whether to soar again in another, even more rhapsodic burst of fake joy or plump for a medium-level expression of delight on 'tic' and hope that this will convince. That's what I go for, but by now I've so distorted my rhythmic sense that I hold on a fraction too long, so it sounds like I'm saying not 'tic' but 'stick' *because this is what I'm actually doing*, 'sticking' embarrassingly to the syllable.

The coda: a few days later I phone her and say that, beautiful though the brooch is, I have quite a few already and could I change it for some earrings. Comes the reply, 'I thought you didn't like it. I could tell from your voice.'

## THE HEARD VOICE

Finding one's voice – speaking out for the first time on a subject that preoccupies or impassions one – is a powerful experience, with the capacity to alter one's view of oneself and one's place in the world. Many people believe that releasing emotions through the voice can be healing.

Being heard can have a similar effect, and can change the actual sound of someone's voice. A 47-year-old man says that, when he's really being listened to, his own voice calms down and gets less angry – it no longer has to take on the world. 'We're all screaming to be heard, and when you occasionally are, it dawns on you, oh

OK, I don't have to scream so much.'[62] A heard voice is rarely an ugly or a whining one, even to its owner, and a heard person is less likely to feel alienated from their voice, or believe they sound like Princess Anne with a cold (unless they actually are).

So powerful is the alchemy of being listened to that it can transform the quality of one's own listening – reciprocity drives the cycle of communication. A 48-year-old scenic artist described a new person she'd just hired. 'He really listened, he heard and he anticipated and that's an unusual quality. That then made me much more attentive to him. It's an interesting thing that, when someone hears you, you listen back much more carefully.'[63]

# 9

## How Our Emotions Shape
## the Sounds We Make
## (and Other People Hear Them)

DAVID BLUNKETT, the former British government minister, is blind, yet in official meetings he would pick up all sorts of extra information, he believes, from changes in the tone of voice of participants. 'Sometimes I can detect what I call a shuffling silence and can tell that people aren't happy.'[1]

Many people will attribute Blunkett's skill to the enhanced hearing that's popularly assumed to accompany blindness and somehow compensate for it, as if nature were so even-handed that for each sensory deficit it bestows a matching sensory gain. But this is hokum:[2] it suggests that the aural only really evolves in the absence of the visual. The myth of the blind person's remarkable innate hearing is just a way for lazy listeners to feel better about our indolence (and other people's disabilities). Blind people don't hear better – they just listen more. And their hearing is easily idealised: attracted to a publisher after hearing her on a BBC radio programme, Blunkett was to find that his affair with her ended disastrously, probably proving that no one – sighted or blind – should initiate a sexual relationship on the basis of the sound of another person's voice.[3]

Aphasics, whose brain damage prevents them from understanding words, are also said to have a superior sensitivity to 'feeling tone'.[4] When a group of aphasics listened to a speech being made by President Reagan, the so-called Great Communicator, they fell about laughing because they detected his histrionics and false cadences. 'They have an infallible ear for every vocal nuance, the tone, the rhythm, the cadences, the music, the subtlest modulations, inflections, intonations, which can give – or remove – verisimilitude to or from a man's voice.'[5] When aphasics lose their understanding of speech, 'Something has gone . . . it is true, but something has come, in its stead, has been immensely enhanced, so that – at least with emotionally laden utterance – the meaning may be grasped even when every word is missed.'[6]

Potentially we each possess the skills of an aphasic. Most of us have an intuitive, post-Freudian sense of the intimate relationship between voice and psyche, embodied in words like tongue-tied, stiff-upper-lip, or lump-in-the-throat. At some level we're aware that the voice acts as an exquisite psychic barometer, sensitive to micro-shifts in feelings, registering what words try and conceal. Often without realising it, expert listeners are attentive to intonation, rhythm, and breath, alert to those moments when there's a 'shortfall' of commitment to what's being said. Though our public and educational institutions may do little to encourage it, an extraordinary array of subtle, skilful voice-reading is practised daily in our work, domestic, and social lives. Since communication, to a great extent, consists of an exchange of vocal cues, being able to grasp the meaning of the modulations of another person's voice and respond appropriately with our own is probably our most important interactive task.

## TRANSFORMED INTO SOUND

How do feelings get turned into sounds? Emotions produce changes in muscle tension, breathing patterns, the brain. Depression, for instance, slows down psychomotor activity.[7] When we're stressed or excited, on the other hand, our laryngeal muscles tense up, making the vocal folds tauter,[8] so that the speaker has to

produce more pressure to force the air through. The vocal musculature is a highly sensitive instrument.[9] The larynx is 'suspended'
between two muscle groups, and undue pressure can disturb the
balance and so modify the sound of the voice.[10]

Emotional states like deception, conflict, and anxiety can change
breathing,[11] which in turn influences the subglottal pressure and so
impacts upon the voice. Breathing faster can alter the tempo of
speech,[12] whereas when we feel powerful we tend to breathe more
deeply, and so our voices become lower. Simply *remembering* an
emotion – a happy event, the shock of an accident – affects the
movement of the diaphragm.[13] A linguist and historian of literature even claimed that he'd managed to authenticate old manu-
scripts by studying his own respiratory changes while reading
aloud poetry and prose. When the respiratory rhythm was different, he concluded that the lines had been written by a different
author.[14]

Changes in facial expressions also affect the pharyngeal muscles. In a grimace, the corners of the mouth are turned down, the
vocal tract shortened and its walls tensed. This helps make the
voice higher, more nasal, and narrower. When a person is appre-
hensive or fearful, their voice 'shrinks' as the mucous membranes
become dry.[15]

A whole array of neuro-physiological structures helps in the
production of the voice, in particular the neocortex, which activates the muscles, and the limbic system, which activates the
autonomous nervous system.[16] On their way to and from the
brain the nerves from the larynx pass through the limbic area –
the so-called 'emotional brain' that controls our feelings, moods,
and drives. It isn't surprising, then, that the voice picks up so many
emotional qualities, or that our emotional state impacts so powerfully upon the voice.[17] As one researcher put it, 'Spontaneous
emotional communication constitutes a conversation between
limbic systems.'[18]

## A SOUND PERSONALITY

Between the 1920s and 1940s a blizzard of studies attempted to prove that listeners could correctly judge a speaker's personality from their voice.[19] Extroverts, for instance, supposedly spoke faster, louder, and with fewer pauses.[20] The very etymology of the word 'personality' seemed to provide encouragement for this approach: coming from the Latin *'per sona'*, meaning to resound, it recognises the intimate connection between the voice and personality.[21]

In the 1950s an American laryngologist even maintained that neuroses had their own, distinctive vocal means of expression, their oral counterpart. 'Neurosis in itself is voice-bound . . . The man who is afraid,' he argued, 'will show it in his voice . . . Voice is the primary expression of the individual, and even through voice alone the neurotic pattern can be discovered.'[22] Purely on the basis of a recording of an adolescent boy's voice, this doctor judged him fearful, cowardly, egocentric, self-conscious, effeminate, intelligent, and gifted. When the boy's Rorschach test was analysed, almost identical conclusions were reached.[23]

Many have tried to read off the psychic state from inflections like this; almost all have failed. Dogged by insignificant results, the research was abandoned in the 1950s,[24] but came back into fashion in the 1980s, although the contemporary studies aren't necessarily any more sophisticated.[25]

## FLATLY DEPRESSED

Researchers these days prefer to investigate how the voice expresses emotions rather than personality, and are trying to prove that different emotions have their own particular acoustic profile. This isn't new. Darwin noted that 'the pitch of the voice bears some relation to certain states of feeling'.[26]

Depression is most easily identified through the voice. Hippocrates believed that the basic symptom of depression was *dimissio animi* –

lowered vital tone, accompanied by taciturnity.[27] By 1921 Emil Kraepelin, the so-called father of modern psychiatry, had observed that depressed people 'speak in a low voice, slowly, hesitatingly, monotonously, sometimes stuttering, whispering, try several times before they bring out a word, become mute in the middle of a sentence. They become silent, monosyllabic, can no longer converse.'[28]

The depressed voice is not only quieter and less inflected but also has a dull, lifeless quality. It trails off at the end of a sentence, as if the speaker is sighing while talking.[29] The writer William Styron recalls, when he went through a major depression, 'the lamentable near-disappearance of my voice. It underwent a strange transformation, becoming at times quite faint, wheezy and spasmodic – a friend observed later that it was the voice of a 90-year-old'.[30] Manic-depressive patients in their manic phase, on the other hand, speak vigorously, with a wide pitch range, lots of glides, and frequent emphases,[31] but revert to flat and halting voices in their depressed state.[32]

The length of speech-pauses is now considered so reliable an indicator of depression that it's used both as a diagnostic tool[33] and, along with an increase in volume, as an objective barometer of improvement after beginning treatment, even identifying the exact moment when improvement begins.[34]

Some even believe that the voice can be used preventively. Thirty years ago a clinical psychologist, interviewing patients in the psychiatric emergency room of a New York hospital, found that the sound of certain voices literally caused the hairs on the back of his neck to rise. The voice, he concluded, contained important psychological information about a person's immediate psychological state, and a dull, lifeless, metallic, hollow sound could act as an early-warning system to alert mental-health workers that a person was seriously considering suicide.[35] A recent study comparing the acoustic properties of the voices of depressed people and suicidal ones managed to distinguish them correctly most of the time.[36] Yet the idea of the voice as an infallible guide to despair remains problematic.

## THE SOUND OF FEELINGS

Depression may be detectable by low pitch, but here's the rub: so too is boredom. Similarly, while anxiety is said to be distinguished by higher pitch, happiness is as well.[37] Contempt, meanwhile, is said to be loud and slow. Anger has high pitch and is faster. Stress is high, loud, and fast[38] (but then confidence is too[39]) while grief is low and full of pauses.[40] Surprise glides, and scorn has an even, descending melody.[41] Only on disgust do the researchers admit defeat, and even here, they suggest that the failure isn't theirs but the speakers': we simply don't encode disgust well, they claim[42], clearly never having spent time in the company of small children or teenagers.

## SOUNDING SCEPTICAL

Work like this has been seized upon as supplying a foolproof way to understanding the hidden clues and cues to human behaviour. Best-selling manuals proclaim that 'Pitch, speed, and volume give away a liar . . . studies show that around 70 per cent of people increase their pitch when lying.'[43] (70 per cent of everyone in the world? Or 70 per cent of those in two or three studies?) Communication has come to be seen as cryptology, the scientific study of codes (preferably with numbers – the fourteen personality types, the six ways of saying 'no', the twelve steps to just about anything), that need to be deciphered if we're to achieve professional and personal success.

So how accurately can specific emotions be distinguished through pitch, volume, and tempo? Even those who believe that the voice is fabulously revealing concede that some emotions are read more easily than others. Anger and sorrow are easy, for instance, but fear is harder.[44] And people differ in how they express their emotions. Most depressed people may have quiet and lifeless voices, but some get louder in depression, and sometimes go higher.[45] There are people who, once they've decided to kill themselves, sound calmer and more settled.[46] Recovery also doesn't always have the same acoustic but is influenced by various factors like age.[47]

One of the things that sank the research on the voice and personality was the realisation that listeners' attributions of extroversion or introversion were based less on actual personality than on theatrical conventions. The same is true about a lot of the work on the voice and the emotions: it tells us not so much about real anger, more about the sort of stage anger that you find in a silent Charlie Chaplin movie.

## UNNATURAL SPEAKER

Peer more closely into many of the more flamboyant claims made about the voice and emotions and you see on what flimsy grounds they rest. In most of the research, for example, a speaker is required to read a nonsensical passage, letters of the alphabet, or even numbers, with different emotions, which judges are later asked to identify.[48] But the speaking voice is different from the reading voice, and simulated emotions aren't the same as spontaneous ones. Some of the studies even use actors rather than ordinary people:[49] one pair claimed extra realism for their study because they used method actors, former members of the Actors Studio![50]

Secondly, this kind of research presumes that the way we say something is unconnected to what we're saying (some studies go to elaborate lengths to filter or mask verbal cues), as though there were pure, abstracted emotions which, although communicated through speech, exist in some kind of non-verbal limbo. They also assume that our emotions are expressed discretely, one at a time, rather than in a cocktail of several simultaneous feelings.[51] And they only seem to recognise a single variant of each emotion, or at best two (like 'hot anger' and 'cold anger'[52]), whereas most of us are Rembrandts of wrath, so many hues of angry voice are at our command.

Finally and perhaps most damning, many studies on the emotions and the voice treat speakers in supreme isolation, completely erasing the person being addressed.[53] Yet most of us speak differently to colleagues and family, or parents and child: how we say it depends, to a very large extent, on to whom we're saying it (as well as where, when, and why). Lose that and you've pretty well eliminated the key aspect of vocal communication.

Of course no research design can reproduce the complexity of messy reality: you have to control some features to allow study of others. But life is not a bipolar scale, and yet most researchers accept without demur the fact that the effects of the individual speaker have been factored out of the analyses as an unwanted source of variance'.[54]

## I READ YOU

So what's the alternative? It was the 'unwanted source of variance' that I was after in my interviews. Simply asking people how they read the voices of those with whom they live, work, or play can reveal some of the complex skills we use to understand those around us and how, *individually*, we learn when to discount and when to give credence at least partly on the basis of the voice.

It's a pretty speedy process. Emotion can be recognised in segments of speech as short as 60 milliseconds (60 thousandths of a second).[55] From their opening telephone 'Hello', daughters can usually recognise their mother's mood and vice versa. And not just mothers and daughters: the voice of a person close to you is as viscerally familiar as the smell of their body, a delinquent tuft of their hair, or the way they chomp their biscuit. These are bits of intimate knowledge we accrue about the people around us, which make those movies where one person is substituted for another so implausible. It's as if we have some internal template of how the other normally sounds – how much warmth and vigour, how free and engaged their voice usually is – and can spot any deviation. We take soundings of each other's voices.

A 48-year-old woman described a row with her 15-year-old daughter. 'I was absolutely furious with her so I adopted a semi-neutral tone which I know in my heart of hearts isn't a neutral tone – it's an accusatory tone disguised as neutral. Which of course was picked up as, "Why are you speaking in a funny voice when you say you're not angry any more?" '[56]

A 54-year-old woman describes her husband:

I almost never listen to the words he uses – I listen to the voice he uses: that tells me more, because if I say, 'Will you do this?' he says, 'Yes, of course,' but you can tell from his voice he really means, 'I really resent this and I'm going to make a row about this in the future.' But you wouldn't know that from the words. Very often there's a dissonance between the words and voice, the words say one thing, the voice another, and you're meant to hear the difference. With me they more usually go together – the words and voice reinforce each other.[57]

She exercises similar interpreting skills with their 24-year-old daughter, currently living in Canada:

You know when she wants you to get off the phone but she's too nice to show it – I can feel her impatience. I also know from her voice when she's telling me that everything is fine but what she really wants you to do is rush in and probe so she can tell you what's wrong. I wonder if you learn to listen to them most when they're teenagers. With a teenager you want to know who's coming down the stairs – Miss Moody or Miss Happy. You learn to listen to the message of your teenage daughter's voice to find out which one it is, and not to say the one word that will trigger an awful day.[58]

Vocal sensitivity is an important dimension of intimate relation-ships. A 39-year-old woman says that, 'When G. conveys with his voice that he's on edge and has had an aggravating day, then oh yes, I speak to him very differently because I don't want to add to it. I'm very careful and tend to back off, or sometimes I'll say, "Tell me what happened." He does the same to me.'[59]

Most of us have at least half-a-dozen experiences like this a day. Voice-reading is, ultimately, a form of empathy: we tune into what another person is thinking and feeling – not always successfully. A 44-year-old woman, after spending the night with a teacher for the first time, found him brisk and businesslike over breakfast and assumed that he was trying to tell her that it was a one-night stand. It was only years later that he revealed that this was because he goes into quick, efficient teacher mode in the mornings. 'Now the name of the school – Rockingham High – has become a sort of code between us, and if either of us sounds cold or is using a very professional, brisk voice when the other feels in need of warmth, we say, "Rockingham High." It's become shorthand for, "Hello, you've disappeared – come back." '[60]

Voice-reading entails a shift in understanding, from one's own style of expressiveness to another's. It invites us to recognise the separateness of listener and speaker – not so much a putting oneself in someone else's shoes, but rather a putting oneself in their voice. Occasionally one can hear something at the threshold of awareness that's just too threatening or disturbing to be admitted into consciousness. A 42-year-old woman recalls:

> W. had just come back from a work-trip to Milan and I was asking him about it and he was talking about a colleague. There was something in his voice that made me anxious for a fraction of a second – it couldn't have been more than that, and I was barely aware of it. I just remember turning and shutting the bathroom door with some tiny speck of nervousness that I didn't allow to surface. Later, when he confessed that he'd been having an affair, I remembered that moment, and wished I'd let myself admit what I'd picked up in his voice, even though I had no grounds for doubting him.[61]

Many of us are able to hear covert emotions in the voices of others, but prefer not to.

Among the people I interviewed, far more women than men acknowledged the importance of reading the voices of their friends and partners, children and colleagues and were able to give examples. This probably reflects the fact that the burden of emotional care-taking in most cultures still falls disproportionately on women's shoulders (a degree of psychological skill – aka 'wiles' – is still considered an intrinsic part of femininity but an optional extra of masculinity), but also that women are socialised into a language of affect – they've developed an ability to translate what they hear non-verbally into words.

## DECODING CHILDREN

Both boys and girls, on the other hand, use voice-reading to steer themselves round the adult world: being finely attuned to nuances in the voices of those people with authority over them like parents and teachers is a form of protection against adult fiat.

One 11-year-old boy, when he hears a certain quality in his mother's voice, heads immediately for a neighbouring room, 'so that I don't get picked on'.[62] His 14-year-old sister confirms that when their mother gets this 'testy, harsh thing in her voice' they know she's angry, while a 13-year-old girl gauges the degree of rage in her parents' voices to decide 'whether to stop arguing with them or whether to go on'.[63] On the other hand the child whose parent says, 'Stop that at once,' but in a tone of voice that fully expects him not to comply, usually doesn't. Parents trying to exert authority over difficult-to-control children, according to one study, often detract from their instructions by using weak voices, and as a result are more likely to be ignored by their children. 'Vocal intonation, which is not easily self-monitored, mirrors an individual's beliefs about his effectiveness as a source of influence.'[64]

Recent research suggests that the ability to read emotional cues in the voices of other children grows in accuracy from infancy to

adolescence,[65] yet remarkably, children who have trouble decoding vocal prosody are judged less popular and lower in social status
by other kids *even before they've left nursery*. Good vocal readers,
on the other hand, are less socially anxious and less fearful of being
criticised.[66]

Perhaps anxiety deafens you to the finer emotional distinctions
in other people's voices; maybe your own anxiety drowns out other
vocal melodies. One study found that very anxious children confused other children's fearful voices with angry ones.[67] Difficulties
in accurately identifying anger in the voice may, one researcher
believes, lead to behavioural problems that are the precursor to
violent or criminal activities.[68] Another piece of research found
that male adolescents who'd been arrested for sexual offences
made more errors in identifying angry voices.[69]

Today there's so much anxiety about children's decoding deficits
that there are books and programmes to help them develop nonverbal skills: these advise parents to teach them rhythm, or encourage the child to practise a sentence in different ways to
communicate different intentions.[70]

Children, in the past, used to learn rhythm from music. And
while chronic difficulties in encoding and decoding feelings in the
voice can cause painful problems of social rejection, most of the
books aren't aimed at the minority with problems but at everyone
else. Titles like *Teaching Your Child the Language of Social
Success* are disturbing, as if it's never too young for the self-
improving makeover. There's even a new term, 'dyssemia', to
mean difficulty using or understanding non-verbal signs and signals, as well as a dyssemia website, and a 'breakthrough programme' for conquering adult dyssemia.[71] Of course you need a
syndrome if you've got a breakthrough to offer. We used to have
other names for 'adult dyssemics' – misanthrope, for example, or
antisocial bugger. But today they have a skill deficit that needs
treatment.

Vocal ability, both encoding and decoding, is a fantastically
useful attribute that develops in response to sensitive parenting and

a benign environment. But skill in using and understanding it is an aspect of the human imagination, rather than a dimension of health and disease.

## LYING THROUGH THEIR TEETH

Can inauthenticity be detected in the voice? Simulated emotion certainly has a different acoustic to a genuine one: within seconds of listening to speech radio, most of us can identify whether it's drama or actuality. The spontaneous voice darts and bounces far less smoothly than we (or over-emoting actors) imagine. As a voice teacher remarked, 'Actors can tend to hold on to an emotion too long. In emotionally demanding scenes, actors tend to stay in one strong affective tone, rather than allow the character's emotional state to be fluid, changeable, and irrational.'[72]

The idea that deception is easily detectable in the voice has been encouraged by the success of the new breed of magician 'mind-readers' like Marc Salem and Derren Brown. Salem argues that ordinary people can tell when someone is lying by developing 'focused listening' and tuning in to momentary vocal wobbles, a rise or fall in pitch or register, 'especially if quickly corrected . . . changing rate of rhythm of speech', a cracking voice, etc.[73]

Brown also maintains that he can spot people who are lying just from their voice.[74] For his 2003 British television 'Russian Roulette' stunt, he picked a volunteer to load a gun and, on the basis of tiny inflections in the volunteer's voice when counting from one to six, Brown claimed that he'd successfully deduced in which chamber the bullet had been placed. Or did he just choose someone suggestible and, through his voice, direct them to a particular chamber?

Whichever the case, such stunts reduce vocal understanding to a party trick, and vastly overestimate its predictive power. For while many of us can tell when a family member or close friend is lying at least to some extent on the basis of their voice, the idea that you can do the same with anyone you might happen to bump into on a train, or that we can confidently generalise about the acoustic of lying, is fallacious. So although some studies have found that liars tend to speak more slowly,[75] in a higher-pitched voice, and with a greater number of pauses[76] and speech errors, typical nonverbal behaviour doesn't exist – there's no vocal equivalent of Pinocchio's nose.[77]

As the burgeoning field of deception studies makes clear, the relationship between voice and behaviour is far more complex, and we all vary in our ability to encode emotions in our own voice and decode those in the voices of others. Good encoders of emotions also tend to be good decoders, but we're not necessarily equally good at encoding and decoding the same emotion – you might be good at showing fear in the voice, for instance, but bad at detecting it in the voices of others.[78] And some people's voices are easier to read than others.[79] 'People behave differently in different situations, and different people behave differently in the same situation.'[80] In fact the detection rates of most professional lie-catchers are modest, and people trained to look and listen for non-verbal clues to deceitful behaviour often do less well than ordinary, untrained listeners.[81]

## YOUR VOICE OR MINE?

Our opinions about other people's voices are shaped by the sound of our own – unwittingly we may be comparing theirs with ours. A comment like 'It's too loud for me', might really mean 'It's too loud in comparison with my own voice'.[82] At the same time, what annoys us in the voices of others may be qualities that we haven't learned to tolerate in ourselves. One woman I interviewed described her irritation at her 8-year-old daughter's 'mouselike voice', but then acknowledged that quiet voices always made her feel aggressive. Her own voice was forceful.

Our personal history also helps shape our reactions to other people's voices. A psychologist, writing in 1931, said, 'Whatever the physical sounds produced by a voice, the effect upon the hearer depends largely on his own past experience . . . in judging a voice, we may – usually unconsciously – be reminded of another earlier voice, significant to us in the past, and our judgement may thus be powerfully influenced.'[83]

A 43-year-old woman who had a ranting father thinks this has led her to overreact to her husband when he rages:

> My radar for an angry voice is highly developed because of being flooded by my father's anger, so I might perceive a higher volume in an angry voice than someone else might because I'm so anxious about it. When my husband first got angry I felt immediately panicky, an immediate reminder of a level of expressed emotion that was out of control, in a way that terrified me. When my husband sounds very angry, I'm transported back in a time machine.[84]

The sound of the human voice has an unrivalled capacity to flood the listener psychologically. A lifeless voice can reproduce the speaker's stagnancy in the hearer. On the other hand, one night in 1956 the playwright Arthur Miller, staying in a motel in Nevada, received a desperate late-night Hollywood phone call from Marilyn Monroe, whom he was to marry later that year. Miller heard 'a new terror' in her. 'I kept trying to reassure her, but she seemed to be sinking where I could not reach, her voice growing fainter. I was losing her, she was slipping away out there . . .'

Miller feared that she might commit suicide, but couldn't think of anyone near by he could summon to help her. Out of breath and dizzy, he slid to the floor and passed out. 'I came to in what was probably a few seconds, her voice still whispering out of the receiver over my head. After a moment I got up and talked her down to earth, and it was over.'[85] Through her voice alone Monroe had managed to project her feelings of overwhelming despair directly into the receptive Miller. Monroe was sinking and Miller sank. A voice can plant feelings deep into the core of another person, even down a crackly telephone line.

But then some voices are more penetrating than others, and some listeners more permeable. A 48-year-old woman finds her husband's voice soothing in much the same way as her late mother's. But, 'When he's angry he yells . . . it's really explosive . . . It makes me feel bludgeoned even if I hear him in the kitchen, shouting, because he's dropped a saucepan. It assaults me, the level of anger in the voice. I don't feel physically threatened by it but it makes me feel unsafe emotionally.'[86] Similarly a 7-year-old girl said of her mother, 'She shouts a lot – it feels like being locked up in a cage and there's a dog barking at me and I can't get out and I can't get the dog's barking out of my head.'[87]

Many of the people I talked to spoke in these terms – as if a raging voice wasn't simply an expression of anger but also had the power to invade and injure them. As a 56-year-old man put it, 'You can be lashed by a person's voice. There was a famous man when I was at officer-cadet school – it was said he could curdle milk at a distance with his voice.'[88] You can turn your back on an angry face, but an angry voice surrounds, saturates and sometimes deluges. A very loud voice can even be a form of abuse: louder than 80 decibels it's potentially destructive to physical tissue and mental processes.[89]

## THE THERAPEUTIC VOICE

Freud may not have been terribly interested in the texture of individual voices, but he created a therapeutic method in which both the analyst and patient's voices played a crucial transformative role. Today nearly every kind of therapy has as its basis two or more people talking together.[90] A psychotherapist argued in 1943:

The voice is a sensitive vector of emotional states and is used by the ego as a vector for neurotic symptoms and defence mechanisms. To hear the voice solely for what it has to say and to overlook the voice itself deprives the analyst of an important avenue leading to emotional conflict . . . since resistances are constantly being acted out by means of the voice (and with much less shame than in other types of acting out), it is doubly important that such behaviour be exposed and analysed.[91]

One American psychiatrist even described himself as an 'aural' therapist, claiming that, purely by tuning in to a patient's vocal cues, he was able to gather all the significant information about them, and judge when to intervene.[92]

Jung believed that his patients used their voice to disconnect from psychic pain and 'speak quite unemotionally about things that have the most intimate significance for them . . . So long as the complex which is under special inhibition does not become conscious, the patients can safely talk about it, they can even "talk it away" in a deliberately light manner. This "talking it away" can sometimes amount to "feeling it away".'[93] On the other hand, sometimes patients lower their voice when saying something important, as though they don't want other people – but also perhaps themselves – to hear it. According to one therapist, 'When they hit a buried piece of psychic shrapnel . . . the voice will drop or the person will pause for a moment and have a hard time finding a word or there's a change in the register . . . there's a shift in the quality, it gets throaty or it gets hesitant or a little cracked . . . you're listening for metaphors, shifts in volume.'[94]

Research carried out in Chicago in the 1960s identified four different vocal styles in patients: 'focused', where they grope towards putting their internal experience into words and move into new territory; 'externalising' – an energetic and expressive voice that talks at someone rather than with them, leaving no space for newness to emerge; 'limited', involving a holding back or withdrawal of energy, as if the speaker were distancing themselves from what they're saying; and 'emotional', where an overflow of feeling distorts or disrupts the speech.[95] Patients might move from style to style within a single session, but the 'focused' style ushered in insight or resolution on the part of the patient, and the more that it was used in the first two sessions, the more favourable (in the opinion of both patients and therapists) was the outcome. When the 'externalising' and 'limited' vocal styles dominated, on the other hand, both parties (in this research) found the sessions relatively unproductive. Trainees, it was suggested, should be taught to listen for those moments when a client's voice slows, softens, and becomes 'focused'. 'Even though the content being discussed may be less than exciting, this is a sign that something here may be alive for the client.'[96]

The therapist's voice is just as important as the patient's in the therapeutic encounter. It can create a bridge between themself and the patient.[97] In psychoanalysis the patient lies on the couch and the analyst sits behind them, out of sight[98] – theirs is like the voice of God, but their voice also envelops like the mother's. Some patients don't want to lie down because they feel it would make them vulnerable, partly because a voice disconnected from a face feels so omnipresent, and deprives them of the information that comes from a face.

Psychotherapists, like other professionals, have their work voices. One recalls how her husband, when he heard her answer the phone, knew from her first sentence whether she was speaking to a friend or a client. 'He could just tell from the tone of my voice and the structure of my voice . . . I become more formal, more businesslike, trying to maintain that kind of blank screen of the voice.'

Her voice plays an especially important role in family sessions:

> Sometimes it can get so heated if a couple begins to argue that . . . I find myself sort of taking a moment and thinking . . . how to modulate my voice because I'm not going to join the fray . . . I've got to speak in the lower register, it's got to be a quieting firmness rather than an escalating one . . . I sit for a moment . . . I don't think how I'm going to speak but I feel how I'm going to speak.[99]

Another practitioner consciously alters her voice with her patients:

Sometimes . . . I use my voice to stimulate a depressed and hopeless one. On . . . other occasions . . . my words are less important than the vocal indication of my presence. Sometimes I remain silent to encourage separation from me. Occasionally my voice backfires on me, as when a patient notices my anger or my anxiety through the sound of my voice.[100]

In analysis the unbearable gets siphoned from the patient's to the analyst's voice, where it's made tolerable, and is no longer expressed manically or lifelessly. Along with the physical boundary of the consulting room and the chronological boundary of the fifty-minute session, the therapist's voice acts as a container for the patient's overwhelming emotions. As one analyst put it, 'The "music and dance" of an interpretation – the poetic or lyrical aspect of language and the emotional tone in which it is spoken – that is transformative to the infantile aspects of the analysand, whatever his/her age, whether or not the literal meaning of the words has been understood.'[101] A 39-year-old woman remarked, 'Someone once told me that after a successful lengthy analysis very often people don't remember any of the words spoken but have a sense of the tone of their analyst's voice, and its rhythm.'[102] A skilful therapist even begins, through their voice, to alert their patient to the imminent ending of the session (and over time the patient learns to understand the meaning of this particular modulation, the therapeutic equivalent of 'last orders').

The very first contact between therapist and client is usually by voice alone, exciting fantasies, hopes, and fears through the medium of the phone. Today some therapists also give sessions by phone. One was surprised at how easy it was to transmit feelings this way, finding that the lack of visual interference produced a 'paradoxical intimacy'.[103] A patient who has telephone sessions twice a week with a therapist in a different city lies down for them in what she calls her 'snoring room'. 'I get to hear my therapist's voice in a far more intimate way than I would in the consulting room, and have learned to read it much better. He feels he's losing out on all the visual, expressive and body cues.'[104]

Tellingly, Freud actually used the analogy of the telephone about the therapeutic process itself: the analyst 'must adjust himself to the patient as a telephone receiver is adjusted to the transmitting microphone'.[105] In effective therapy, the patient not only internalises the psychotherapist's voice but also, perhaps, in some sense learns to copy it, so developing the capacity to soothe or stimulate themselves. In the process the patient's own voice can change – become lighter, or slower, or less grating. Together, the therapist and patient's voices can make a vital, if often unacknowledged, contribution to the patient's growth.[106]

## THE THIRD EAR

Freud advised the analyst to turn their 'own unconscious like a receptive organ towards the transmitting unconscious of the patient'.[107] One of his first students called his 1949 book about his experience as an analyst *Listening with the Third Ear*. In it he said that 'he who listens with a third ear hears also what is expressed almost noiselessly, what is said *pianissimo*'.[108] Ordinary people do this too.

As communication and service industries displace manufacturing ones, and social mobility increases, along with the number of divorced parents whose main contact with their children is by phone, the ability to interpret voices becomes an increasingly prized skill – too often unacknowledged. It atrophies with disuse, but can be regained with practice. A mixture of concentration and relaxation is involved:[109] You have to direct and focus the ear while allowing it to remain loose and receptive. Athletes and yogis, musicians and writers do something similar with their instrument.

# 10

# Male and Female Voices: Stereotyped or Different?

WHEN THE New York-based Barbie Liberation Organisation (BLO) discovered that Barbies and GI Joes used the same voice-box parts, it bought up 300 of them, switched their voice chips, and then surreptitiously returned the dolls to the stores in time for Christmas 1989. Unsuspecting customers found that they'd brought home testosterone-pumped GI Joes who trilled, 'Will we ever have enough clothes?', 'Math class is tough,' and, 'Let's plan our dream wedding!', while the pinker-than-pink Barbies barked, 'Attack!', 'Vengeance is mine!', and perhaps most fittingly, 'Dead men tell no lies.'

So vital is the voice in gendering our identity that the BLO couldn't have provoked greater outrage if they'd given the dolls a sex-change and not just a voice-change operation. The way that men and women speak (and how we think they ought to) not only reveals our changing ideas of masculinity and femininity, but also helps to create them: since our voice ties us into our social role, sounding like a man or a woman is a crucial factor in being accepted as one.

## ANATOMY OR STEREOTYPE?

Men's average pitch is about 120 Hz, and women's 225 Hz,[1] with men's usually lower than women's because they have a larger larynx and longer, thicker vocal folds as a result of physiological change at puberty.[2] Puberty changes the voice of both sexes, but while the boy's larynx grows about 1 centimetre during puberty to form the Adam's apple, girls' vocal folds grow only 3 to 4 millimetres;[3] the speaking voice of boys can drop a full octave, but that of girls only a third[4] or half an octave.[5] So the changes in boys' voices are more dramatic, with some boys experiencing a sudden break in the voice – an involuntary change in pitch and quality that occurs because secretions of the sex hormone testosterone have made their larynx grow so fast that they have difficulty coordinating it. The breaking voice has thus become an emblem of male puberty.[6]

But geography, culture and race can influence the arrival of puberty – it happens earlier in the children of city-dwellers and more educated children,[7] for instance, but later in northern countries.[8] And what happens at puberty can't explain all the differences between men and women's voices. Every culture establishes contrasting norms and conventions for the sexes that go beyond the biological differences.[9] The differences in secondary sex characteristics are much smaller in men and women than in the male and female of most other species.[10] Indeed much of our nonverbal behaviour, far from being natural, has been developed to accentuate and draw attention to sex differences, rather than just reflect them.[11] We could sound more similar to one another, as one sociolinguist put it, but we choose not to.[12] In practice, the opposite takes place: 'Men seem to be under some kind of social and psychological pressure to make their voices sound as different as possible from women (and, perhaps, vice versa).'[13]

As a result men and women really do seem to be speaking from different places. Social codes penetrate so deep into the body that men often breathe more from their abdomen than women (producing the characteristic 'belly laugh'). One speech therapist trains her male-to-female transsexual clients to talk more from the head than from the chest, thereby avoiding the 'foghorn' effect that comes from men's greater chest cavities. This also helps them reproduce women's 'lighter' sound that emanates from smaller bodily cavities like the upper portion of the voice-box around the throat and head.[14]

Pitch has become a weapon in the gender wars. You'd expect, for example, pre-adolescent boys and girls of the same height and weight (who therefore have similar-sized larynxes) to speak in the same pitch regardless of sex, and yet a celebrated study found that the sex of children could be identified from their voices long before puberty, and that the average acoustic differences between boys and girls are greater than they would be if anatomy were the sole determining factor. What's surprising is how early these differences begin to appear. One study found that, while vocal range for both sexes increased from 85 Hz to a maximum of 97 Hz during the first six months of life, *by the end of the first year* vocal range continued to increase to 110 Hz for the girls, but dropped back for boys to 80 Hz.[15] By the age of 5 or 6, boys and girls can be distinguished fairly easily through recordings of their voices, even though their pitches still overlap a lot.[16]

So why? Children learn early what is culturally appropriate for their sex.[17] Mothers might have talked to them (in motherese) differently from fathers; fathers give their children more commands than mothers.[18] Pre-adolescents may be trying to match their parents' pitch.[19] Children as young as 4 or 5 have been heard adopting a lower pitch when they play at being Father or doctor, and higher to signal Mother or nurse.[20] Another reason might be that, at around 7 to 8, boys begin to restrict their intonational range.[21] Is it pure coincidence that, between the ages of 7 to 10, children also become more aware of prosodic rules?[22] By then boys have realised that speaking in a monotone is a male thing. Cool male icons, like Clint Eastwood, use this restricted range too: vocal restraint is part of what makes them strong and silent.[23]

Children might also be picking up quite specific labial habits.

Rounding the lips lengthens the vocal tract and lowers pitch: pre-pubertal boys may have already picked up that habit to sound masculine. Spreading the lips, on the other hand, shortens the vocal tract and raises pitch – the way some women talk and smile at the same time produces this effect.[24] The net result by adulthood is that 'men talk as though they were bigger, and women as though they were smaller, than they actually may be'.[25]

So have women and men simply learned contrasting vocal tunes, with most of the difference attributable to convention and upbringing? Or has biology provided them with different voices, and then culture evaluated them differently, prizing one and deriding the other?

## STRIDENT PREJUDICE

Taboos against women's voices have a long history. According to the second-century Babylonian Talmud (where rabbis interpreted Jewish law), 'the voice of a woman is nakedness'.[26] Men were prohibited from reciting the Shema, one of the most important Jewish prayers, while hearing a woman's voice because it was so seductive that it might distract them with impure thoughts.[27] St Paul followed suit with his edict, 'Let your women keep silent in the churches, for they are not permitted to speak . . . For it is shameful for a woman to speak in church.'[28]

The literature extolling silence in women is voluminous. According to Aristotle, 'Silence is a woman's glory.'[29] Sophocles' Ajax declared, 'Silence gives the proper grace to women.'[30] As a sixteenth-century writer on rhetoric put it, 'What becometh a woman best, and first of all: Silence. What seconde: Silence. What third: Silence. What fourth: Silence.'[31]

Mute nymphs and voiceless maidens also figure prominently in Greek myths and other fables. Echo, the talkative nymph, is punished by Juno with the loss of her voice: all she's able to do is repeat other people's words.[32] King Tereus of Thrace rapes Philomela, daughter of the King of Athens, and then, to prevent her telling anyone about his crime, cuts out her tongue.[33] Hans Christian Andersen's Little Mermaid is prepared to forfeit the use of her voice in order to live a human life: she trades in her voice to win herself legs.[34]

When women did speak, men drew on a thesaurus of contempt to describe their voices.[35] A New England preacher proclaimed in 1619 that 'the *tongue* is a *witch*'.[36] As late as the eighteenth and nineteenth centuries it was argued that, if women persisted in speaking in public, their uteruses would dry up.[37] In 1906 *Harper's Bazaar* said of the American woman: 'She sometimes spoke through her nose, she twanged, she whiffled, she snuffled, she whined, she whinnied,'[38] while Henry James compared the female voice to the 'moo of the cow, the bray of the ass or the bark of the dog'.[39]

## CURB THEIR TONGUE

The invention of the megaphone, loudspeaker, and microphone did nothing to change the common belief that women made poor orators because their voices weren't powerful enough. In fact the history of women's exclusion from broadcasting represents perhaps the most blatant example of prejudice against women's voices: according to Bell Laboratories, 'The speech characteristics of women, when changed to electrical impulses, do not blend with the electrical characteristics of our present-day radio equipment,'[40] the fault lying with the women rather than the equipment.

Belief in the unsuitability of women's voices for announcing began in the early days of radio, in both the US and Britain. According to the *Daily Express*, 'Many hardened listeners-in maintain that . . . Adam has a more natural broadcasting voice than Eve. Some listeners-in go so far as to say that a woman's voice becomes monotonous after a time, that her high notes are sharp, and resemble the filing of steel, while her low notes often sound like groans.'[41]

Many different reasons for denying women access to the British

and American airwaves were advanced in the 1920s. One newspaper reported that 'the general opinion is that there is only one
woman in about 10,000 who is sufficiently educated in the general
problems of the day to be able to announce news items as they
should be spoken', and then went on to quote an official saying
that 'women would no doubt get flustered in the rushing from one
studio to another'.[42]

The female timbre was singled out for particular opprobrium.
The wireless correspondent of the *Evening Standard* suggested that
women's high-pitched voices irritated many listeners, and that they
talked too rapidly, over-emphasised unimportant words, or tried to
impress listeners by talking beautifully.[43] High voice in women was
associated with demureness, and low voice with sexuality, so that –
in a Catch 22 – the voice that escaped accusations of promiscuity
wasn't considered authoritative enough for serious broadcasting.

Women were also indicted both for conveying too much personality through their voices ('Critics consider that women have
never been able to achieve the "impersonal" touch. When there
was triumph or disaster to report, they were apt to reflect it in the
tone of their voices'[44]) and too little ('For some reason, a man . . .
can express personality better by voice alone than can a woman'[45]). America, too, threw up similar complaints about lack
('Few women have voices with distinct personality,' according to
the manager at a Pittsburg radio station[46]) and excess ('Perhaps the
best reason suggested for the unpopularity of the woman's voice
over the radio is that it usually has too much personality'[47]).

In 1933 the BBC finally caved in and, in an 'experiment', hired a
Mrs Giles Borrett to announce not the news but, daringly, 'This is
the *National* programme from London. The tea-time music today
comes from the Hotel Metropole, London.' The barricades had
been breached, the fortress stormed. Or, in the words of the next
day's newspapers, £500 A YEAR FOR A FEW WORDS A DAY: BBC
DEBUT OF WOMAN WITH GOLDEN VOICE', adding, for good measure, 'HER BABY LISTENS IN'.[48]

Borrett's voice was reviewed by the *News Chronicle's* music critic: '[She had] good, clear vocalisation, correctly pitched, pleasing in its cadency, yet free from pedantic exaggeration', leaving the paper's radio critic to report on the technical, or perhaps electrical, side of the story – the reaction of her 15-month-old son ('He gurgled with pleasure when he recognised the voice of his mother').[49]

On 21 August 1933 Mrs Giles Borrett advanced further, reading the BBC six o'clock evening news bulletin for the first time, although two months later BBC officials declared that the experiment had failed, not for personal but once again for 'technical' reasons.[50] Elsie Janis, Mrs Borrett's American counterpart, appointed as first female announcer in 1935, met almost exactly the same fate. Her NBC employer soon declared that he was not 'quite sure what type of program her hoarse voice is best suited for, but he is certain she will read no more Press Radio news bulletins. Listeners complained that a woman's voice was inappropriate.'[51]

In the end, it was the war that created change – if only temporarily: as the men were called up into the army, the women began to be called up into the BBC. In 1939 the BBC appointed its second full-time female radio announcer and by 1941 she'd been joined by seven others.[52] Sometimes these announcers were on duty for forty-eight hours at a stretch, but there are no reports of them being flustered. Yet scarcely was the war over than back came the view that 'women have never been able to achieve the "impersonal" touch',[53] while the BBC's report on the employment of women found that many female applicants 'who have clear voices and good pronunciation are entirely lacking in "mike personality"' – adding, a tad gratuitously, that this was 'a most elusive quality'.[54] The arrival of television brought more of the same – a woman announcer fired because she 'was told [she] had too much personality'.[55] In 1975 the only woman employed in a policy-making job at BBC Radio I ruled against a female daytime DJ 'because men are more impersonal'.[56]

How different is it today? In 2000, writing at a time when BBC radio, at least, seemed to be full of 'babes' and 'ladettes', I argued that 'as cultural shifts go, it's harder to imagine a greater one than that which has befallen women and radio' in under three decades.[57] But it's also easy to overstate, as I did, the extent of the change, both in employment and attitudes. When, in the 1990s, sixty years after the Borrett saga, male commercial radio broadcasters were asked to explain the paucity of female DJs on their stations, they raised exactly the same objections: that 'people prefer to listen to a man's voice on the radio rather than a woman's voice',[58] and the old favourite: 'People are sensitive to voice . . . and if a woman's voice sounds grating or high . . . shrill, then that will switch them off,'[59] once again placing the responsibility for switching the radio off on to women's voices themselves.

## YOU DON'T SAY

Among the other faults commonly ascribed to women is that they talk too loudly and too much. And yet in reality there appears to be no difference between male and female volume[60] – if anything, men tend to speak louder.[61] Loudness certainly seems to be judged differently depending on the sex of the speaker.[62] Talking loudly is considered an act of aggression in women, but in men as no more than they're entitled to. Sit in a restaurant for any length of time, and pretty soon roars and guffaws gust over from a nearby table of men. Groups of women and girls seem to have got much noisier in public over the past five to ten years, yet men drinking and eating together still dominate public spaces with their voices. Interestingly, after male-to-female transsexuals had undergone vocal surgery in a German clinic, they found that their voices didn't sound as loud as before. They didn't mind, though, because they associated speaking loudly with maleness.[63]

Women are supposedly also voluble. A British Telecom advertising campaign encouraging men to talk more was amusingly mocked by Michael Frayn. 'A characteristic man's telephone conversation [British Telecom says] runs like this: "Meet you down the pub, all right? See you there." They find this "abrupt".

I find it distinctly garrulous. "Meet you there . . . see you there" –
the poor fellow's saying everything twice.'[64]

In fact men talk more.[65] When 100 public seminars were
analysed, men, it was found, dominated the discussion time in
all but seven.[66] In American classrooms, according to one study,
boys spoke on average three times as much as girls, were eight
times more likely than girls to call out answers, and *teachers
accepted such answers from boys but reprimanded girls for calling
out.*[67]

So enshrined in popular belief is the notion that women speak
more than men that a popular rhyme learnt by small children goes,
'The daddies on the bus go read, read, read . . . the mummies on
the bus go chatter, chatter, chatter.' Yet in a famous experiment,
where men and women were separately shown three paintings by
Albrecht Dürer, and asked to speak about them for as long as they
wanted, the average time of the female descriptions was 3.17
minutes, and of the male ones 13 minutes. (These statistics aren't
entirely accurate, because three of the men simply talked until the
cassettes, which had a recording time of 30 minutes, ran out.)[68]

Interruptions are interesting.[69] Although there aren't clear-cut
gender differences, women seem to be more likely to be interrupted
when they smile than men – and women smile significantly more
when they begin to speak. A woman's smile seems to serve as an
invitation to men to interrupt them, whereas a man's smile has
precisely the opposite effect, inhibiting women from interrupting.[70]

So how do we square the fact that women are almost universally
disparaged as chatterboxes with the proven reality that it's men
who dominate conversation? Perhaps women's talk is evaluated
differently from men's? When two actresses spoke dialogue of
exactly the same length of time, listeners recognised that fact (same
with a pair of actors). But when the roles were played by a man and
a woman, the women were judged – by both men and women – to
be talking more.[71] The feminist Dale Spender famously argued in
1980 that women seemed excessively talkative not, as had been
assumed, in comparison to men but rather as compared with

silence. In other words, if silence is the ideal for women, 'then any talk in which a woman engages can be too much'.[72] In 1975 the presentation editor of BBC Radio 4 defended the scarcity of women continuity announcers with the argument, 'Women are still relatively rare in radio; if you have two on – it sounds a lot.'[73]

## I HESITATE TO SAY

Some early feminist writing on the voice seems not only to confirm stereotypes about women's voices, but to add to them. Women's voices are more expressive, emotional,[74] and cheerful than men's.[75] They tend to sound tentative, indecisive, and deferential, their statements often taking the form of questions or requests. So lacking in a sense of her own authority and power is a woman that she even answers a question to which only she knows the answer, like 'When will dinner be ready?' with an 'Oh . . . around six o'clock?'.[76] Men, on the other hand, 'consistently avoid certain intonation levels or patterns: they very rarely, if ever, use the highest level of pitch that women use . . . most men have only three contrastive levels of intonation, while many women have at least four. Men avoid final patterns that do not terminate at the lowest level of pitch, and use a final, short upstep only for special effects' – to sound deliberate, for instance, or interrogative.[77]

Beguiling though these theories were – and they were widely repeated – there was almost no evidence to support them. On the contrary, when they were tested, few of them stood up.[78] Women, for instance, didn't use more question intonation than men[79] (today, of course, everyone is using it), and it was never indisputably proved that their intonation arched and swooped more than men's.[80]

And yet it can't be denied that we use our voices to establish certain socially admired characteristics of our gender. Today feminists are still finding differences in the way that men and women use their voices, but their arguments are more nuanced. So, for example, they've suggested that, when women do use a greater pitch range, they're trying to show how warm and friendly they are,[81] or they feel that they've got to win listeners' (particularly male listeners') attention. Similarly, by speaking with shoulders rounded, chest collapsed, and without taking a full breath, are women making sure that they don't take up too much space?[82] As a voice teacher observed, 'The deeper ranges of the voice connect with the self at a fundamental level of power and many women avoid the feeling and expression of power because they do not want to dominate.'[83]

Yet if some of the stereotypes have fallen away, new ones have arrived to take their place. For instance, it's been argued that when girls reach adolescence they move from the full speaking voice they had as children, via half-voice, into silence[84] – outspokenness gives way to circumspection, confidence to compliance.[85] Really? Teenage girls today, in revolt against the stereotypes, now rival boys in noisy outspokenness. Their role models in popular culture are gabby. Perhaps the school of feminism that sees girls as vocally suppressed at adolescence, written as it is by women now in their 50s and 60s, is actually more of a commentary about their own (1940s and 1950s) girlhoods. In truth there's greater pressure on teenage girls today to be loud than retiringly quiet, to be feisty and opinionated rather than tentative or introvert.

And let's not forget that men too, are stereotyped vocally. Even if the stereotypes governing the male voice are more positive ones, connoting authority and power, men too, are expected to suppress 'unmasculine' qualities in their voice.

## WHO'S NORMAL NOW?

One way in which the stereotypes have been challenged is by reversing them. 'If either sex has to have the prejudicial label of abnormality attached to it, the stack of evidence points to male speech.'[86] Girls, for instance, vocalise more than boys: they learn to speak earlier, develop vocabularly faster, have higher comprehension at 17 months, and surpass boys in speech abilities at about the age of 11. Boys have more speech disorders than girls. Girls don't use the pitch range of which they're capable, perhaps because they're rarely praised for it. (If they hit its upper reaches, they're accused of being 'shrill'.)

Yet linguistic theory has traditionally treated the male as normal, the female deviant. One linguist declared, for example, that the Bengali initial 'l' is often pronounced as 'n' by women, children, and the uneducated classes. But since women, children, and the uneducated constitute the majority of speakers of Bengali they (as a female commentator later remarked) should serve as the norm.[87]

Today it's no longer fashionable to describe men as dominating women through their voices – the talk is all about complementary social roles and different psychological styles. The misconception that women speak more is given a sympathetic gloss: men believe it, goes this argument, because they experience women 'as talking at times when they would be less likely to talk themselves, and about matters about which men would be less likely to choose to talk about themselves'.[88] The best-seller *You Just Don't Understand*, first published in 1991,[89] popularised this approach (which reached its apogee in John Gray's *Men Are from Mars, Women Are from Venus*[90]). Early feminist theorising about gender and speech may have generalised too sweepingly, but today all communication problems between men and women seem to have been redefined – at least in popular culture – as differences in style and misunderstandings between social equals, entirely obliterating issues of power and inequality in the process.[91]

## WHO YOU TALKING TO?

What further complicates our understanding of the differences between men and women's voices is that our voices change depending on who we're talking to.[92] Women, for instance, adopt a kind of baby talk – high-pitched with wide-ranging intonation – to express affection to their boyfriends.[93] Both men and women talk louder when speaking to someone of the opposite sex,[94] women making 'masculine-like vocal adjustments' to avoid feeling at a power disadvantage.[95] The average pitch of women is usually lower in formal situations than in informal: when they turn to serious topics women suppress the high pitches they use in casual conversational style, perhaps because it's so disparaged.[96] On the other hand a woman attending a course on presenting a positive image was shocked, when a recording was played back, to hear herself talking in a confident mid-pitch to a group of women until a man walked through the door, whereupon she immediately switched to a little-girly voice.[97]

The way that mothers and babies change their voices depending on the listener is even more striking. A 10-month-old boy babbled at 390 Hz when playing with his mother, but dropped to 340 Hz when with his father. Similarly a 13-month-old girl veered between 390 and 290 Hz depending on which parent she was interacting with.[98] When they talk to girls of 10–14 months, mothers make language-like sounds, but to boys of that age they make non-language sounds like car-noises. They ask more questions of 2-year-old girls, but tell 2-year-old boys what they want them to do.[99] These differences seem to emerge between 3 and 6 months, but the question remains: are mothers responding to the different interactional styles of boys and girls themselves, or to stereotypes?[100]

## WHO SAID THAT?

Perhaps, instead of focusing on speakers and apparent differences in male and female style, we should be concentrating more on listeners, for the human voice is so saturated with social beliefs that we actually hear and interpret men and women differently. A baby's cry was interpreted as anger when listeners were told that the infant was a boy, and fear when they were told it was a girl.[101] Verbal fluency has been evaluated negatively in women and positively in men.[102] Class too is heard differently – girls are identified as middle-class more often than they actually are, and boys as working-class, resulting in problems 'placing' working-class girls and middle-class boys.[103]

And in a classic 1968 monograph, listeners evaluated the same personality attributes differently, depending on whether the speaker was a male or female. 'Throatiness' in men, for example, suggested an older, more realistic, mature, and well-adjusted person. The throaty woman, on the other hand, was perceived as being less intelligent, more masculine, unemotional, ugly, sickly, careless, naïve, humble, neurotic, quiet, uninteresting, and apathetic.[104]

In one fascinating study, conducted in 1988, Americans of both sexes described the ideal voice in almost exactly the same terms as the male American voice, so that an American man could possess an ideal and an ideal male voice at the same time. It wasn't the same for an American woman, though: she couldn't have a voice that was both perfectly female and matched the American ideal – she had to choose.[105]

Is this still true today? Is it now possible, for instance, to sound feminine and authoritative at the same time? Perhaps there's been enough social change for some female politicians – like Condoleezza Rice, Hillary Clinton, and Patricia Hewitt – to have begun to pull this off. On the other hand, listeners' prejudices have proved remarkably resilient, shape-shifting in response to new conditions. If you ask the talking computers installed into next-generation BMW cars in Japan to give you directions, for example, you'll be answered by a synthesised deep male voice. This, according to a Stanford University professor researching people's reactions to computer voices, is because, 'Our studies have shown that directions from a female voice are perceived as less accurate than those from a male voice, even when the voices are reading exactly the same directions.'[106]

Indeed so powerful is the prejudice against women's voices that it even distorts the fact that men and women are heard differently. In a 2005 experiment British researchers found that men and women's voices activated different regions in the male brain, perhaps because the frequencies in a woman's voice are more 'complex' to process.[107] Not much more than this was stated with certainty, since the researchers hadn't yet conducted the complementary study to find out how women's brains in turn process male and female voices. And yet all round the world headlines claiming that men's brains weren't designed to listen to women's voices rang out ('Can't hear you, dear . . . blame my brain,'[108] 'Why Men Don't Listen to Women',[109] etc). Truly we hear only what we want to.

## MOUTHING DESIRE

In Greek mythology the Sirens, creatures with the head of a woman but the body of a bird, lured passing sailors with sweet singing until their ships came so close to the rocks that they crashed. To drown out their songs Orpheus plugged his ears with beeswax, played his lyre, and sang even more sweetly himself. Women have been accused of seducing men with their speaking and singing voices ever since.

A beautiful or erotic voice, of course, can excite sexual desire – Susan Sontag's husband said that, the moment he heard her voice, he knew he was going to marry her – but this is as much the case with men's voices as with women's. For both sexes a body and breath are required to make a voice, yet again and again women's voices are reduced to pure body, while men's are treated as disembodied. The German philosopher Theodor Adorno articulated this idea when he wrote:

> Male voices can be reproduced better than female voices. The female voice easily sounds shrill . . . in order to become unfettered, the female voice requires the physical appearance of the body that carries it . . . Wherever sound is separated from the body . . . or wherever it requires the body as complement – as is the case with the female voice – gramophonic reproduction becomes problematic.[110]

Our feelings for the earliest female voice we encounter, our mother's, clearly have something to do with this attitude: since we hear that first voice *in utero*, perhaps it's not surprising there's so much of the body associated with woman's voice.[111] Certainly, in our journey from foetus to adult, something curious happens to the status of the female voice. Although the mother's is in some sense the first voice-over that we ever hear,[112] in both cinema and television voice-over, narrators are predominantly male – the female voice has been stripped of its social and public authority. Forever associated with matters internal, subjective, and corporeal, the mother's voice must be repudiated.[113]

Perhaps sexualising women's voices is one way of doing this (since it leaves them only with the power to ensnare). There's a logic to sexualising the voice: it comes out of the mouth, an emblem of female sexuality that can be contorted into provocative shapes like the pout. And since women's mouths today are outlined, glossed, stained, plumped up with collagen, or tricked out to look natural, *they're not what they seem* – duplicity surrounds the very chamber of the woman's voice. If the shrill voice turns listeners off, the sexy one even more reprehensibly risks turning them on.[114] It's not surprising, then, that the hijab or face-veil worn by some Muslim women hides their mouth, literally muffling their voice in the process.[115]

The case of breathy voice is particularly interesting. In sexual intimacy, because hormonal factors change the copiousness and consistency of lubricating mucus in the larynx, making it vibrate less efficiently, the voices of both men and women become breathy.[116] Breathy voice has therefore come to be associated with sexiness.[117] At the same time, since it's less efficient than ordinary voice, breathiness is seen as a sign of vocal problems. But breathiness is also a characteristic of normal female voices. Sexy, sick, and female – all three go together. The 'little-girl' or 'baby-doll' voice, with a high degree of breathiness, was an essential component of the dumb blonde stereotype[118] epitomised by Marilyn Monroe, although since her death Monroe's considerable intelligence has become apparent. What she did, or was required to do, to her mind was similar to what, at least in public, she did to her voice – the girly, innocently sexual persona diminished and constricted them both.

A study over twenty years ago found that British women were deliberately (though not necessarily consciously) adopting a style of speaking that made their voices less efficient and more monotonous, sounding (Monroe-fashion) as if they were sexually aroused (even when they weren't).[119] So sexuality isn't only projected on to women's voices by listeners but is also actively used in speaking by women themselves. There's nothing biological about it because Spanish women's voices are no breathier than Spanish men's.[120] It's an example of how the human voice has to try and accommodate often conflicting cultural demands.

## THE HORMONAL EARTHQUAKE

If some doctors are to be believed, it's all down to hormones. Hormones certainly affect the human voice. Eastern bloc female athletes developed deeper voices in the 1980s after taking steroids containing androgens. Some hormonal treatments for endometriosis have a similar effect.[121] Ovulation can also change women's voices:[122] according to some laryngologists, oestrogen and progesterone alter the structure of the laryngeal mucus just before ovulation, causing decreased range, loss of power, loss of harmonics, and a flat, colourless timbre[123] (although the connection between the voice and ovulation isn't undisputed,[124] except in the case of women suffering from premenstrual tension).[125] Female professional singers sometimes develop hoarse voice and vocal fatigue just before their periods – in 1968 the National Theatre in Prague found one-third of its female singers to be suffering from 'menstrual dysphonia' (voice problems).[126] La Scala, Milan used to include 'grace days' in its contracts, obliging singers not to perform during the premenstrual period and while menstruating (though they were still paid).[127]

What's fascinating is why researchers focus on the effects of menstruation on women's voices rather than, say, of sexual activity on a man's voice, even though a singing teacher at the Royal Opera House says, 'At Covent Garden I can always tell what a tenor has been doing the night before. They always have difficulty with the top notes.'[128] Such different reactions to men and women's voices reveal a lot about the state of gender relations. Implicated in every corner of our personal and social lives, the human voice resonates with our anxieties and values.

Consider the contrasting attitudes to the ageing voice. Compared to women's voices, male voices deteriorate significantly from the age of 50.[129] As they age their voices get higher – 'His big manly voice / Turning again towards childish treble, pipes / And whistles in his sound.'[130] Women's, on the other hand, get deeper.[131] Men's vocal folds also become less flexible as they age, leading to an even more reduced pitch range.[132] And yet, to add to premenstrual vocal syndrome, women are now graced with yet another syndrome, menopausal vocal syndrome, while the more dramatic ageing of male voices is simply regarded as a normal physiological process.

The signs of menopausal vocal syndrome are apparently a quieter speaking and singing voice, vocal fatigue, a decreased range, and loss of timbre, although there's no evidence of large numbers of women suffering from this or being disabled by it. Yet that doesn't prevent some from consulting a doctor about their worry that 'the hormonal earthquake caused by the menopause was upsetting the entire balance of their emotional lives and vocal careers'.[133] If laryngologists continue to publish papers about the disagreeable symptoms of menopausal vocal syndrome, then perhaps this is hardly surprising.

These doctors regard the child's voice as supple, but in their view the vocal condition of even a perfectly healthy woman deteriorates thereafter. During the child-bearing progesterone years, they contend, one-third of women suffer from 'vocal premenstrual syndrome' and can only achieve vocal stability through oral contraceptives[134], while during the menopause, these laryngologists recommend hormone replacement therapy 'to avoid the development of a male voice'. Thus almost the entire female vocal cycle is found deficient or pathological. It never seems to occur to these specialists that, after a lifetime of being called shrill, some women may positively relish their deeper post-menopausal voices.

# 11

# How Men and Women's Voices
# Are Changing (and Why)

MEN EVERYWHERE DON'T sound alike, any more than women. Polish men use a higher pitch than American men,[1] who've been found to speak in a much lower pitch than German men.[2] Are these differences caused by language or culture? And can we read into the fact that Mexicans expect the male voice to be much louder than Americans do[3] something about the different ways in which Mexicans and Americans define masculinity, or how assertive they're expected to be? Cultural differences in men and women's voices are profoundly revealing of a society's doctrines and desires. But they also enable us to track social change: the transformation in women's lives over the past forty years has had striking oral consequences.

## HISTORICAL WOMEN

Just as the Second World War made it possible for women to find work on radio, so too did it allow them to sound different on screen. The 'fast-talking dames' of the 1940s were a flamboyant contrast to the quiescent female stars who'd preceded them: the Barbara Stanwycks talked snappily with no hint of gentility, Hepburn's voice dripped with assurance and mockery. Carole

Lombard, Jean Harlow, Claudette Colbert, Ginger Rogers, and Bette Davis were anything but dumb (and rarely blonde). The characters they played excelled at repartee,[4] and refused to be silenced or derailed.[5] Exuberant mistresses of screwball comedy, they challenged the simplistic idea that women merely simpered, flattered, or remained silent until 1970s feminism gave them a voice.

Of course their voices, like their lines, were shaped by men. Howard Hawks famously made Lauren Bacall deepen her voice with months of punishing exercises until it attained the degree of huskiness he thought it needed for the 1944 film *To Have and Have Not*. Was this to get an older, sexually experienced woman's voice emanating from a 20-year-old woman's body?[6] Or did he prefer women's voices in the lower registers because there they harmonised more easily and unobtrusively with men's?[7] If Bacall – who, in the film, even went by the nickname of Steve – was to be the sultry foil to a much older Humphrey Bogart, did she have to match him vocally, pitch for pitch?

For all the acoustic liberation of the war years, by 1948 women were being urged to practise vocal exercises in front of the mirror to ensure that words that began with a 'v' didn't transfer their lipstick to their teeth. The result was a pouting style of speech, common to the film stars of the era.[8] In the 1950s, when gender roles were so precisely differentiated, so too were Western men and women's voices,[9] although these were to undergo change again as the women's movement began to affect women's daily lives.

## THE JAPANESE PEAK

Women in almost every culture speak in deeper voices than Japanese women. American women's voices are lower than Japanese women's, Swedish women's are lower than American's, and Dutch women's are lower than Swedish women's.[10] Vocal difference is one way of expressing social difference, so that in Dutch society, which doesn't differentiate much between its image of the ideal male and of the ideal female, there are few differences between male and female voice. The Dutch also find medium and low pitch more attractive than high pitch.

Japanese women, by contrast, adopt a very high pitch to distinguish themselves acoustically from Japanese males.[11] When Japanese women are being polite, they can reach an abnormally high peak of 450 Hz, while English women in one study never exceeded 320 Hz. Similarly Japanese men speak in a lower pitch than their English counterparts, although the range of which they're capable is hardly different. Sometimes, the same study found, the English men even reached the top pitch used by English-women; this was never the case with the Japanese men.[12]

Perhaps Japanese women's extremely high voices come from their smaller physical size, or from differences in the language.[13] But then why aren't these shared by Japanese men? More likely the large difference in pitch reflects the more rigid sexual and social roles that have existed in Japanese society until recently. Through the way they control their larynx, Japanese women are displaying their femininity,[14] using the voice to give an impression of power-lessness[15] in a culture where modesty, innocence, subservience and helplessness are much more highly prized as female attributes than in the West.[16] Japanese women's high pitch has even been compared to the Chinese practice of foot-binding.

## BIGGING IT UP

How do the differences between the voices of Japanese and English women square with evolutionary theories based on animal behaviour? Darwinian ideas of sexual selection, in trying to explain why male toads have deeper-pitched croaks than females and why female toads seek out the males with the deepest croaks,[17] suggest that there's a relationship between size, sound, and attracting a mate. The larger a creature is, the noisier and deeper the sound it makes. Aggressive or dominant humans, according to this argument, use their low pitch not only to signal their body size to rivals but also to intimidate them. Submissive humans, on the other hand, make high-pitched sounds to give the impression of being as small and non-threatening as possible.[18] Add in sex differences to this argument, and 'sound symbolism' (high equals submissive, low equals aggressive) now corresponds with, and is produced by, anatomical differences in men and women.[19] Volume and low pitch, it follows, are a way of demonstrating and ensuring male dominance (a premise shared, interestingly, by both evolutionary biologists and some feminists), and winning a female.

Yet human beings today live in highly evolved, complex social structures[20]: as falling birth rates, easily available contraception, and large numbers of child-free lesbians and gay men testify, mating is no longer the same irresistible human imperative. Nor is a deep voice always equated with sexual potency – nobody thinks David Beckham, of the skilful right foot and squeaky voice, any less virile for his lack of an aggrandising growl.

On the other hand, if men's voices are deep in order to convey the size and mastery essential to triumphing over their rivals, shouldn't we expect the voices of women, as they enter the competitive commercial world in greater numbers, to deepen accordingly? It may be entirely coincidental, and not so much about dominance as having to fit in with male values and standards, but this is exactly what's happened.

## PLEASE LOWER YOUR VOICE

Women's voices have deepened significantly over the past fifty years. When recordings of 18–25-year-old women made in 1945 were compared with similar recordings made in 1993, the average pitch of the later batch was 23 Hz lower.[21] Some of this deepening might have been conscious: women working in the Australian media have admitted lowering their voices at the suggestion of voice coaches or sound engineers.[22]

The pressure to speak huskily had already been observed back in the 1970s. Women, one researcher remarked, were tending to use an average pitch-level lower than advisable – around two-thirds of an octave higher than men's, rather than the more usual one octave. Under the influence of actresses and TV personalities, women were restricting the range of their voices, making them less expressive, and risking injury.[23]

Even in Japan. When TV announcer Etsuko Komiya joined a serious news programme from a lighter daytime show, her male co-host urged her to lower her voice, to 'speak, not squeak'.[24] And so she did, soon abandoning her natural high-pitched voice. 'I feel so embarrassed when I watch my early tapes because I sound as if I was speaking from the top of my head,' she said in a later interview. 'When reading hard news, it's important to sound credible and comprehensible. I trained hard to get a lower voice for that purpose.'[25] Monitoring this descent, a professor in a department of engineering found that her voice had fallen from an average of 223.4 Hz in 1992 to 202.6 Hz in 1995. Komiya's strategy seems to have been successful: she ended up as the programme's sole anchor.

The trend for abnormally deep female voices has caused concern, which Komiya shares. 'Naturally, my voice resonates around the nose and mouth. By lowering the tone of my voice, I can feel that I am straining my vocal cords.' Other female Japanese newscasters' voices are also deepening.[26] 'Although low-pitched voices sound stable and might relax listeners, it can weigh down the overall impression of the programme if announcers read all the stories, including cheerful pieces, in a low voice,' said another female announcer at the station. In a country where a high-pitched voice used to be considered the apotheosis of femininity, some Japanese women are even undergoing surgery in their search for a deeper voice.[27]

This rejection of the 'nightingale' voice isn't confined to media workers, but has spread to other industries. According to a Japanese voice specialist, in a male-oriented society men's voices sound trustworthy, so to gain trust women have to deepen their voices. A lower voice also connotes maturity.[28]

Comparable changes have taken place in the West. The British broadcaster Jon Snow believes that 'a woman without bass registers in her voice would find it very hard to get on in broadcasting unless she was exceptionally beautiful'.[29] An online poll conducted by an American centre for disorders to discover the 'Best and Worst Voices in America' found that the worst voices were all higher than normal for their gender, and that Americans preferred melodious, low-pitched voices.[30] Americans' idea of the ideal female broadcasting voice resembles more and more the ideal male one.[31] A 70-year-old woman I interviewed admitted:

> I probably force my voice to be lower than it should be. I think its natural pitch is probably a little higher, but of course that's not acceptable – when I was in theatre taking acting lessons, my acting teacher told me to drop the pitch down. Now if you go to the Supreme Court of the United States and listen to Justice Sandra Day O'Connor, she has this little high squeaky voice, so it obviously isn't an authority issue.[32]

In pursuit of social and economic equality, have women traded in one vocal convention for another? Instead of liberating women, and men, to use the full range of intonation and colour, has the new tyranny of huskiness made women with higher voices feel inadequate? Professor Higgins famously asked, 'Why can't a woman be more like a man?' Well, now she increasingly is.

Yet perhaps, in contrast to the 1950s, ours is an era of 'gender-neutral' voices.[33] The deepening of women's voices might be an expression of lack of confidence, of the need to emulate men, but it might equally be an indication of greater confidence. Already in 1975 one study found that feminist wives spoke longer than their husbands, and more often had the last word.[34]

## THE MUTE MAN

Most of the people I interviewed revealed something different, however: that conflicts in the home almost invariably followed the same pattern – female anger, male silence, provoking even greater female anger. In all but two couples this sequence of rage and silence seemed not only gendered but almost to have been inscribed in the family DNA.

'B. is not a big fighter,' said a 37-year-old woman, 'but I yell at home all the time. On the whole he's silent, which makes me yell more. I want more engagement – I will push him for a response . . . I think, oh boy, that's exactly like your father – no wonder your mother threw plates. I'm definitely like his mother.'[35]

A 57-year-old man admits:

> I go very quiet in conflict and L. gets louder. Sometimes I join in the shouting match and then realise, just like my father, that's not anything I'd win so pull out of it . . . Things get resolved by a bit of dumb insolence on my part. I think that she goes too strident too quickly, as if to pre-empt a more rational tone . . . so I say I do object to being shrieked at, and in a metallic tone . . . I'm sure that sets her teeth on edge, but it certainly doesn't modify her voice.[36]

A 60-year-old man says his wife 'has got a louder voice. If she's minded to shout it's awesome, and all I can literally do is just retire. I try to shout but she always prevails – she's totally uncontrollable if she gets cross . . . I walk out. I'd like to come back but I don't dare because I know I'd be losing the argument and would have to apologise.'[37]

And finally, a 54-year-old man. 'I'm silent a lot of the time, using silence as a weapon as my father did, because silence on my part also means I can't think of what to say. When she gets agitated, I always say it's not what you're saying that irritates me, it's your hectoring tone. And this drives her really mad.'[38]

By withdrawing from conflict, it's usually argued, men are exercising power. Male silence makes their behaviour seem un-emotional and rational. 'To not say anything in this situation is to say something very important indeed: that the battle is to be fought by my rules and when I choose to fight.[39] Perhaps. All these men are certainly trying to control the women, but they're withdrawing not because of any sense of power but precisely its opposite – what they experience as powerlessness. In contrast to public life where, as we've seen, they tend to talk louder and longer, in the private realm they find themselves drowned out: when the strategies for control they use at work are no longer effective, they feel at a disadvantage.

The women here may be silenced by men in public, but within their own home they metamorphose into uninhibited yellers. The men, on the other hand, silence themselves at home, in the belief that they can't prevail over their shouting partners. What they hear in the women's timbre – the whine of helplessness, a loss of control – is anathema. They're also often the recipients of women's contempt and disappointment, expressed at full volume, which makes them retreat instinctively to the safety of their own lower, more level pitch or to silence, their very inexpressiveness enraging the women even more.

Is the mute male the product of an earlier culture? Teenage boys today seem (except to their parents) anything but taciturn. They can hold their own in telephone marathons with teenage girls, and happily converse at interminable length about daily trivia. The mobile phone has helped make talk cool. According to a recent survey, the arrival of the mobile phone has caused men to spend more time on the phone than women.[40] With the Internet permitting controlled self-disclosure and self-invention, perhaps the new technologies have carved out a new space in which men feel more comfortable speaking. Or maybe men are developing new ways of speech that, mediated by technology, don't seem so girly. Either way, it sometimes seems as if boys are the new girls.

## HIS MASTERFUL VOICE

If women, as we've seen, are under pressure to masculinise their voices, then men are increasingly expected to adopt more traditionally female ways of speaking. As service industries grow and technology takes over many of the occupations that used to depend on brawn, social skills become more vital.[41] And just as women's voices have been compared to men's, so men's voices are now being compared to women's – and are being found wanting. An era that exalts an expressive style of communication is beginning to demand the feminisation of the male voice.

This poses a dilemma for men, who are stigmatised when they speak like women.[42] Women who talk like men gain in status, but men who talk like women risk ridicule. Men with higher-pitched or swoopy voices are mocked as effeminate, camp, or gay, and labelled simpering or mincing.[43] Imitating a man in a woman's voice is a way of insulting him, but the reverse isn't the case.[44] A 43-year-old man said that he found pronouncing French with a good accent impossible because the amount of labial mobility it required felt 'cissy'.[45]

Men today have to find a way of making their voices more expressive, while at the same time 'achieving' masculinity in the voice (or avoiding sounding female). In effect they're being asked to develop voices that are at once masculine and feminine. And yet, for some, inexpressiveness is the very foundation of masculinity. As one 63-year-old man said:

> Our father's voice was the one me and my brothers modelled ours on. It didn't have a great range – my father had great difficulty in expressing emotion. If there was any trace of emotion in one's voice, you could be attacked by him – he thought women were at the mercy of emotion, and the goal in life was to avoid it. So, when we talked, we tried to drain our voices of emotion – every trace had to be expunged. The result is that, when broadcasting and presenting programmes, I've had to really try and put back what normal people have.[46]

A successful male broadcaster today would be expected to have a far greater expressive range.

According to an American voice teacher, 'White American men hold their voices in a very tight, small place. If they use even one ounce more of expression than they think they should be using, all kinds of issues about their masculinity and sexuality come up for them. If you show too much emotion in the voice, you run the risk of sounding feminised in some way.' On pitch too, they operate within a strictly circumscribed range. 'I've heard a few men who literally have some kind of frozen basso thing about them. But most of the guys I encounter are capable of two and three more octaves vocally than they actually use.'[47] Black American men, on the other hand, use a much wider pitch range.[48]

So, despite major vocal changes affecting both men and women, most of us are still trapped in vocal worlds not of our making – women's mouths are still regarded as a gateway, and men's as a sentinel. The full vocal range isn't yet available to us all, male or female. The human voice can tell us, with remarkable clarity and resonance, how far our gendered lives have changed. In the voices of men and women we can hear their evolving roles and relative power.

# 12

# Cultural Differences in the Voice

EVEN BEFORE I open my mouth to speak, the culture into which I've been born has entered and suffused it. My place of birth and the country where I've been raised, along with my mother tongue, all help regulate the setting of my jaw, the laxity of my lips, my most comfortable pitch. They play a part in dictating how fast I speak, how soft or loud, how much silence I can tolerate. My vocal habits mark me out as urban or rural, upper or working-class, while the rules about turn-taking that I follow or flout also place me. I speak with my voice, but my culture speaks through me.

For a long time anthropologists tended to treat speech purely as a route through which cultural values were communicated. But then, in the 1960s and early 1970s, came the explosion of interest in the 'ethnography of speaking'. Now the rules of speech in different countries, regions, ethnic groups, and social classes themselves became the object of study.[1] Why were some clans so much more voluble than others,[2] and what positive or negative social values were associated with being taciturn?[3] Researchers chronicled the use of the voice for sacred purposes – chants and invocations – and rhetorical ones (political oration, funeral speeches, etc). They understood that in every culture, no matter how sophisticated, a kind of tribalism is expressed through the voice – we use it to sniff out similarity and difference, us and not us – and to signal where in the social hierarchy we position ourselves (or are positioned by others). Through a society's precepts about the voice you can trace its ideas about ancestry and lineage, rank and kin – even its metaphysics.

## MAGIC SPEAKING

In traditional societies the voice is one of the single most important components of ritual. A pioneering piece of twentieth-century anthropological field-work, Bronislaw Malinowski's 1935 account of magic among the Trobriand Islanders near New Guinea, showed how it could also be a vector of magical beliefs. The magician, Malinowski observed:

> prepares a sort of large receptacle for his voice – a voice-trap, we might almost call it. He lays the mixture on a mat and covers this with another mat so that his voice may be caught and imprisoned between them. During the recitation he holds his head close to the aperture and carefully sees to it that no portion of the herbs shall remain unaffected by the breath of his voice . . . When you watch the magician at work . . . then you realise how serious is the belief that the magic is in the breath and that the breath is the magic.[4]

In ritual, voice and breath often connect a speaker with spirits handed down by their forebears. Young chanters of the Ata Tana Ai of eastern Indonesia, when they 'receive the tongue and take the voice' of their ancestors' knowledge, acquire, supposedly in a flash, historical and linguistic understanding.[5] The Kalui people of Papua New Guinea make no distinction between spirit voices, the weeping and singing voice, and the sounds of rainforest birds.[6] The Dogon people of Upper Niger in North-West Africa don't distinguish between literal and spiritual voice either. To them the voice carries the life-force, so it's hardly surprising that individuals are vulnerable to each other's bad voices. A nasal voice, producing speech caught between the nose and the throat, they call 'decayed': like a stagnant miasma, it gives off a bad odour that penetrates into the listener's very body.[7] Until recently Westerners tended to regard such views not as evidence of the underdevelopment of their own vocal attitudes but as primitive.

## THIS IS MY PITCH

Tone of voice is like a language, and different cultures use it differently. The Kaingang of Brazil communicate degree and intensity by changing pitch, as well as by facial expression and posture, whereas we're more likely to do it through words.[8] The Christian habit of addressing God in hushed tones would strike the Sioux as bizarre, since they use high pitch and loud voice to communicate the solemnity of religious ceremonies, addressing the Great Spirit in shrieks and loud cries.[9]

But there are also cross-cultural similarities in the way we use our voice. An Italian with no knowledge of English can distinguish between 'When can you do it?' asked courteously and when asked curtly – in both languages ending the question on the very bottom pitch rather than a rising one introduces an element of insistence.[10]

Of course no culture is static, and speaking styles constantly evolve. The more complex and rule-laden a society becomes, it's been suggested, the more sharp sounds take over from lax ones: there's a higher level of muscular tension in countries (and individuals) that prize conformity, discipline, and self-control.[11] In India and Pakistan, on the other hand, the jaws are held rather inertly and loosely apart.[12]

## TALK ABOUT TALK

In other cultures differences in prosody are not only noticed but also openly discussed. For instance Japanese *aizuchi* – backchannel cues in the form of agreeing sounds or affirmative noises made by the listener while someone else is speaking – can be words ('*hai*' [yes], '*so*' [that's true]), echoes of one of the speaker's key words, or even grunts. They maintain harmony and tell the speaker that the listener is paying attention.[13]

*Aizuchi* require a keen sensitivity to the voice, for they're anything but random. Listeners interject them at particular stages in a speech – at a low-pitch point, for example, or when the speaker slows down, or raises their volume or pitch. It's as if, alongside his or her actual words, a speaker is transmitting non-verbal vocal instructions to the listener about how to react. 'Prosody alone is sometimes enough to tell you what to say and when to say it.'[14]

The Japanese refer to *aizuchi* in everyday conversation, discussing how a particular *aizuchi* was read, or how someone else's *aizuchi* made another person feel.[15] Chapter 1 showed how the Tzeltal speakers of Tenejapa in Mexico had many words to describe the voice. The Japanese too, it seems, are not only acutely responsive to vocal cues, but have also developed a language in which to discuss them.

## EMPHATICALLY VERBAL

Even though Bronislaw Malinowski helped awaken generations of anthropologists to the connection between the voice and magic, one of his most important concepts is curiously deaf to the vocal dimension. It was Malinowski who coined the word 'phatic' to describe the kind of conversation that has no intrinsic meaning but is simply a kind of small talk – 'a type of speech in which ties of union are created by a mere exchange of words',[16] where the very fact of saying something – anything – is significant, irrespective of the verbal content. 'The breaking of silence, the communion of words, is the first act to establish links of fellowship . . . The modern English expression, "Nice day today" or the Melanesian phrase, "Whence comest thou?" are needed to get over the strange and unpleasant tension which men feel when facing each other in silence.'[17]

Curiously, nowhere does Malinowski specifically acknowledge or even mention the voice. Yet phatic communion's power lies precisely in what's communicated nonverbally, especially vocally – that's why the words aren't important. In phatic speech words are simply carriers of the voice – excuses for it, opportunities for it to be heard. When another person speaks they're implicitly volunteering information about themselves: through vocal self-exposure, purely by using their bodies to produce a sound, they've made themselves less threatening. Phatic speech is essentially phatic voice.

## A DIFFERENT CLASS OF VOICE

We express through our voice not only friendliness but also social rank; we use different registers at different times, for 'no human being talks the same way all the time'.[18] In Burundi a peasant farmer, no matter how naturally eloquent, is obliged to stammer, shout or generally make a rhetorical fool of himself when his adversary is a herder or other superior. But put him in a council in the role of judge and he'll speak with grace and dignity.[19]

In Senegal low-ranked Wolof speak loudly, quickly, in a high pitch, and with a wide pitch-range. In political meetings their voices rise to a falsetto, reaching more than 300 syllables a minute. High-ranking Wolof nobles, on the other hand, are expected to sound breathy and talk in a soft, low voice that can drop to a basso profundo, sometimes mumbling at 60 syllables a minute or less:

> Power is ... displayed in this form of talk – the power to command the audience's attention. The people who speak in the most extreme versions of this style are those whose high status and authority are unambiguous. Although they speak little, their right to the floor is unquestionable ... When the high noble does begin to say something, an immediate hush falls ... Everyone leans forward, straining to make out what is being said.[20]

The Wolof, of course, aren't born with these contrasting vocal styles – they develop them as they take their social place.

Nor are such contrasts peculiar to the developing world – they're also present in industrialised countries. The setting of the mouth, according to the French sociologist Pierre Bourdieu, is shaped by class, with the working-class way of eating and speaking dominated by the refusal of 'airs and graces'.[21] A working-class English-speaking Glaswegian will have a different degree of openness about the jaw and place their tongue differently from a middle-class English-speaking Glaswegian, and those settings will be modified again by age and gender.[22] In fact, quite fine distinctions of status are discernible through the voice, *even to listeners who don't understand the language of the speakers.* When non-French-speaking Canadian students listened to a tape of French Canadians reading a short passage, they recognised their social status accurately purely on the basis of their vocal properties.[23]

Though this may be depressing, it isn't surprising. Bourdieu has coined the words 'habitus' and 'bodily hexis' to describe the ways in which class is inscribed in the body. 'Habitus' refers to the way that childhood learning is mapped on to the body so that it becomes 'a living memory pad', while 'bodily hexis' (from the Greek word for habit) describes how social status comes to be embodied in our way of standing and speaking,[24] in our very mouths. In other words, the social body and the physical body enjoy an intimate relationship.

One of the earliest, most influential (and enjoyable) studies on status and speech was conducted by William Labov in three large New York City department stores in 1962. Labov had noticed that New Yorkers of different social classes pronounced 'r' when not followed by a vowel (as in 'card', and 'four') differently, and decided to test this out in Saks Fifth Avenue, Macy's, and S. Klein (Manhattan stores at the top, middle, and bottom of the price and fashion scale). Assuming that sales staff, even if they didn't themselves share the social status of the stores' clientele, borrowed the customers' mannerisms, Labov adopted the splendid research method of approaching sales people in the store and asking for directions to a particular department located on the fourth floor. The answer usually would be, 'Fourth floor,' whereupon he'd lean forward and say, 'Excuse me?', so eliciting another, more forceful, 'Fourth floor.' Labov conducted as many such interviews in as many aisles as possible until people began to notice. Then he proceeded to the fourth floor to ask, 'Excuse me, what floor is this?'

His interviews revealed that the majority of Saks's employees, half of Macy's, but only one-fifth of Klein's sounded the 'r' clearly, bearing out his theory that a more emphasised 'r' was now the more prestigious pronunciation, and that social class is inscribed in the pronunciation of even a single sound.[25]

## CROSSTALK

Because our voices are so imbued with our culture, the potential for cross-cultural misunderstanding is enormous. Pitch or volume can mean one thing in one country but something quite different in another. The Chinese, for instance, lower their pitch or volume to draw attention to the seriousness of the subject or the intensity of their feeling about it. Since Americans use high pitch and volume to communicate this, they often find the Chinese over-deferential and insufficiently forceful.[26]

On the other hand, West Indians, to emphasise a point, might suddenly and dramatically change their pitch in the middle of a conversation, and start speaking louder. To other ethnic groups this seems like an expression of anger or aggression and so they react as if the speaker has been rude, leaving the West Indian feeling that they've come up against an example of racial discrimination.[27] As we have seen, volume can also cause problems in American–Arab conversation: Arabs sound loud to Americans, and Americans too quiet and insincere to Arabs.[28]

Silence is perhaps the most culturally bound aspect of communication. A Japanese speaker, in the absence of *aizuchi* from a non-Japanese listener, can get the impression that they haven't been understood[29] – the Japanese give backchannel feedback twice as often as the English. But they also regard halting speakers much more positively than fluent ones, as their proverbs ('The mouth is the source of calamity', 'Talkativeness is a mouth's fart', 'Honey in the mouth, a dagger in the belly') testify. The Americans and the British, by contrast, are discomfited by silence. So while the average length of pause between two Americans is .74 seconds, in an all-Japanese meeting it's 5.15 seconds and can even extend to 8.5 seconds.[30] An American trying to be polite to a Japanese colleague might end up reducing his or her credibility. Similarly the Japanese tend to find American taxi-drivers talkative, while Americans can be thrown by what they see as the tremendous reticence of Japanese taxi-drivers.[31]

Even the placing of pauses can cause problems. White American children are taught to pause at the beginning of clauses or before conjunctions, whereas black children pause wherever a significant change in pitch occurs, even if it's within a clause. White children, as a result, sometimes see black children's speech as less grammatical than their own, which forces black ones, when talking to white ones, to code-switch into the white way of speaking to avoid communicative problems.[32]

The consequences of these differences in vocal style can be serious. Alaska Native Americans in the 1970s received on average 20 per cent longer jail sentences than non-Native Americans because, to show deference to authority when talking to the police, they slowed down the rhythm of their speech and paused for longer than usual before speaking. Their hesitation was interpreted by non-Native police as a sign of antagonism rather than respect. Again, when the Native Americans didn't respond to questions in court with the expected speed, officials thought they were expressing hostility and gave them stiffer sentences.[33]

Countries' different vocal melodies can create powerful, and sometimes misleading, impressions. A senior lecturer in German studies at a British university observed, 'When my students visit Germany they often think German people have been rude, but it's just the different intonation and phrasing.'[34] These conflicting tunes result partly from linguistic differences, but they can also occur between countries that speak the same language. Since Americans, when they ask a question, are far more prone than the British to use the high-rising tone (which sounds lighter than the falling tone), Americans sound casual to the British, while the British (with their preponderance of falling tones) sound formal to the Americans,[35] even though American society in many ways is the more formal.

Cultural differences in the use of pitch can, as we have seen, make foreigners seem impolite or tentative. Thanking someone for an invitation to dinner, English men will raise their pitch, but Japanese men won't: to the English this sounds unsociable, cool, or downright rude.[36] Indian accents have a musical quality, but their inflections can sometimes be wrongly interpreted as questioning and uncertain, and the voice treated as lacking authority.[37]

And finally tempo, too, has a cultural dimension. New Yorkers regard quick talk as evidence of quick thinking, so that a Mid-western professor's habit of pausing before speaking and then speaking slowly is seen not as a sign of being thoughtful but of being dim, which is how societies or regions of faster talkers have always regarded slower ones.[38] Even the Finns, long characterised as silent, distinguish between the faster talkers of Carelia in the south-east of the country, and the slow-speaking residents of south-west Finland. Every country, it seems, needs to point the finger at someone slower than themselves, although in reality tempi may have less to do with national characteristics than with how urban a country is. Rural speakers are usually slower and quieter than urban ones – who, after all, have to make themselves heard over the din of the city. (A farmer complained about the city people who walk through his woods. 'They talk so loudly. No wonder they never see any wild life.')[39] Finland, until the 1960s, had a famously dispersed rural population. If the Finns were silent, it's probably because they didn't have anyone around to talk to.[40]

What accounts for the enormous growth in interest in cross-cultural misunderstandings, to the extent that one British bank has even made it the subject of its television and poster advertising campaign? Partly it reflects the general preoccupation with communication: making oneself understood is no longer taken for granted, especially today when we're increasingly required to commune beyond national boundaries. A lot of this material is also fun – anthropology for beginners, saloon-bar sociolinguistics. And yet some examples of cross-cultural miscommunication are just a whisper away from talk of 'national character' and can easily lapse into stereotypes, even if they're legitimated by linguistics.

Many of the cross-cultural comparisons involve the Americans and Japanese. Is this because of those nations' financial and corporate dominance? Or because the Japanese, to the British, Americans and other Westerners, constitute a kind of instant 'other', a new exotic, a negative American – high-tech enough to be part of the same modernist universe, and yet socially distinctive. In a sense they've come to represent cultural difference.

Much of the time, human beings manage to converse through the voice quite successfully, yet this perspective seems to suggest that human communication is riddled with misunderstanding, that we're divided by voices rather than connected through them. Speaking about the Japanese like this also draws covertly on old, war-engendered prejudice: stereotyping them as cold automatons turns the Japanese into almost high-tech machines themselves – even though, as we've seen, their society (like most others) is in flux.


## THE GLOBALISED VOICE

Globalisation plays a part in our obsession with inter-country misunderstanding: our interest is driven partly by anxiety about its impact on trade. Oddly, we've become interested in vocal difference at the very moment that it's beginning to disappear. It's not only the distinctive voices of Japanese women that are beginning to resemble their Western counterparts. Other vocal fashions are sweeping across continents, indifferent to local prosody. The high-rising terminal (HRT), for instance – in which the intonation of questions is applied to statements – seems to have begun in New Zealand,[41] moved over to Australia, migrated to American teenagers (especially female),[42] and eventually colonised Europe.

HRTs have been cited as another example of 'cultural cringe', Australians' low self-esteem, and traced back to Australian soap operas like *Neighbours* and *Home and Away*. TV, others say, rarely disseminates phonological change – only word-of-mouth can do that.[43] The increase in international travel certainly exports not only germs from one part of the world to another but also vocal patterns: intonation can be contagious.

In the beginning, HRTs were also seen as a marker of casualness, and a characteristically female expression of doubt.[44] So are we all tentative, or female, now? Or perhaps they're the intonation of rapport. Differing views about their meaning began to surface after the publication in the *New York Times* in 1993 of a humorous piece by a professor of journalism, who'd noticed them spreading among his students and coined the phrase 'uptalk'. 'Uptalk' is a sign not of deference or self-doubt, according to another commentator, but of 'identity and group affiliation'.[45] HRTs had mutated from tentative to cool – so much so that when, in 1993, the US's National Public Radio examined the phenomenon by phoning professional men and women in their 20s and 30s and listening to their answering-machines, it found messages like this: 'This is Bob? I'm not in right now? Leave a message at the beep?'[46]

HRTs remain the lingua franca of Australian soaps and American teen movies, and can still be detected in young voices. But now that they've spread to all age groups their life-span is surely limited.

But the most compelling example of how globalisation changes the human voice is the call centre. Increasingly today the letter and the body have been replaced by the voice: all call-centre transactions are conducted through the medium of the voice, now the sole link between customer and company or service. The way call-centre operatives have had to transform their voices seems at first just another striking example of how we alter our voices to fit in with other people's.[47] A Scottish study found that staff answering telephone enquiries will say 'Aye' instead of 'Yes' if they are speaking to a person with a distinctly Scottish accent, or drop their 'h's if the caller has a cockney accent.[48]

But the location of call centres tells another story. More than half the world's top 500 companies now outsource either IT or business processes to India, where over 170,000 people work in call centres. They must first complete a spell of 'accent training' or so-called 'accent neutralisation' (although there's nothing neutral, of course, about the accents that the workers are trained to use). Students repeat 'Peter Piper picked a peck of pickled pepper' while holding a sheet of paper ten inches from their face: they have to pronounce the 'p's so that the paper blows away from it.[49] They also learn to say 'villain' rather than 'willian', utter 'extraordinary' as one word,[50] and repeat 'can't' with a long 'a' after watching Rex Harrison in *My Fair Lady*.[51]

Excellence in Indian call centres is now almost synonymous with anglicising the voice. A British chief executive, visiting India to check out potential call centres, said admiringly, 'In two operations the agents had virtually no Indian accent.'[52] In a nice irony, the Indian call-centre workers are expected to use an accent, with its perfectly pronounced 't's and 'p's, that many of their non-Received Pronunciation (RP) British counterparts can't achieve. Decades after 'How now brown cow' deformed the elocution of generations of Brits, Indians are still required to practise it.[53]

There are other vocal demands on call-centre workers. The need to 'build rapport', to sound helpful and confident talking to people with strange accents and bizarre figures of speech, all while under the pressure of the clock, puts immense strain on the voice. The tensions in international capitalism are played out in the voices of call-centre workers – no wonder so many of them lose theirs.[54] As one 27-year-old engineering graduate put it, 'You don't realise how much working in a foreign language takes out of you. You try not to miss a single word of what people are saying. They expect you to be familiar with their culture, but they don't care a damn about yours.'[55]

Now staff are quitting after racist abuse from British callers, who've detected from their accents that they're not English. They tell of callers barking, 'Get on with it. I don't understand a word you're saying.'[56] According to the training vice-president of one Mumbai-based company, 'Usually they won't use abusive language but you can tell from the tone of their voice that they're angry.' Another former call-centre worker admitted, 'I found it difficult to work for British clients. They wouldn't call you names but you could hear the hostility in their voices.'[57] Sometimes Indian employees hit the mute button and scream in Hindi, before returning to a polite, ingratiating persona.[58]

A vocal power struggle is taking place. For most workers, says a head of training, 'It is very challenging to unlearn their natural manner of speech.'[59] Of course the concept of 'neutral accent' or 'English correction' is a shifty one. Those who use it know that the idea of coercing workers to adopt new patterns of pronunciation is controversial, so they rephrase it in more acceptable terms. 'We're not trying to create accents, we're just trying to influence the way he or she speaks,'[60] says one.

According to another, 'I don't know that we're doing elocution as such . . . not necessarily altering vowel sounds, but perhaps working on pitch and delivery.'[61] Says a third, 'A lot of times our trainees say, "Why do I need to sound British? . . . I'm an Indian and this is my accent and I want to sound like this . . . I don't want to lose my identity." What we tell them is . . . we want to eradicate any chance of the customer not being comfortable, so we need to enhance your accent just to be a great customer-service agent.'[62] In reality, most Indian call-centre training colleges remove anyone not using an acceptable accent after training.[63]

The experience of call-centre workers shows how misleading is the term 'globalisation'. Although it evokes some McLuhanesque global village, an egalitarian nirvana where all have equal access to the world's bounty, in reality there's a strongly colonial aspect to call centres, with the same old countries in the roles of colonisers and colonised. On the other hand it's ironic that the British, having destroyed India's own industries and obliged Indians to speak their language, are now losing some of their own call-centre jobs to India.[64]

With the spread of call centres, the growing interest in cross-cultural vocal misunderstanding, and the way that class and dominance are expressed and detected paralinguistically, the idea that the voice has diminished in importance in modern industrial societies has never seemed more unfounded.

PART THREE

## 13

# From Oral to Literate Society

L IVING IN A CULTURE so saturated with images and books, it's extraordinarily hard for us to imagine ourselves back into one without print, where all beliefs, values, and facts are communicated face-to-face, stored only in human memory.[1] Try to conjure up a world without any means of preserving on paper or recording digitally: instantly what's possible both contracts dramatically and expands in unexpected ways. The relationship between a citizen and the law, a citizen and their government, is acoustic. 'A communication system of this sort is an echo system, light as air and as fleeting.'[2]

Without books the meaning of words is pinned down by context and the inflection in which they're spoken, rather than by the dictionary.[3] (Inflections were more important in oral culture, and it was more important to understand them – more depended on it.) A whole array of mnemonic devices – rhythm, repetition, antithesis, alliteration and assonance, epigrams, proverbs and formulaic expressions – were developed to fix words and ideas in the mind of both speaker and hearer. (You only have to read transcripts of interviews even today to see how differently the voice travels when the only requirements it has to satisfy are those of the now.)

No wonder oral cultures regard hearing as so powerful: phonetics and sound are their chief means of communicating information, a fact echoed in the metaphors they use to describe knowledge. The African Basotho – and modern-day New Yorkers – consider 'I hear' to be equivalent to 'I understand',[4] while among the Ommura people of New Guinea, the verb 'iero' means both 'to hear' and 'to know'. The word used by the pre-classical Greek philosopher, Heraclitus, to signify 'know' (ksuniemi) originally meant 'to know by hearing'; by the early fourth century BC Plato still equated individuals who knew a lot with individuals who heard a lot. Religious texts, too, put the aural in prime position: in Scripture, Calvin observed, the phrase 'to hear' was virtually synonymous with 'to believe',[5] and the Israelites believed that God made the universe by speaking ('And God said, "Let there be light," and there was light').[6] It wasn't until the Renaissance that God was portrayed visually: before then he was conceived as sound or vibration.[7] All this allows us to understand why traditional societies, as we've seen, so often attribute magical properties to the voice and spoken word. Once the word was written down, it began to lose some of its magic power.

## LOSING THE VOICE

It's only when you stop to think that you realise how stupefying a human leap it is to use a graphic symbol or letter to represent a sound that makes up part of a word rather than an entire object. In fact human society became literate relatively late in its history.[8] Not until the fifth and sixth centuries BC in the city states of Greece and Ionia were there societies that could be called literate. The *Iliad* and the *Odyssey* were basically oral creations.[9]

The spoken, as compared with the written, word 'must have one or other intonation or tone of voice . . . it is impossible to speak a word orally without intonation'.[10] Literacy, however, downgraded intonation and our sensitivity to it, preferring to use grammar to help establish meaning.[11]

Literate societies tend to treat written words as labels attached to objects.[12] When words are written, they become part of the visual world. 'Like most of the elements of the visual world, they become static things and lose . . . the dynamism which is so characteristic of the auditory world in general, and of the spoken world in particular.'[13] While sounds, paradoxically, come into existence only when they're going out of existence – aurally, not all of a word can exist at the same time – the alphabet makes a word seem like a thing (rather than an event), and one that's present all at once.[14] Phonetic writing divides ear from eye, giving humans an eye for an ear.[15] The implications of such a shift are simply endless.

In oral cultures the role of the voice was pre-eminent: as a communicative tool it had no rival. Now it has several. Today, even with the development of digital media, official status can only be conferred by print – stamped with the authority of the written.[16] In many professional or legal environments the voice and spoken word count for little: they rank as an appendage. The arrival of printing and literacy changed the voice's status – decentred it from official life. The voice knows that it no longer has to communicate everything.

On a personal level, memory down the generations has atrophied: we're able no more to memorise huge chunks of narrative because we don't need to – unless, that is, we've got a so-called 'photographic memory', in other words, one based on the eye rather than the ear.[17] This change has happened alarmingly fast: people schooled in the 1940s and early 1950s were expected to memorise long pieces of poetry that many of them can still recite. But by my late 1950s and early 1960s education, that requirement was no more.

Indeed as oral exams diminish in importance (except in the case of foreign languages, as though sound only matters when it's alien), students can leave university without ever having had to deliver a formal speech.[18]

The differences between speaking and writing are profound. Once released, the voice or spoken word can't be recovered – these pigeons never return home. Written words, on the other hand, are infinitely correctible. 'To make yourself clear without gesture, without facial expression, without a real hearer, you have to foresee circumspectly all possible meanings a statement may have for a possible reader in any possible situation.'[19] No wonder, this author added with feeling, writing is so agonising: we have to do it without the colours of spoken intonation.

On the other hand many people now have traces of writing in their voice: in the way we structure our thoughts and arguments, we've increasingly begun to talk as if we were writing. Those whose speech owes more to oral cultures meet with prejudice. Anyone moving their lips when reading – the body-trace of an earlier age that celebrated reading aloud – is sneered at as a poor reader. The so-called 'restricted code' of speaking identified by Basil Bernstein was oral in origin, while the (more highly prized) 'elaborated' one was based on reading and writing.[20] Ironically, although written cultures are more reflexive and self-conscious, oral cultures have an entire lexicon to describe the voice and speech that we can only listen to with envy.

## STILL TALKING AFTER ALL THESE YEARS

In celebrating oral culture, though, there's a real danger of idealising it ('Oral cultures encourage fluency, fulsomeness, vo-lubility'.[21] 'The eye has none of the delicacy of the ear'[22]). People in auditory societies are often described in terms also used for small babies, as though they lived in a pure world of sound that somehow existed beyond the purview of society. In fact oral cultures, necessarily slower to change than literate ones, can be conformist and controlling places where power relations are rigid.[23]

There are also risks in polarising Europeans as people for whom 'seeing is believing', and rural Africans for whom 'reality seems to reside far more in what is heard and what is said'.[24] It's a circular argument and a self-fulfilling prophecy: once you start seeing modernity as built on 'the inexorable rise of Newtonian sight', not only do you romanticise hearing but you also write off Western culture's capacity to integrate ear and eye.[25]

In fact the arrival of printing didn't eclipse the voice. Although silent reading was rare until the advent of printing – before then, books were mainly used for reading to others – reading aloud remained a common practice throughout the nineteenth century. The very look of early printed books owed more to speaking and reading aloud than to our present sense of textuality: printed title pages of sixteenth-century books appear comic to us, often dividing up major words, spilling them over to the next line, and altering typeface within sentences without any notion of inconsistency. The text was fashioned not by our print conventions but by the needs of the ear, with some of the eccentricities of font size and placement acting as a reminder of the individual voice. 'The early age of print still felt . . . [reading was] primarily a listening process, simply set in motion by sight.'[26]

Even the most resolutely visual and literate societies retain traces or residues[27] of the oral tradition. In sixteenth- and seventeenth-century New England there was still a lot of speaking about speaking – the power of talk was a favourite topic of conversation. Words and deeds retained a close relationship, and preachers' voices were in some sense supposed to echo that of God himself.[28] The writing of James Joyce and Gerard Manley Hopkins, with its spoken rhythms, is another example of oral residue,[29] as is the defence of the doctoral dissertation in some universities, in person and orally, and the way that children like to hear story cassettes over and over again.[30]

But are these just residues? Although most of our official lives are recorded on paper or, today, hard disk, we use our voices in daily life just as much as our ancestors did. What's changed is that we don't recognise the fact. Some of the historians of literacy have contributed to this by suggesting, for example, that rhythm, which had been essential to support oral memory, is now no longer needed.[31] In fact, as chapters 6 and 7 show, it remains the bedrock of vocal interaction and learning. By concentrating so hard on

those quarters where print has supplanted speech, these writers
sometimes risk taking their agenda from literacy rather than
orality. Of course the starting-point for all of us is our own society,
but by casting back from the vantagepoint of the literacy that *in
some respects* supplanted orality, we risk regarding oral societies as
ones that existed before writing rather than without it.

It's practically impossible not to treat literacy as inevitable, and
oral societies as the precursor of literate ones. If nothing else,
chronological bias encourages us to think of the former as being
supplanted by the latter. But this kind of either/or-ism isn't helpful,
because literacy didn't replace orality, only supplemented it. Indeed
in certain quarters the voice has become more and not less
important – to American presidents, for instance, as the next
chapter shows. Between parents and children, couples and friends
it also remains, as I've tried to show, central.

The experience of childhood is the nearest we come to understanding what pre-literate, oral societies were like – in this sense we
all develop from our own personal primary orality. Yet although
children live in an aural culture, it's not long before they realise that
their elders are bi-cultural. But just as Western society hasn't lost its
dependence on the voice even as it's developed other forms of
communication, so too do children to some extent hold on to their
own orality. Even fluent child readers love reading aloud.[32]

## DOWN WITH WRITING (THEY WRITE)

Studies of orality seem to acknowledge print and writing's more
sublime creations sometimes only as an afterthought, and yet the
idea of a schism between eye and ear, the written and spoken word,
is hard to sustain when you think of the oral treasures that
Shakespeare produced through the medium of writing. Though
they obviously differ from Homer's, they're no less glorious. Yet to
the historians of literacy, print sometimes seems like an insensitive,
marauding coloniser, with the oral reduced to disenfranchised
native.

Marshall McLuhan, who wasn't just the epigrammatist of the ephemeral (as he's now usually characterised) but also a serious scholar, was especially dismissive of print, which he compared unfavourably with orality – either 'primary orality', or the 'secondary orality' he believed was ushered in by the electronic age that succeeded the 500-year-long typographic era.[33]

The philosopher Jacques Derrida took up the opposite position, complaining that Western thought regarded speech as superior to writing, a prejudice he called phonocentrism. Its origin, he claimed, lay in the fact that both speaker and listener (unlike reader and writer) are joined by the unity of time and space. 'My words are "alive" because they seem not to leave me: not to fall outside me, outside my breath, at the visible distance . . .'[34]

Both camps tend to treat sight and sound as rivals: the propagandists for the ear have turned into the detractors of the eye, and vice versa. Isn't it time to make peace between them?

## LITTLE VOICE

All this talk of orality remains surprisingly mute about the real, embodied voice. Despite eloquent descriptions of its transitory qualities (see chapter 1), the scholars of orality still equate voice mostly with speech, the acoustic with the spoken word. In one of the most important texts[35] 'voice' simply doesn't appear in the index, as if orality were a phenomenon without agency or medium. These scholars, all too often, wed voice to language: while it's hard to divorce them, we limit our understanding of the voice if we're unable, even temporarily, to think about it without immediately focusing also on its language partner. And when these scholars finally attend to acoustic matters, they usually concentrate on how sound shapes speech and thought, as if voice and sound were interchangeable.

In a peculiar way the orality debates have marginalised the voice almost as much as contemporary visual culture does. One writer even makes a point of insisting that theories of orality can't and shouldn't deal with the expressive, impermanent aspects of orality, i.e., 'a performance of a person's mouth, addressing another person's ear and hearing with his own personal ear the spontaneous personal reply'.[36]

Why not? Because, apparently, 'oralist theory has to come to terms with communication . . . as it is preserved in lasting form'.[37] What a contradiction! If these theorists lose contact with the fleeting nature of the physical voice, if its tones and cadences aren't ringing in their ears, how can they hope fully to comprehend the extent of the shift that's occurred, how can the voice they track be anything more than a simulacrum, a model as inert and flattened as they feel the text to be?

Part of the difficulty, of course, lies in finding ways of talking about the voice – that slippery, elusive (and exquisite) thing. Without speech and words, the voice seems to consist purely of sound – the grunts, cries, and laughs that are the body's own soundtrack. Yet although the voice is made of sound, it's also more than sound – it's charged sound, revved by a private, bodily engine. But music and abstract art have developed their own lexicon, so why not the voice too?

## ILLUMINATING THE VOICE

The 'eye's eclipse' of the ear'[38] hasn't gone unchallenged. It's the written word and not the spoken one, some say, that 'is a fragile thing' – just think of how the library of Alexandria went up in smoke.[39] On the other hand, even in cultures steeped in print for centuries, spoken knowledge is still a major form of human exchange. Science, for example, depends as much on the lecture, discussion group, lab work, conversation among friends, as on reading and publication.[40]

The philosopher Michel Foucault has argued that at the beginning of the nineteenth century the relationship between the invisible and the visible underwent a radical change.[41] A host of new inventions enabled the inside of the body to be seen for the first time. These instruments of the eye were themselves made possible and inevitable by an emerging new world-view – a belief in the primacy of sight, in the credo that seeing leads inexorably to knowing (and thence to believing). This new conviction held that, exposed to the effulgence of the medical gaze, obscurantism would dissipate. Each fresh invention widened and deepened the certainty that the human body need never be opaque again. Samuel Johnson had defined 'to enlighten' not only as 'to illuminate . . . instruct, to furnish with increase of knowledge' but also 'to supply with sight, to quicken in the faculty of vision',[42] and these new instruments of the gaze, it's sometimes argued, helped to marginalise the voice by downgrading the faculties of speaking and hearing.

Yet, paradoxically, the invention of a whole array of instruments of observation produced greater knowledge about how those aural faculties operated. The Englightenment, according to one cultural historian, advanced not only the study of optics but also of acoustics, encouraging a wide and fluid interest in all the senses, even if 'the new experimental philosophy sought at a number of levels to make sound intelligible by rendering it manifest to the eye'.[43] As well as the Englightenment, another scholar of sound has suggested, there was also an Ensoniment – a series of ideas, institutions and practices that made the world audible in a new way. The sense of hearing, in particular, was 'measured, objectified, isolated, and simulated'.[44]

The desire to view the organs of speech didn't originate in any case in the Englightenment. Already in the early sixteenth century the anatomy textbook of Leonardo Da Vinci included several drawings of the larynx. Dissecting corpses in hospitals in search of new information about the anatomy, physiology, and pathology of the human voice, Da Vinci attempted, as befitted a painter, to visualise the organs that produced it.[45]

Though the Englightenment, too, tried to visualise the voice, it also generated an astonishing amount of talk. Taverns, salons, clubs, theatres – metropolitan venues like these thrived. The explosion of print – 56,000 titles were published in Britain in the

1790s[46] – and the multitude of newspapers and pamphlets, far from silencing the voice, only gave rise to more public places, like the coffee house, in which it could be exercised.

In the coffee house there was no seating according to class or status:

> Distinctions of rank were temporarily suspended; anyone sitting in the coffee house had a right to talk to anyone else, to enter into any conversation, whether he knew the other people or not, whether he was bidden to speak or not. It was bad form to even touch on the social origins of other persons when talking to them in the coffee house, because the free flow of talk might then be impeded . . . Tone of voice, elocution . . . might be noticeable, but the whole point was not to notice.[47]

It therefore marked a certain democratisation of the voice – 'certain' because women were completely excluded.[48]

Beyond the metropolis, too, the voice was the main medium for leisure activities. Poor country people passed the long winter nights by reading aloud. And the advance of print culture sparked a corresponding interest in what remained of traditional oral culture, creating a drive to 'pickle and preserve folklore, songs and sayings'.[49]

## THE VOICE OF PANIC

So have the voice and hearing been displaced by the printed word and sight, or is the acoustic just as omnipresent as ever? There's no doubt that contemporary culture graces the visual with a special status. Expressions like 'I see what you mean', 'I'll see to it', 'Point of view', 'Seeing is believing', demonstrate how we equate sight with understanding, doing, and reality, fetishising the graphic. And yet the uncritical acceptance of the idea that ours is a visual culture has seduced us into believing that the voice has been somehow superseded, even though new technologies, as we'll see, have enhanced its importance rather than diminished it. In fact it's as much the *propaganda* of print culture as its reality that has marginalised the voice: the voice hasn't lost its centrality, only its representation – it's a citizen stripped of the vote. The voice is still the connective tissue between humans, even if the Stalinists of vision have tried to erase this from the public record. Let's not be deluded by optic's own illusion.

## MR SEVEN PER CENT

There's a coda to this story, one that demonstrates that, if under-estimating the role of the voice is dangerous, so too is exaggerating it. In the 1960s and '70s, just at the point where Marshall McLuhan and other historians of orality were drawing attention to the differences between the ear and the eye, another researcher was attempting to measure the relative importance of the voice, the face, and the word in communication. The way in which UCLA psychology professor Albert Mehrabian's limited (and frankly tenuous) conclusions were taken up by the public imagination and distorted by a klaxon of publicity is a fascinating case-study in the popularisation of science, as well as a dispiriting example of the human desire for certainty.

Those who've never heard of Mehrabian will almost certainly be aware of his conclusions. Bandied about in TV advertisements for cars or banks, or simply part of common currency, they claim that only 7 per cent of meaning in human communication comes from words, the rest from the voice and face.

In the first of Mehrabian's studies, three women had to say the word 'maybe' three times, to express like, neutrality, and dislike, and then seventeen other women were asked to imagine that the speaker was saying the word to someone else, and guess the speaker's attitude to the addressee.[50] The second study, to test the decoding of inconsistent communications, was more ambitious – two women read out eight words ('honey', 'thanks', 'dear', 'maybe', 'really', 'don't', 'brute', and 'terrible') in positive, neutral, and negative tones, and ten people were asked to imagine that the speakers were talking to someone else, and to judge the information given by their tone. The study concluded that 'the tonal component makes a disproportionately greater contribution to the interpretation of the total message than does the content component'.[51] On the basis of the results of both pieces of research Mehrabian tentatively touted the idea that the verbal content communicated 7 per cent of the meaning, the vocal 38 per cent, and the facial 55 per cent.[52] By the time he came to write *Nonverbal Communication*, published in 1972, he was less hesitant in expounding his 'simple linear model' of 'Total feeling = 7 per cent verbal feeling + 38 per cent vocal feeling + 55 per cent facial feeling'.[53]

So an entire theory about human communication was based on three women saying one word, followed by two women reading eight words – reading, not even spontaneously saying – and two dozen listeners 'imagining' these words being spoken to someone else, and speculating about what feeling was being expressed! My primary-school daughter is currently studying what constitutes a fair test: I doubt these two would pass muster.

According to the philosopher Karl Popper, a proposition is unscientific if it's falsifiable by observation or experience.[54] Imagine, then, that I ask someone the way to the bus stop for the no 24 bus at nine in the morning. Were I only to pay 7 per cent attention to the verbal content of their reply, I might well still be wandering around looking for the bus at nine in the evening. It's palpably absurd to suggest that in human conversation words play such an insignificant role, and the face such a major one: it clearly depends on who is talking to whom, when, where, and why. And yet Mehrabian's theory has been disseminated all round the world, gathering such facticity on its way that it's now often referred to as 'the 7 per cent rule'. Seminars on public speaking are based upon it; it's the mantra of management-training and communication-skills workshops; and it underpins New Age therapies like neurolinguistic programming. Mehrabian's percentages have entered the culture as if they somehow contained information about the genetic basis of effective communication.

To be fair, Mehrabian can't necessarily be blamed for the ways in which his work was popularised, or perhaps even bowdlerised. Although research like this is really no more than a 'Let's pretend', a simulation that's one step up from rank speculation, it goes largely unchallenged, as most scientific work does (except by other scientists). For, despite the development of more critical attitudes to science over the past two decades, we still tend to accord it a privileged status, as though it were thought with the thinkers removed, discoveries just waiting to be discovered.[55] When an American psychologist did the promotional media circuit for her new book on anger, she noticed that she was rarely asked by reporters how she knew what she knew. Most of them couldn't evaluate her work, and simply seized on it for its value as controversy.[56]

Mehrabian seems not a little embarrassed by the independent life his 'research' has acquired. 'My findings are often misquoted,' he said. 'Please remember that all of my findings on inconsistent or redundant communications dealt with communications and attitudes. This is the realm in which they are applicable. Clearly it's absurd to imply or suggest that the verbal portion of all communication constitutes only 7 per cent of the message.'[57] Even when his studies were originally published, Mehrabian envisaged them chiefly being applied to inconsistent communications in the families of schizophrenics.[58]

On the other hand, Mehrabian has appeared to encourage the application of his findings to all sorts of everyday situations, as when promoting his later volume Silent Messages on his website. And a man who also wrote a book purporting to give 'scientifically based information for choosing baby names', with chapters on names that connote success, or sound masculine and feminine,[59] is clearly no enemy to popularisation.

But what's more interesting is why the findings from two such plainly minor studies spread like a particularly virulent infectious disease. It's especially intriguing since Mehrabian wasn't the first person to try and quantify the role played by the voice in communication. In 1955 a linguist called Lotz maintained that only 1 per cent of the acoustic signals emitted by the adult human voice were of linguistic use, the remainder being vocal or phonic.[60] Yet Lotz's claims were never, to my knowledge, seized upon with such enthusiasm by the extra-linguistic world. The reason for Mehrabian's fame as compared to Lotz's is surely that, in the decade or so that separated them, communication had come to occupy a much more vexed place in social and community life – it was now a problem for which solutions were sought. The point about Mehrabian's work is that it carried the tang of scientific exactitude: it offered the promise that the ingredients of successful human communication could be precisely calibrated. Of course you can more easily pin the tail on the donkey than fix a percentage to the role played by the voice in conveying meaning: if Mehrabian's studies measure anything, it's the extent of our desperation to find a formula to help us communicate.

The overselling of the voice is to some extent a consequence of our failure to develop a shared language of the voice. In that vacuum exaggerated claims and unfounded fears have flourished. Lacking a lexicon of nuance, we've developed one of hyperbole instead. The challenge is to acknowledge and celebrate the role of the voice in both public and private life without distorting it.

# 14

# *The Public Voice*

DESPITE PROFOUND HISTORICAL changes in public-speaking styles, the voice has remained a vital instrument for inspiring, influencing, and convincing – a weapon of mass persuasion. Commanding and controlling it is just as crucial to the political leader in the modern digital world as it was in the Athenian polis. The voice hasn't been displaced from the centre of public life; if anything it's gained in importance. Tracking transformations in the public voice can tell us a lot about the evolving nature of public life – what we demand from politicians, actors, and other public figures now and in the past.

Ideas about the voice that developed in Ancient Greece have been extraordinarily influential, enduring right up until the 1960s. Although Aristotle described the voice in spiritual terms ('Voice is a kind of sound characteristic of what has soul in it; nothing that is without soul utters voice'[1]), he also conceived of it as a major oratorical tool.[2] With written texts only just beginning to be widely available, rhetoric ('*rhetor*' was the Greek word for 'orator') was critical to the art of swaying others.

In Greek cities public speaking had become so pivotal by 450 BC that 'it was positively dangerous to neglect it'.[3] So obsessed were the Athenians with the improvement of the voice that they employed three different classes of teachers for the purpose: the *vociferaii* to strengthen the voice, the *phonasci* to make it more sonorous, and the *vocales*, the finishing masters, in charge of intonation and inflection.[4]

Greek physicians were also fascinated by the voice. Hippocrates realised that the lungs and trachea played a role in its production, and the lips and tongue in articulation.[5] He tried to label vocal states like clarity, hoarseness, and shrillness, and use them for diagnostic purposes. Galen, known as the founder of laryngology,[6] declared that 'the glottis is the principal organ of the voice', and was the first to describe workings of the vocal apparatus.[7] Finally, the writers of 'physiognomics', who claimed to be able to deduce character from physical qualities, expatiated on the relationship supposedly existing between certain kinds of voices and people.[8] Between them, in this most oral of cultures, these specialists had at their disposal a rich vocabulary to describe the characteristics of the speaking voice.

Rhetoric was essential to the Roman Empire too. Although Cicero's was so florid that he's been accused of contributing to the degeneration of eloquence,[9] his writings on the psychological dimensions of the voice now seem very modern. 'For every emotion of the mind has from nature its own peculiar . . . tone . . . and the variations of his voice . . . sound like strings in a musical instrument, just as they are moved by the affections of the mind.'[10] Cicero's complete works were rediscovered in 1421 and exerted an enormous influence over the Renaissance. His style of oratory provided the model for public speaking until the spread of electronic media: when Harvard was founded in 1636, the first criterion for admission was to demonstrate an understanding of Cicero.[11]

But Quintilian, another Roman writer who outlined 'the art of speaking well' (both effectively and virtuously),[12] was classical rhetoric's most distinguished analyst. 'Every human being,' he argued, 'possesses a distinctive voice of his own, which is as easily distinguished by the ear as are facial characteristics by the eye.' When it rings with passion but is also controlled as a result of vocal exercises, 'the voice, which is the intermediary between ourselves and our hearers, will then produce precisely the same emotion in the judge that we have put into it. For it is the index of the mind, and is capable of expressing all its varieties of feeling.'[13] The study of rhetoric remained a central part of formal education for around two and a half millennia, until the arrival of the Romantic movement in the nineteenth century.[14] The University of Oxford still has its own public orator, but oratory today is usually equated with pomposity and demagoguery, and despised for it.[15]

## SOUNDING THEATRICAL

Theatre compels us to listen to the voice, but the actor's voice has changed enormously over time. Greek theatre included highly rhythmical recited or chanted texts, for which actors trained carefully, even dieting and fasting to keep their vocal instrument in perfect condition.[16] Clarity and projection were vital. Sophocles, it's said, had to give up acting in his own plays because of his weak voice.[17]

Tragedy and comedy both relied on masks, which meant that the actor's face had no role to play in creating effects – all was voice and movement. Masks fortified vocal power in another way because the space between them and the actor's head provided an extra resonating chamber for the voice, helping the speaker control its direction and volume as well as rhythm and tone. From behind the mask he'd release a whole array of ritual laments associated with funerals. Greek and Roman amphitheatres were huge, and actors were expected to be vocally powerful enough to reach every corner, although they were helped by the superb acoustics – a modern writer has confirmed that in Epidaurus, even today, you can hear a pin drop distinctly in every one of its 14,000 seats.[18]

In the eighteenth century, acting began its divorce from oratory. The British actor, David Garrick, seemed to typify the new fashion: to most of his contemporaries he seemed entirely natural.[19] And yet the declamatory style of performing survived for a long time.[20]

Nineteenth-century poets and writers, like nineteenth-century politicians, recited slowly and to modern ears over-melodiously, as if
their voices were imprisoned in a relentless sing-song they were
powerless to escape. Rhythm seemed to monotonise rather than
enliven.[21]

On the British stage, heroic acting continued well past the middle
of the twentieth century. There, voices like John Gielgud's still
twirled and paraded. Gielgud's sonorities epitomised the voice
beautiful: they've been described as 'full of velvet and self-
esteem'.[22] At the same time whole schools of theatre were developing around a freer use of the voice and its connection with
the personality. Alfred Wolfsohn, a Jewish refugee from Nazi
Germany, believed that the range of the human voice could be
extended to seven or even nine octaves if actors realised that it was
produced by many different parts of the body and resonated
throughout it.[23] His successor, Roy Hart, drew on a similarly
uninhibited and cathartic wide range of vocal sounds.[24]

Regional accents were heard in British theatre and movies for the
first time in the 1950s, ushered in by John Osborne's *Look Back in
Anger* and the new wave of film. The voices of so-called angry
young men, instead of aspiring towards 'proper speaking', were
propelled by the energy of dissent, their very acoustic a snub to the
social order.[25]

Film liberated the actor from the need to project. Fluency was no
longer the ideal – actors now tried to hesitate, falter and stumble in
the style of the marginalised, those who could have been a con-
tender. Too much vocal clarity had come to signal acting, rather
than good acting. The Method school became associated with
stuttering and incomprehensibility and Marlon Brando caricatured
as the Great Mumbler. He insisted, 'I played many roles in which I
didn't mumble a single syllable, but in others I did because it is the
way people speak in ordinary life . . . In ordinary life people
seldom know what they're going to say when they open their
mouths . . . They pause for an instant to find the right word, search
their minds to compose a sentence, then express it.'[26]

The eighteenth- and nineteenth-century actor's voice had made no attempt to conceal that it was repeating the dramatist's lines; by the mid-twentieth century film actors tried to sound as if they'd dreamt them up themselves. The most acclaimed film stars of the time – Brando, James Dean, Montgomery Clift – gave off a brooding intensity that suggested psychic struggle, as though the voice were just the outward flickerings of an essentially interior process.[27]

Actors were responding to broader changes in the vocal culture. Male and female ways of speaking have always been shaped by the social values of the time, and even by its clothing. In the nineteenth century the boned corset limited the amount of breath that the 'well-dressed' middle- or upper-class Victorian woman could take in, severely constricting the sound that she could make. 'The speaking voice which resulted would have been the breathy, thin, high and unresonant, rather childlike voices we associate with some of Dickens' female characters.'[28]

If you listen to voices from the first half of the twentieth century today, they sound almost somnolent. George Eliot observed that in the nineteenth-century British drawing-room guests were stilled 'by the deep-piled carpet and by the high English breeding that subdues all voice'.[29] A 46-year-old man remembers that the voices of his youth 'were more muted, an indication of a more general mutedeness in feeling and expression',[30] while a 63-year-old man recalls, 'We had a far less frenetic voice – you have to raise your voice to be heard today.'[31]

As the pace of life quickened so did the voices. The rapid riffs of bebop – Charlie Parker's soaring sax, with its fresh phrasing and harmonies – provided a new acoustic, while the beat poets (not for nothing was Allen Ginsberg's most famous poem called 'Howl'), hippies and, later, rap music played their part in dissolving vocal inhibitions, leading to an audible generational divide.

There was always a place, though, for traditional rousing oratory. Martin Luther King's 'I have a dream' speech, delivered at the Lincoln Memorial in Washington DC on 28 August 1963, is often singled out as the most powerful example of twentieth-century public speaking. A formal speech that brilliantly evoked the American dream and how far it had yet to be fulfilled in black Americans' lives, it started slowly and built in pace, volume, and urgency. King used his voice in a musical way, in places almost singing: by rhythmically repeating phrases like 'Now is the time' he worked the crowd. As one witness declared:

> Three centuries of the rhetoric of the South were pulled together in one exalted outburst. Every device ever contrived by every preacher of the South, black or white, was put to use, until his huge audience, black and white, had been carried beyond itself, no longer merely the sum of its members. He bit into the gathering 'Amens,' the answering 'Yeah! Yeah!', the thundering applause, for they were not to be allowed to rest, but were to be carried to a higher pitch with each ejaculation.[32]

By the 1980s the sound of Woody Allen's kvetch, Richard Pryor's falsetto and Bob Geldof's Live Aid roar were markers of just how unbuttoned the public voice had become. Today's voices are far less clipped and fluting, their vowels much less strangulated and their consonants less precisely articulated than those of even a half-century ago. At the same time the attenuation of public speaking produced far less resonant voices than those of previous generations. By the early 1990s contemporary speaking voices were already lighter than a decade earlier, and few could fill a space without being electrically boosted by a microphone.[33]

Although the postmodern experiments of performance artists like Laurie Anderson treated the voice like another instrument, splicing together snatches of monologue with music and sound effects, it could still act as a bridge for characters (like Samuel Beckett's) otherwise divided by their misunderstandings, bodies and despair. In Beckett's Not I,[34] first staged in 1972, a woman sits on a darkened stage, the only thing visible her red mouth – as if the voice were the final embers of life.

## LISTENING TO HISTORY

Today, with the preservation of the speeches of many major twentieth-century figures,[35] the afterglow of the voice has been infinitely extended. The story of twentieth-century politicians is in a sense a story of the voice, of successful leaders able to inspire a nation through vocal skill.

## FRANKLIN DELANO ROOSEVELT

By the time Americans sat down beside their radio sets on Sunday evening, 12 March 1933, to listen to President Roosevelt, the Great Depression had put a quarter of them out of work, and every bank in America had been closed for at least eight days. In his 'fireside chat' (the first of thirty-one), FDR announced that the banks would reopen the next day, and that most of their deposits would be guaranteed by the federal government.

Although Roosevelt's administration didn't call it a fireside chat – that term was coined by CBS's Harry Butcher in a press release before the second, on the New Deal – the description stuck and soon the President was using it himself. It served to emphasise their colloquial nature:[36] though FDR wasn't sitting by the fireside himself, only his listeners, the President made it sound by his intimate delivery as if he were there too.[37] When he got before a microphone, his secretary of labour recalled, 'His head would nod and his hands would move in simple, natural, comfortable gestures. His face would smile and light up as though he were actually sitting on the front porch or in the parlor. People felt this, and it bound them to him in affection.'

Roosevelt was one of the earliest to understand the difference between broadcasting and other public speaking. He didn't orate.[38] Instead, a friend recalled, he 'made any speech that he delivered so much his own that it was what he might say in conversation. He never seemed to be reading to an audience. Neither did he seem to be reciting'.[39] He appeared, said another observer, to be 'talking and toasting marshmallows at the same time.'[40]

FDR was also the first American president to use radio to project his ideas and personality directly into American homes, and bypass newspapers that were critical of him.[41] As the political commentator David Halberstam has pointed out:

> He was the first great American radio voice. For most Americans of this generation, their first memory of politics would be sitting by a radio and hearing that voice, strong, confident, totally at ease. If he was going to speak, the idea of doing something else was unthinkable. If they did not yet have a radio, they walked the requisite several hundred yards to the home of a more fortunate neighbor who did . . . Most Americans in the previous 160 years had never even seen a president; now almost all of them were hearing him, in their own homes. It was literally and figuratively electrifying.[42]

Radio allowed FDR to conceal his handicap: his disembodied broadcast voice could be transformed in the listener's imagination into the emanation of someone able-bodied, with the ability to travel unimpeded.[43]

Roosevelt's urbane and self-assured voice has been compared to a bell in the darkness. As it developed from patrician to paternal, it had a containing effect on listeners, and helped give the nation hope in the Depression, just as Churchill's voice would during the Second World War. 'Roosevelt's voice,' said the philosopher T. V. Smith, 'knew how to articulate only the everlasting yea.' Its measured pace and level tone seemed to guarantee Americans safety, no matter how volcanic the events of the world.[44]

## HITLER

Hitler's voice – aggressive, resolute, and staccato – incarnated Nazi ideology. He used it to galvanise himself and the public. It was an instrument for the incitement of mutual frenzy.

This ability, when he first became aware of it in a beer-hall in 1919, thrilled him. As he wrote in *Mein Kampf*, 'What before I had simply felt within me, without in any way knowing it, was now proved by reality: I could speak!'[45] Hitler preferred the spoken to the written word, believing it had magic power.[46] He tried to refine his oratorical skills by studying the techniques used by Weiss Ferdl (a popular Munich comedian) to capture the attention of noisy beer-hall crowds before beginning his act. Hitler also scrutinised the Munich beer-halls' acoustics, adjusting the pitch of his voice to suit each one.[47]

What Hitler said was unoriginal.[48] His voice was metallic, its timbre harsh compared with Goebbels', which was beautiful. He spoke for too long, and was often repetitive and verbose. 'These shortcomings, however, mattered little beside the extraordinary impression of force, the immediacy of passion, the intensity of hatred, fury and menace conveyed by the sound of his voice alone.'[49]

Today Hitler is seen as a ranting, mob-inciting, demonic figure, carried away by the hysterical tenor of his own oratory, a view created largely by short clips – now iconic images – from 1930s' newsreels. In reality he made sophisticated, calculating use of public-speaking techniques. Speaking from rough notes – mainly a series of jotted headings with key words underlined – rather than reading his speeches, he conveyed an impression of spontaneity and freshness. He could deal expertly with hecklers (although so could his police).[50] He understood the exact moment to resume speaking through applause just after it had started to drop off.

'He did not rant and rave all the time – a physical impossibility for a man who spoke normally two hours or more – but addressed his audience in quiet tones at first, even hesitantly.'[51] The whole performance was carefully choreographed. Hitler would delay his entry into the hall. Once there, he'd begin with a pause 'that seemed to become utterly unbearable'.[52] Then he'd start speaking softly and slowly, a low-key opening designed to allow tension to mount. Gradually he'd grow louder or, taking up a more fighting tone in response to a catcall, insert biting sarcasm. Deliberately repeating words like 'smash', 'force', 'ruthless' and 'hatred', he'd lash himself up 'to a pitch of near-hysteria in which he would scream and spit out his resentment'.[53] His voice became hoarse, sometimes croaking, as his tirades reached a climax. But for all his seeming abandon, he never lost control.[54]

When an American professor of the time analysed Hitler's voice, he found that it had a typical frequency of 228 vibrations per second, compared with the usual frequency for anger of 200 vibrations per second. 'It is this high pitch and its accompanying emotion that puts the people in a passive state,' he maintained. 'He stuns them with his words in much the same fashion as we are stunned by an auto horn.'[55] So aggressive were Hitler's harangues that they left listeners only one option: if they were not to become the object of his attack, they had to identify with the aggressor.

His gatherings resembled revivalist meetings. Men would groan or hiss and women sob involuntarily 'if only to relieve the tension'.[56] The German writer Joachim Fest has compared Hitler's speeches and the public response to a collective orgy. 'The sound recordings of the period clearly convey the peculiarly obscene, copulatory character of mass meetings: the silence at the beginning, as of a whole multitude holding its breath; the short, shrill yappings; the minor climaxes and first sounds of liberation on the part of the crowd; finally the frenzy, more climaxes, and then the ecstasies released by the finally unblocked oratorical orgasms.'[57]

According to Traudl Junger, Hitler's infatuated secretary, 'The same man who made speeches . . . with that rolling "r" and that roaring – I never heard him speak like that in private. He would speak in such a flattering, such a moderate tone. In his private life he had that gentle Austrian intonation too.'[58] And yet Hitler himself admitted, 'I must have a crowd when I speak. In a small intimate circle I never know what to say.'[59]

His relationship with large crowds was symbiotic: he communicated an excitement to them that in turn provided fresh impetus to his voice.[60] Hitler used his speeches to make himself feel alive and intoxicate himself, 'whipping himself and his audience into anger and exultation by the sound of his voice'.[61]

His giant gatherings also used hypnotic drum-rolls to make the crowds more receptive and breathe in unison.[62] Applause-manipulation facilities were built into the very design of the Nuremberg stadium. Strategically positioned microphones were wired to amplifiers hidden behind the rostrum so that the cheers and chants of 'Heil Hitler' could be amplified but also played back at the crowd. The artificial source of the increased fervour was invisible to the newsreel cameras.[63]

Radio carried Hitler's voice not only across Germany but throughout the world. In September 1938 Virginia and Leonard Woolf heard the Nuremberg rally live on the radio. She wrote in her diary, 'Hitler boasted and boomed but shot no solid bolt, mere violent rant, & then broke off. We listened to the end. A savage howl like a person excruciated; then howls from the audience; then a more spaced and measured sentence. Then another bark . . . the voice was frightening.'[64] Because of the time difference, Hitler's voice was often heard on American radio in the morning just as listeners were going to work. The very sound of his voice, some later recalled, convinced them that danger lay ahead.[65]

So important was Hitler's voice to his leadership that one of the dafter ideas to issue from the Office of Strategic Services, the American intelligence agency, was a plan to slip female sex hormones into his meals so as to raise the pitch of his voice.[66]

Hitler's raucous style of speaking had become so identified with tyranny and genocide that never again would it find favour. Except perhaps for some Eastern European/Soviet bloc despots, no post-war politician would thunder and bellow themselves into a frenzy. Hitler made demagoguery suspect. Today we expect the persuasive voice to be smooth and cajoling, to caress and flatter the listener rather than harangue them into acquiescence. We expect it to sound more female.

## CHURCHILL

'He who enjoys the gift of oratory,' wrote a 21-year-old Winston Churchill, 'wields a power more durable than that of a great king.'[67] So the young Winston set about cultivating it, despite the inhibitions caused by a slight stammer and lisp. He analysed Disraeli's speeches and copied their rhetorical devices – short syllables and punchy, dramatic, cathartic endings,[68] sometimes spending more than ten hours composing a single speech.

Compared with the standard, high-pitched voices of the British upper classes in the 1930s and '40s Churchill's timbre was deep, warm, and fruity.[69] 'The people of Britain had the lion's heart,' he said. 'I had the luck to give the roar.' And roar he certainly did – one historian has described 'the growls, the sudden leonine roars, the lyrical sentences, the cigar-and-brandy-toned voice, the sheer defiance coming straight from the viscera insisting upon no surrender in a war to the death'.[70]

A recent BBC radio experiment compared a recording of an extract from Churchill's 'Finest Hour' 1940 address to the House of Commons with one of the same speech made by the broadcaster Melvyn Bragg. Bragg's version lasted 60 seconds, as against Churchill's 90.[71] Churchill not only spoke slower but also inserted dramatic pauses into almost every prosodic phrase. As for pitch-range, both Bragg and Churchill spanned an octave, but Churchill's voice was three semitones lower than Bragg's and his stresses idiosyncratic throughout. Churchill's low pitch made him sound like some elemental force. In his 'Battle of Britain' speech, by so dramatically dropping his voice on the words 'but if we fail',[72] he evoked the abyss of a new dark age.

## REAGAN

Ronald Reagan was the first actor to become President – proof, in some people's eyes, that in modern America the two roles were now indistinguishable – and his voice played an important part in his popularity. He was known as 'the Great Communicator', although some think he just read autocue better than any president before or since.[73] His was a reassuring voice – soft and folksy, that put people at ease. One observer commented that Reagan's voice 'recedes at the right moments, turning mellow at points of intensity. When it wishes to be most persuasive, it hovers barely above a whisper so as to win you over by intimacy, if not by substance . . . He likes his voice, treats it like a guest. He makes you part of the hospitality. It was that voice that carried him out of Dixon and away from the Depression.'[74]

Listeners responded to his self-deprecating wit and the relish with which he told stories. Through the warmth of his voice and his narrative skills he seemed to breach the distance between president and citizen. And yet Reagan, an experienced film actor, made skilful use of the feminine, expressive dimension of his voice. Recalling the death of American soldiers in the bombing in Lebanon, he had a catch in his voice and a tear in his eye, pursing his lips as if to control strong feelings[75] – the 'acted sincerity' techniques that Tony Blair would come to use later, albeit less convincingly. Reagan spoke informally and sentimentally: he had a picket-fence, small-town-decent-American-values kind of voice.[76]

Reagan mastered television the way Roosevelt mastered radio. His stint as radio announcer had honed his vocal skills, but he also understood the difference between the two media.[77] He was one of the earliest users of the autocue, which came to be called his 'sincerity machine'.[78] It enabled him to look directly at his audience instead of having to refer to notes, so creating the impression that he was speaking impromptu rather than delivering speeches crafted for him by others.

And yet Clinton, as much as Reagan, deserves the title of Great Communicator. Though Clinton's voice was often hoarse, it electrified people, one describing it as like molasses dripping through cornbread.[79] Through his voice he established a rapport with many different kinds of Americans, expressing a much-parodied 'I feel your pain' empathy.

## THATCHER

The surprising thing about the young Margaret Thatcher's voice is
how light and charming it sounds compared with her later one.[80]
Encountering the prejudice against women's voices and their
supposed shrillness, emotionalism, and lack of authority examined
in Part Two, Thatcher effected what was probably the most
significant change in any modern politician's voice.

Her retinue of advisers had already identified her supposed vocal
failings. Tim Bell, the advertising executive she brought in to run
the Conservative Party's election-publicity campaigns, recalls:

Physically she had a problem in that she spoke from the top of her
chest . . . had a slightly stressed larynx, and was prone to coughs
and colds. Her voice would sound strangulated when she was
tired, and the voice would get tired sooner than anything else . . .
[such as] the legs. When she first became leader of the opposition
she had a schoolmarmish, very slightly bossy, slightly hectoring
voice. It was a voice from the 1950s that was long gone.[81]

Britain had started to proletarianise and we wanted to reach
out to . . . ordinary people rather than the upper-middle classes.
So we – Gordon Reece, her publicity adviser, and me, head of
Saatchi and Saatchi – wanted her voice to deliver a message that
was simple and to the point and lightweight, e.g., you can't spend
more than you earn. If you deliver that in an upper-class voice
you sound patronising. If you deliver it in a B1, CD or Estuary
voice, you don't. The BBC voice she spoke in has been completely
discredited – her voice was grand and rather bossy and heavy-
weight . . . We knew that she'd developed the ability to project
her voice in Parliament and meetings but she'd do it too on TV
and it doesn't work – TV is chatty. She . . . [didn't] do small talk
and chatty.[82]

In 1978, before she became Prime Minister, Bell and Reece took
Thatcher to Laurence Olivier for vocal advice:

We sat in his place in Hampstead for four hours while he talked about how he had invented the theatre and acting and how he had made Shakespeare come to life. When, after four hours, she went ... [to the toilet] Gordon said, 'Can we talk about Margaret Thatcher's voice?' and Larry said, 'Absolutely typical of a politician – all they want to do is talk about themselves.' He gave her half an hour's advice on how to project her voice – mostly about learning to speak from down here rather than from the top and to talk to the person at the back of the hall in a public meeting because this helps you project.[83]

In spite of this, certain venues remained challenging for her throughout her tenure as Prime Minister: according to someone who knew her well, she had particular difficulty in the House of Commons, where she faced interruptions 'and the acoustics were curious'.[84] The playwright Ronald Millar, another adviser on communication issues, remarked that 'the selling of Margaret Thatcher had been put back two years by the mass broadcasting of Prime Minister's Question Time, as she had to be at her shrillest to be heard over the din'.[85] Millar taught her that lowering the voice brought the speed down to a steadier rate, and recommended holding to a steady and equable tone at Question Time eventually to drive through, not over or under, the noise.[86]

Tim Bell also tried to shape Thatcher's broadcast voice to her message:

When she did a party political broadcast we used to use a simple technique. We'd have two drinks ready – ice-cold water, and warm water with lemon and honey ... If she was coming to a point where she was trying to sound sympathetic and sensitive, we'd give her the honey-and-lemon water to soften and relax her voice. If she wanted to sound forceful we'd give her the ice-cold water.[87]

That Thatcher received voice training has become part of political lore, and yet it's dismissed by Bell, who worked closely with both Thatcher and Reece, neither of whom mentioned any such training to him. 'I'd be amazed if anyone popped up and said, "I gave her coaching lessons" – though they'd attract gales of publicity. So she probably just chose to lower it.'[88] She was certainly conscious of pitch: scribbled next to some of her pre-1979 speeches are notes on their ideal delivery.[89]

On the other hand, most people believe that it was Gordon Reece himself who directed the vocal change. As one journalist put it, she 'had high notes dangerous to passing sparrows' and, because her accent had changed when she was at Oxford, the effect was 'a bad case of stage posh'.[90] Reece advised her to speak more slowly and, on radio, closer to the microphone, which made her sound huskier. In 1977, within two months of him working with her, her voice was so much softer and lower that one interviewer unwisely asked her, 'Have you got a cold?'[91] In a BBC tape Reece can be heard coaching her to say the words, 'The socialists must learn that enough is enough,' and trying to encourage her away from her more derided duchess tones.[92]

Women's pitch, as we've seen, has dropped over the past four decades; Margaret Thatcher lowered hers in just one. When a linguist analysed recordings of Thatcher's voice over a span of ten years, she found that it had been artificially lowered by 60 Hz, or about half the normal difference between a female and a male voice. 'Not only was she speaking in an acoustic "no-person's-land", but there is a strong possibility that she was committing vocal abuse which could lead to the more serious pathological condition of vocal nodules'.[93] The news, many years after she left office, that Thatcher's doctors had advised her to do no more public speaking, seems to bear out this prediction.

Thatcher's new voice aimed for a forceful but caring sound; in reality it appeared strained, abnormally low, and contemptuous, making her sound as if she had the tiresome task of talking to an uncomprehending small child.[94] (The writer Keith Waterhouse declared, 'I cannot bring myself to vote for a woman who has been voice-trained to speak to me as though my dog had just died.'[95]) Not only did her new voice end up more caricatured than her old one, but it also gave the impression that Thatcher was trying to emulate men – that in order to sound authoritative she'd had to make herself sound male.

Her new pitch may have had another unintended consequence. When television interviews with Thatcher were compared with those of another British prime minister, Jim Callaghan, the surprising finding was that Thatcher was interrupted almost twice as often as Callaghan. Surprising because Thatcher seemed much more domineering in interviews than Callaghan, and usually tried to talk through interruptions to finish what she was saying.[96]

Was she interrupted more than Callaghan because she was a woman and in these interviews questioned by a man? Another explanation is that, although falling intonation usually provides a cue to when one speaker has finished talking and another may begin, Thatcher's intonation fell even when she hadn't finished a clause and wasn't ready to give way. By artificially changing her voice, Margaret Thatcher may have interfered with the paralinguistic cues that help smooth switches between speakers.[97]

## TONY BLAIR

Tony Blair's voice marked a significant break with those of previous British politicians. Like Clinton's, it brimmed with emotion, both of them using their voices to show empathy. At the same time they also had to avoid sounding effeminate. Sounding 'feminine' is a taboo for both male and female politicians yet, paradoxically, Clinton and Blair's voices signal the feminisation of the male public voice.

This was achieved partly by language,[98] but also through the voice. Unable to use the female glissando (slide in pitch), Blair falls back on a cracked-voice register break. Only immense self-control, this seems to be saying, enabled him to contain his overwhelming identification with other people's pain. The voice he used after the death of Princess Diana has been employed so routinely ever since that it now sounds as if the poor chap is almost perpetually choked with feeling, constantly on the verge of tears. How can he be moved by so many different things, so emotionally incontinent?

In fact, since Blair possesses the fluency of most educated modern politicians, he gives the impression of having deliberately injected hesitancy into his voice in order to create a sense of authenticity. And by over-using the device, he's drawn attention to it. This has happened over a relatively short period of time: if you listen to Tony Blair talking in 1997, he sounds not only much younger but also much less prone to the voice-crack. In the years since then, he's begun to impersonate himself.

Added to this, he sometimes roughs up his articulation and loosens his vowels: in certain circumstance the Blair 'I' emerges as an 'ah', or an Estuary-style mini-glottal stop suggestive of the common touch is inserted. Perhaps this too is a way of masculinising his voice, of restoring a sense of blokeishness and ensuring that the emoting hasn't over-feminised him or made him cissy (for the wobble in his voice must never give the impression of a wobble in policy or resolve).

Together these mannerisms have backfired, making Blair sound tremendously contrived. A 1999 Internet survey (though it may have polled the self-selected) found only 31 per cent of people considering the Prime Minister's voice trustworthy.[99] In 2003, after one parliamentary Question Time had been broadcast, a radio listener emailed the BBC to rail against Blair's 'smug, lisping, slightly effeminate, pseudo-Geordie, pseudo-Estuary, pseudo-Middle English, pseudo-public-school voice that, like Tony, tries to be all things to all people'.[100] How, increasingly, we warm to the unspun voice.

Yet Tony Blair's voice isn't simply a personal confection – it's also an expression of the new intimacy. Presentation in politics has become paramount, with the voice now a central part of impression-management (see chapter 17), leading to a corresponding cynicism about its manipulation. Blair, many voters now believe, 'seems to base his policies on trying to charm the nation'.[101]

But then never before was being a nice bloke such an essential attribute for political leaders (effectiveness was). Today, by contrast, one of the prime requirements of a politician is the ability to act sincere.[102] Yet should the effort of it, the art of it, become audible, the loss in credibility is enormous.

This marks a singular change in the history of public life. Before 1868, natural expression lay outside the public realm, and the language of politics was at one remove from intimate life. Since then, and particularly over the past decade, ours has become a culture of personality,[103] which has to be laid out for public perusal. See how perfectly the following description of the end of public culture, first published in 1977 when Tony Blair was barely out of university, fits him:

> It became logical for people to think of those who could actively display their emotions in public, whether as artists or politicians, as being men of special and superior personality. These men were to control, rather than interact with, the audience in front of whom they appeared . . . now what matters is not what you have done but how you feel about it . . . The modern charismatic leader destroys any distance between his own sentiments and impulses and those of his audience, and so, focusing his followers on his motivations, deflects them from measuring him in terms of his acts . . . It is uncivilised for a society to make its citizens feel a leader is believable because he can dramatise his own motivations. Leadership on these terms is a form of seduction.[104]

Tony Blair conducts his campaign of seduction primarily through his voice.

## BUSH, GORE, AND KERRY

Geniality, as a result, is no longer just an asset but an essential prerequisite for successful public office, one that the American politician can demonstrate through the voice. Whatever his intellectual superiority, Al Gore's stiff, sanctimonious monotone – often compared to that of a robot or metronome – put him at an enormous disadvantage beside George Bush's vocal affability. Even towards the end of the 2000 presidential campaign, when the body-language specialists had gone to work on him, trying to defrost Gore's timbre and limber his inflections, there was little suppleness of pitch or tone-colour.

The Howard Dean story, by contrast, is one of vocal excess. When, in 2004, the Vermont governor hoping to be selected as Democratic presidential candidate came a poor third in the Iowa caucuses, he made a raucous shriek of a concession speech, straightaway dubbed his 'I have a scream' speech, that immediately ruled him out of contention. If he couldn't control his voice, went the thinking, then how could he possibly keep the country or world in check?

Like Gore, John Kerry suffered from a lack of vocal charisma. His flat, stentorian speaking style[105] belonged to another era. His language and voice separated him from voters rather than bringing him closer: if his voice was to be believed, Kerry couldn't articulate their pain and certainly couldn't feel it. Voters never felt they were glimpsing the man within – Kerry offered none of the new intimacy.

The 2004 presidential election was distinguished by the number of voice coaches publicly analysing the candidates' delivery and offering advice. They berated Kerry's lack of vocal range, arguing – naturally – that he needed decent voice training. 'Kerry's got that deep, deliberate voice . . . He isn't the sort of person you want to sit down and have a drink with, necessarily . . . He is somebody whose speech was formed in boarding schools,' offered a Stanford University linguist.[106]

Critics faulted him for the lifelessness of his voice, its lack of dynamism and emotion,[107] which marked him out as aloof and patrician. He needed to de-Brahminise his delivery,[108] and take frequent dramatic pauses, varying tempo and register, volume and pitch.[109] Again and again commentators and journalists referred to the voice, sometimes even finding voters who regarded it as critical. 'I think a quality as seemingly trivial as vocal tone will play a factor in swaying as yet undecided voters.'[110] The playwright Arthur Miller said that political leaders everywhere have come to understand that to govern they must learn how to act.[111] Kerry, even though he was apparently assisted by a coach who graduated from the Yale Drama School, never did.[112]

George W. Bush, on the other hand, sold himself not just as the defender of the free world but also as the kind of bloke you'd have round for a barbecue. His voice positioned him as a regular guy, though in reality he was the scion of a powerful family. He constructed a persona of ordinariness that, combined with an ideology of conviction and the language of westerns, counted for more than his gaffes and malapropisms. As one commentator put it, 'Mr Kerry has a problem with rhetoric. He doesn't have his own sound. You may hate Mr Bush's sound but it's his, and a lot of people like it. He sounds normal, which for all its pluses and minuses as a style does tend to underscore the idea that he *is* normal.'[113] Indeed, Bush's many stumbles may have even endeared him to Middle America, which remains suspicious of East Coast fluency and uncynical about folksiness.

Among the other factors determining the outcome of the 2000 and 2004 American presidential elections, Gore's and Kerry's voices and personalities were a major element. An ability to communicate over the airwaves, which in a politician like Roosevelt had seemed like a serendipitous personal gift, is now mandatory. The modern political voice, compared with the ancient one, has been demoticised. Rhetoric is dead: the new orator must sound like a buddy.

# 15
## How Technology Has Transformed the Voice

WITHIN THE SPACE of twenty years at the end of the nineteenth century, three major new machines of sound reproduction emerged. They developed out of, and brought with them, new ways of thinking about the relationship between body and voice. Through them the voice was distributed, preserved, and diversified. The advent of the talkies in 1927 brought further change, and seemed to reunite body with voice, a process later accelerated by television. Yet despite this renewed intimacy between ear and eye, in little more than 100 years ideas about the voice had undergone a profound metamorphosis. The number and range of voices to which most of us are exposed also multiplied astonishingly in what was becoming a noisy new world. Technology didn't kill off the voice, as some feared and others believe – in many respects, its importance was enhanced – but the new technologies did help transform it.

## UNBODIED

Arriving in 1876, between the development of the telegraph and the phonograph, the telephone was in some sense just waiting to be invented. Commonly described as the first technology to disembody the voice[1] – to transport someone's voice without the accompaniment of their body[2] – the telephone extended the reach of the ear in an unprecedented way.[3]

Before its arrival, hearing voices when the speaker wasn't present was seen as a sign of either mysticism or insanity.[4] Now those qualities were projected on to the telephone itself, which seemed like 'a kind of extra-sensory perception',[5] 'our sixth-and-a-half sense'.[6] So disturbing was the apparent rupture between body and voice that it inspired not just awe and fear, but also contempt. The president of the Western Union turned down the chance to buy Bell's patents for $100,000, famously saying, 'What use could this company make of an electrical toy?',[7] and the chief engineer of the British Post Office sneered that the Americans needed phones because they lacked servants. 'If I want to send a message – I . . . employ a boy to take it.'[8]

The newspapers were full of foreboding. 'It is difficult,' wrote the *Providence Press*, 'to really resist the notion that the powers of darkness are in league with it.' The *Boston Advertiser* described a 'weirdness' never felt before in the city, while in *Scientific American*, an anonymous reporter recorded how disorientated the new development made him feel:

My own material existence I am reasonably assured of. I can imagine my friend at the other end of the line. But between us there is an airy nowhere, inhabited by voices and nothing else – Helloland, I should call it. The vocal inhabitants of this strange region have an amazing vanishing quality. Even while you are talking casually to one or another of them, you may become aware that you have been unaccountably 'cut off' . . . The telephone seems to have no visible agency.[9]

The idea that the voice could be canned, just like beef and milk,[10] seemed to denude it of its human quality, transforming it into just another commercial commodity. The voice of the female phone operator was employed to assuage these fears. Although the first commercial switchboard that opened in New Haven, Connecticut in 1878 hired boys, within a decade they'd been almost totally replaced by women, who were seen as more patient, polite, and pliable.[11] 'The dulcet tones of the feminine voice seem to exercise a soothing and calming effect upon the masculine mind, subduing irritation and suggesting gentleness of speech and demeanor; thereby avoiding unnecessary friction.'[12] If the telephone made people anxious because it gave alien voices direct access to the listener's ear, then the employment of women acting as intermediaries between the public and private worlds was seen as an antidote. Brought in to domesticate the telephone, they helped shift it in the public's mind from technological intruder to a medium through which social contact could be maintained by talk, even if the association of the phone with women soon opened it to male derision.

Today the phone's ability to connect people's voices across countries and continents is its most cherished function. As an Australian woman observed, 'The telephone is more personal than the post. What I want to know is what my friends are feeling, and that I can hear on the phone.' Another regarded the telephone as 'the instrument which enables women to build up their psychological neighbourhood'.[13]

The phone's spread is easily overstated – eight out of every ten people in the world have never made a phone call, and by 2001 there were more phones in New York than in all of rural Asia.[14] It's also had a contradictory effect on the voice, on the one hand helping to disperse people and so reducing the opportunity for face-to-face speech, and on the other compensating by reconnecting them (an example of how 'a technological device is eventually used in solving a problem it helped create').[15] While Bell predicted some of its effects – 'I believe in the future . . . a man in one part of the country may communicate by word of mouth with another in a different place'[16] – he was convinced that its chief function would be for news and entertainment, rather than speech. Indeed in the early days Philadelphia operators would give callers summaries of the news, while in Budapest in 1898 a telephone-newspaper was started up – as if the telephone were simply the radio-in-waiting.

## THE VOICE PRESERVED

Although the phonograph (precursor of the gramophone) or 'speaking machine' arrived only a year after the telephone, in the press at least it was treated almost as its successor. According to the *New York Times*, 'The telephone was justly regarded as an ingenious invention when it was first brought before the public, but it is destined to be entirely eclipsed by the new invention of the phonograph. The former transmitted sound. The latter bottles it up for future use.'[17]

*Harper's Weekly* was similarly fickle, extolling the phonograph's egalitarianism:

The telephone, which created such a sensation a short time ago by demonstrating the possibility of transmitting vocal sounds by telegraph, is now eclipsed by a new wonder called the phonograph. This little instrument records the utterances of the human voice, and like a faithless confidante repeats every secret confided to it whenever requested to do so. It will talk, sing, whistle, cough, sneeze, or perform any other acoustic feat. With charming impartiality it will express itself in the divine strains of a lyric goddess, or use the startling vernacular of a street Arab.[18]

The phonograph excited not only marvel but also, because of its associations with ghostliness, anxiety. As the press secretary of Thomas Edison, the phonograph's inventor, wrote, 'Whoever has spoken into the mouthpiece of the phonograph and whose words are recorded by it has the assurance that his speech may be reproduced audibly in his own tones long after he himself has turned to dust. The possibility is simply startling . . . Speech has become, as it were, immortal.'[19] Slipping from science to séance was all too easy.[20]

Indeed the phonograph, while it preserved speech, seemed also to possess the power to efface the speaker,[21] a classic ingredient of sci-fi horror. *Scientific American* made it sound almost as though the human voice were being entombed. 'The voices of . . . singers . . . will not die with them, but will remain as long as the metal in which they may be embodied will last.'[22]

Like the telephone, the phonograph seemed to represent the industrialisation of the voice. The *New York Times* worried about the demise of reading but also the arrival of 'bottled orations',[23] while the composer John Philip Sousa feared that the new machine would somehow render the voice redundant, declaring that 'with the phonograph vocal exercises will be out of vogue! Then what of the national throat? Will it not weaken? What of the national chest? Will it not shrink?'[24]

## THE IDEAS IN THE MACHINES

The idea that the new acoustic technologies brought unprecedented changes to the human voice is seductive, but leads all too easily to idealised fantasies of a pre-technological marriage of body and voice.[25] It can also give rise to a crude technological determinism: by looking at the telephone and phonograph's past from the vantage of their future, we make the route between them seem inevitable.

In fact, as we've seen, Bell originally thought that the telephone would emit as well as transmit. As for the phonograph, this had two needles at first – one for recording and another for playback. Indeed, because of its poor sound quality, Edison envisaged it mainly as a dictating machine, with its recording function uppermost in his mind. So the eventual path followed by the telephone and phonograph, and the social changes they produced, weren't foretold by the machines themselves, but were only one among many different tangents they could have taken. Edison even imagined that the phonograph might also be used as a 'family record', a kind of phono-album or 'registry of sayings, reminiscences, etc, by members of a family in their own voices, and of the last words of dying persons'.[26] Indeed, of the ten ways in which he thought his invention could 'benefit mankind', all but two were concerned with speech and the spoken voice rather than music, and yet the phonograph, and its commercial successor, Emile Berliner's gramophone (patented in 1887), gained their commercial success – once the early interest in speaking records had subsided – almost entirely from music and the singing voice.

On the other hand, the route eventually followed by the devices wasn't entirely accidental. Traditional histories of technology often mistake the causes of the new sound-reproduction machines for their effects. Technologies don't emerge out of nowhere but from the genius of great men: they're designed by human beings with certain ideas in mind that have evolved from the zeitgeist. As one perceptive cultural historian put it:

> Our most cherished pieties about sound-reproductive technologies – for instance, that they separated sounds from their source or that sound recording allows us to hear the voices of the dead – were not and are not innocent empirical descriptions of the technologies' impact. They were wishes that people grafted onto sound-reproduction technologies – wishes that became programs for innovation and use.[27]

In fact, many different cultural and economic shifts had helped to shape the new voice-machines, in particular the idea of the voice as a personal attribute, and the development of markets ready to embrace the new products, which itself was made possible by the growth of corporate capitalism, and American households' emerging sense of private space as they opened themselves up to consumerism.[28]

## MORE WAVES

Like Bell and Edison before him, Guglielmo Marconi didn't foresee what would emerge from his patent – in this case the development of broadcasting.[29] Radio, like the telephone and phonograph, elicited hope, but also nervousness. While the telephone voice travelled through wires, and the phonograph and gramophone voices were pressed on to foil or cylinders – material forms all – the medium of radio was the vast unbounded ether with its disturbing, uncanny connotations of the supernatural, telepathy, and clairvoyance.[30] And whereas the phonograph offered a means of preserving the voice, the radio only emphasised its transience and insubstantiality.[31]

The BBC was established in 1922, but two years earlier the first commercial radio station, KDKA, had opened in Pittsburgh. Between the 1930s and mid '50s, thanks partly to the mass production of cheap radio sets, radio was the dominant mass medium in the United States, with more Americans owning a radio than a telephone or phonograph.[32] With the introduction of the telephone, the phonograph and then radio 'there was a revolution in . . . [the] aural environment that prompted a major perceptual and cognitive shift, with a new emphasis on hearing'.[33] As one commentator in the late 1940s observed, 'After two decades of radio and sound movies, Americans are becoming more auditory-minded than visual-minded.'[34]

The spread of radio didn't only draw attention to the ear, but also brought new styles of speaking into vogue. As we have seen, it marked the beginning of the end of classical oratory: the orotund voice fell out of favour and an anti-oratorical sound became coveted, with the microphone favouring those who didn't boom into it as if addressing a mass public meeting. Writing about radio in 1936, the psychologist Rudolf Arnheim said that 'if a man is speaking *before* others, the most natural thing would be that he should also speak *to* them . . . as if they were sitting in front of him and could even answer him'.[35]

Already by 1933 an innovative BBC talks producer was arguing that the 'holy voice', the clerical intonation supposed to carry well in large echoing churches, and the poetry or 'elocution' voice, were both unsuited to broadcasting. Radio, she suggested, was bringing a new consciousness of speech, and broadcasters should think about cadence and rhythm:

> Why is it that some people, with voices like corncrakes or like sparrows, can hold the breathless attention of a vast audience? These successful voice and personality projectors seem to possess a particular range of personal qualities – they are human, sincere, unaffected and vital . . . much of the personality is revealed in characteristic cadences, hesitations, stresses, change of pace and general vocal gesture . . . within the limits of intelligibility, speakers' idiosyncrasies of voice should be left to speak for themselves.[36]

Yet it would be decades before her aspirations would be realised. In the meantime, radio speech remained mostly the province of men using a 'voice of authority' that purported to be the disinterested voice of truth[37] but was shaped partly by newsreels like *The March of Time*. 'Time Marches On!'[38] its narrator, Westbrook Van Voorhis, would thunder (first on radio but from 1935 also in the cinema), his melodramatic delivery and portentous script later parodied by Orson Welles in *Citizen Kane* ('Then, last week, as it must to all men, death came to Charles Foster Kane'). In one way or another, this style of commentary would persist until the 1950s.

In Britain during the 1930s radio was all but closed to working-class people: they were listeners rather than broadcasters. On the rare occasions they were interviewed, the recordings were taken away, transcribed and polished, and later read out from a script by the interviewee – no wonder they sounded so wooden. When, unusually, a group of Durham miners was put in the studio and told to talk unscripted, they swore so liberally that a young studio assistant was dispatched into the studio with a giant sign bearing the words 'Do Not Say Bloody and Bugger'. This so silenced the men that the assistant had to return to the studio with another sign saying 'As You Were'.[39]

Interviews were confined to the studio partly because the recording equipment of the time was so bulky: the arrival of more mobile vans allowed working-class people to be interviewed on their home turf, bringing their accents, rhythms and cadences on to the airwaves.

## VOICES OF WAR

The Second World War was a war of voices. It was the war that tipped radio into a prominent place in British life. Many Britons first knew that war had been declared, not from reading it in the newspaper (as with previous conflicts), but through hearing the sombre radio announcement of Neville Chamberlain 'speaking to you from the Cabinet Room of Number 10 Downing Street'. So grimly resonant was his address that most Britons over a certain age (and quite a few below) can still reproduce its intonation and tempo.

Radio comedy programmes like 'ITMA' created a kind of audio home front. Listeners came to recognise the characters' voices – Tommy Handley's nasal German spy Funf, Dorothy Summers' mumsy Mrs Mopp – and their catchphrases became enormously reassuring, providing an aural anchor in turbulent times. Whatever else was raging in the world, British listeners knew that they could tune in at eight-thirty on Thursday evenings and be sure to hear Colonel Chinstrap say, 'I don't mind if I do,' the lugubrious Mona Lott declare, 'It's being so cheerful as keeps me going,' Ali Oop repeat, 'I go – I come back,' and Mrs Mopp ask, 'Can I do you now, sir?'[40] The audience soon anticipated these catchphrases, felt gratified when they were eventually delivered, and quoted them so liberally that the radio voice became, in a sense, their own.

Radio news was a new phenomenon[41] and its style constrained, with the exception of the first eyewitness American broadcast of a major disaster, the crash of the Hindenburg in 1937. When the large German-built Zeppelin aircraft, powered by hydrogen, combusted in New Jersey after its transatlantic flight, reporter Herbert Morrison began to scream, 'It's broken into flames! It's flashing –

flashing! It's flashing terrible! . . . This is one of the worst cata-strophes in the world . . . Oh the humanity and all the passengers!', adding amid sobs, 'I'm going to have to stop for a moment because I've lost my voice. I can't talk. This is the worst thing I've ever witnessed.'[42] Listeners got a vivid sense, through the sound of Morrison overcome with emotion, of being there themselves.[43]

Orson Welles' infamous *War of the Worlds*, based on H. G. Wells' novel about the invasion of Martians, drew uncannily on the tenor of Morrison's report. It caused mass panic, but although the naivety of the listeners of the time is often mocked (and nothing dates so much as old radio voices with their clipped articulation), some parts of *War of the Worlds* remain genuinely disturbing today, especially the reporter's mounting hysteria in the face of the encroaching horror.[44]

The audience's over-reaction partly reflected the enormous credibility that, over a relatively short period, the radio voice had acquired.[45] Radio, for Americans, had become a major source of news. Pre-war dispatches from CBC reporters stationed in Europe had made a deep impression on American listeners, who came to trust these voices from over the ocean. Hearing the New York announcer's voice exhorting the ether with 'America calling Berlin; come in London' brought to the radio some of the qualities of the telephone, and made it seem as if the announcer's voice embodied the city itself.[46] The constant breaks in the flow – 'We interrupt our regularly scheduled broadcast' – together with the familiar voices and unexaggerated, conversational style of the reporters (Bob Trout, Ed Murrow, and William Shirer) soon established these programmes as a source of truth.[47] So when Welles mimicked those very same conventions – the interrupting of the scheduled live broadcast of dance music, emergency news flashes, horrified eyewitness reports from an intrepid radio repor-ter in the field – he was mobilising (and simultaneously destroying) all the credibility of the voice of radio news.[48]

War changed the radio voice in Britain too. At the suggestion of the Minister of Information, anxious that the Germans could easily imitate the voices of upper-class newsreaders, the BBC hired Wilfred Pickles, a popular radio presenter from the North of England, to read the news. To a modern listener Pickles' vowels bear only faint traces of his northern origins, but the British public was so finely attuned to accent that the merest hint of a flattened vowel on air was enough to cause outrage,[49] and Pickles' appointment launched a heated national debate about accent and dialect.[50] As with wartime women announcers, the experiment didn't last long. Pickles soon returned north, and the 'BBC voice' remained the yardstick for broadcasters for two more decades at least.

There were exceptions, like Charles Hill, the Radio Doctor and voice of the wartime Home Front – the first British national broadcaster to demonstrate a real affinity with the audience. Through his colloquial vocabulary and style of address, Hill developed the art of plain speaking.[51] Until the Radio Doctor and the *Brains Trust* programme, 'serious' broadcasting had consisted of turgid talks by stentorian speakers, as if the radio talk were simply a lecture, addressed to a mass, but Hill spoke to the audience as a constellation of individuals positioned in families, gathered round the hearth.

## FRESH TALK

Yet once the war had ended, the British radio voice reverted to type, until its deferential style was detonated by *The Goon Show*. Running from 1951 to 1960, this slapstick comedy series not only blew a raspberry at sobriety but also introduced a style of clowning and cartoon radio never heard on the air before. The Goons parodied the class-streaked British voices of the day, from the posh cad to the cockney idiot, and created an extraordinary profusion of vocal sounds – a falsetto giggle, the deep-voiced Miss Throat, unaccountable Indian accents. Endlessly imitated, these character voices entered the public's own conversations, bringing a new acoustic into British social speech. It was the nearest Britain had come to the Marx brothers, or the aural style of Eddie Cantor's

1930s' American radio shows, or George Burns and Gracie Allen.[52]

The slow death of the BBC voice was also helped by the birth of British pirate-radio stations in 1964, speaking in a tone and language that young people wanted to use themselves. The DJ John Peel recalled:

> When I started doing radio programmes on a pirate ship in 1967 . . . one of the things that was seen as astonishing – and the same thing was true when Radio 1 started – was that I used to speak in what was then my normal voice, and I didn't have one of those mid-Atlantic, 'Hi there, great to see you, the John Peel Show' [voices] which is what everybody seemed to be doing. Just using my ordinary voice was seen as rather exciting.[53]

Peel's voice, with its low, uninflected, and apparently working-class Liverpudlian burr, became one of his distinguishing features (even if he was actually the product of the British public-school system and had deliberately roughened his accent).[54]

The BBC responded to the pirates by establishing its own pop-music network, Radio 1, in 1967. Never before had so many people sounded so cheerful for so long, or displayed such 'pseudo-proletarian spontaneity',[55] the relentless mateyness spoofed twenty-five years later by Harry Enfield and Paul Whitehouse's bland, glib, ageing DJs, 'Smashie and Nicey'.

A new audio world was emerging. Phone-ins, making their British debut in 1968 on BBC Radio Nottingham, and on national radio two years later, provided an unprecedented opportunity for the voice of the ordinary person to be heard – not exactly unmediated, for presenters still had the power to question and interrupt them, but at least spontaneously on air. As one phone-in presenter put it, 'Today there is nothing special about hearing the voice of the public on the radio . . . [people] like to hear fellow human beings talking, even if the talk is a load of rubbish.'[56]

The truly upper-class voice, once de rigueur for broadcasting, would eventually fall completely out of favour. Patricia Hughes, the announcer whose crisp, ringing tones had been the very embodiment of BBC Radio 3 in the era of 'elocution', lamented her eclipse: 'I've got a voice that nobody wants.'[57] Channel 4 News presenter Jon Snow says his original voice now embarrasses him, so much has his accent changed. 'I can't believe that I ever spoke like that . . . People who sound like public-school toffs would have a great deal of difficulty in journalism now.'[58] Snow didn't deliberately demoticise his accent – it just developed out of the new culture of broadcasting.

Yet despite the diversity of voices to be heard on local radio, networked BBC radio still hasn't a single newsreader with a regional accent. In America the TV news anchor Dan Rather recalls that, when he started in broadcasting, he tried to reduce his Texan accent because, 'I was told, "The Midwestern accent is the least identifiable accent and the most acceptable." Until, deep in the 1950s, you did hear regional accents on radio and television [but] some time in the late 1950s, beginning and early '60s . . . the pasteurisation or homogenisation of voices that you've begun to hear happened.'[59]

## ALTERNATIVE VOICES

If the phone-in brought the voices of ordinary listeners on to the airwaves, it did so on the professionals' terms. Callers-in were still subject to an editorial selection process, and had to conform to producers' ideas of what counted as fluent and articulate. Speakers who were marginalised or ignored by the mainstream media turned instead to community radio, often described as giving a voice to the voiceless. But community radio did this literally as well, by using non-professional broadcasters from a wide range of backgrounds, whose voices skipped to a tune very different from those on commercial or even public stations. On community radio you could hear languages, dialects and accents that were almost entirely absent from network radio. KPFA, established in Berkeley, California in 1949, was the first in the world to be financed by subscriptions from listeners.[60] Over fifty years later, its voices still startle, some sounding so low-key and making so few concessions to broadcasting that listening becomes a form of eavesdropping. The explosion of 'free radio' in the 1970s bridged the gap between the radio voice and the off-air one. In the Kwa language of Yoruba, there are two words for radio: '*ghohun-gbohun*' (snatcher of voices), and '*a-s'oro ma gb'esi*' (that which speaks without pausing for reply).[61] To these features of broadcasting community radio provided a radical alternative.

## IN SYNC

Al Jolson was mobbed outside the Warner Theatre on Broadway and 52nd Street after *The Jazz Singer* premiered on 6 October 1927.[62] Silent screen stars, thereafter, were required for the first time to take a voice test. One actor recalled a door bursting open and a man running out yelling, 'Walter Beery has a voice! Walter Beery has a voice!'[63] When she heard her voice test played back, Jean Arthur cried out in dismay, 'A foghorn!'[64] When Mary Pickford heard hers, she was horrified. 'That's a pipsqueak voice. It's impossible. I sound like I'm twelve or thirteen. Oh, it's horrible.'[65] (The clumsiness of the early, non-directional mikes didn't help. Pickford had what Douglas Fairbanks Jr. called a 'small, tight voice', one that needed caressing by a microphone close to her. Instead, she was required to strain her voice to reach one far away.[66])

The careers of silent movie stars were now on the line. When Clara Bow, who became Paramount's most prominent casualty,[67] saw her test, she screamed, 'How can I play . . . with a voice like that?' She was 'stunned and helpless', recalled Louise Brooks. 'She already knew that she was finished.'

Even a star as big as Garbo had to take a sound test, and was nervous. 'I feel like an unborn child,' she told a friend beforehand. The studios were concerned, not about her timbre but her Swedish accent. 'Garbo Talks!' proclaimed the eventual publicity for *Anna*

*Christie* (1930), her first talkie, as if she'd been silent before then, off the screen as well as on. When, thirty-four minutes in, she finally spoke her first words, 'Give me a whisky . . . ginger ale on the side . . . and don't be stingy, baby,' audiences cheered, and critics compared her voice to wine, velvet, a cello, mahogany.[68]

Not everyone was so fortunate. Suddenly every vocal blemish mattered: even Stan Laurel became concerned about his childhood lisp. Voice coaches were now in demand. Sound-men became the new Hollywood aristocracy. 'Esther, see that man up there in the booth?' Richard Dix asked Esther Ralston when they were filming *The Wheel of Life* in 1929. 'You mean the soundman?' 'Yes, and you'd better be nice to him.' 'Oh. Why?' 'Because he can make a baritone out of you and a soprano out of me.'[69]

But what was important was not so much a perfect voice as one that matched the actor's already established screen personality. The most infamous mismatch belonged to John Gilbert, and the discrepancy killed off his career. Those who'd worked with Gilbert before had never thought that his light baritone voice was particularly high-pitched, although it was breathy, nasal and lacking in chest tones. It was a voice that might have fitted Gary Cooper or David Niven but not Gilbert's swashbuckling image: his once admired love scenes now seemed comic and made audiences laugh. When the cinema had been silent, the public could project its own vocal fantasies, based on the actor's appearance, on to the stars:[70] radio and silent cinema both allowed the public to 'dream of the harmony of the whole'.[71] Once sound was introduced they were often disappointed. Within a few years Gilbert's confidence was shattered and he drank himself to death. As an MGM production manager later said, 'It was a miscarriage of justice. Today, his voice problem would have been rectified in five minutes.'[72]

## SEEING VOICES

Yet once the novelty of cinema sound wore off it became, in a sense, inaudible. Critics concentrated on the visual image, rarely referring to the soundtrack (which anyway described music and effects more than voice), except as an afterthought, something added to the visual image rather than its equal partner.[73] Theorists, with a few notable exceptions, deconstructed 'the male gaze' but paid very little attention to the dominance of the male voice.[74] The voice was treated as a natural, unmediated aspect of film as compared with the artfulness of the image. Movie credits themselves expressed the industry's hierarchy – in some of them (*Master and Commander*, for instance) still, the sound designer's name comes after that of the hairstylist.

It wasn't until 1980 that the status of the voice in the Hollywood soundtrack began to be debated,[75] and even then there was a tendency to idealise the cinema voice as somehow purer than the image,[76] even though it was always recorded from a particular point of space or field, and given artificial priority over other sounds.[77] American movies, for instance, usually put each speaking character on a separate soundtrack, in a kind of auditory close-up[78]. One analyst has talked of 'the sound film's fundamental lie: the implication that the sound is produced by the image when in fact it remains independent of it'.[79] When Jean-Luc Godard put an omni-directional mike in a French café and picked up all the sounds equally without giving special hierarchical place to voice, it was not only strange and shocking but also revealed just what a sonic fiction Hollywood sound had created.[80] With its layering of 24-tracks and overlapping dialogue, Robert Altman's 1975 *Nashville*, too, made the vocal conventions of most Hollywood movies seem absurdly thin and artificial.[81]

Since then, Hollywood has entered multi-track nirvana, with voices perfectible through a multitude of post-production techniques, including Dolby sound, computerised mixing, and digital formats. In amongst all the consummate sounds, the juice of the voice no longer bubbles through as it once did. Audio has become over-processed.

Yet in one respect the voice has become more important to American film: major stars are now brought in to voice-animation movies. They're cast, curiously, not just for their vocal image but also for their visual one, with the characters' faces recognisable cartoon versions of the stars' own. Watching a cartoon today, the viewer calls up their memory of the star's features, and reunites the actor's body with the voice.

One voice teacher believes that voice has actually declined in importance in American cinema. 'There's been a big shift. Most Americans can do vocal impressions of the great stars of the Hollywood studio era. Everybody can do an impression of Jimmy Stewart or Bette Davis or Katherine Hepburn, but I defy anyone to do their impression of Brad Pitt's voice or Ben Affleck's voice,'[82] although today's film actors might just be more vocally versatile and less mannered than their predecessors. With such vocal homogeneity, it's the character roles where the flashy vocal technique is to be found – Meryl Streep's (sometimes self-conscious) versatility with accent; Brando's much-imitated rasping Don Corleone voice for The Godfather (1972)[83]; and Mercedes McCambridge's voice of the demon inside Linda Blair in The Exorcist (1973).[84]

## WHO DO YOU HEAR?

Once voice and body had been severed so dramatically through the silent film, there was always some anxiety about how well they could be sutured back together. Techniques like post-synchronisation raised the possibility that strange, alien voices might be substituted for an actor's real one.[85] Voice-overs posed the unsettling question of where the speaker was actually situated – beyond the screen, in the cinema, or perhaps in some altogether fictional space, which itself disturbed the assumption that talking cinema had restored body and voice.[86] The arrival of sound also introduced the vexed question of language. Just as children, once they start to speak, lose some of the Esperanto of nonverbal communication and become confined like most adults to their own language community, so too did film, once sound was added, lose its claim to universality, its ability to cross language frontiers with ease. What's more, when sound came, as a British observer found, 'a problem of accent and class . . . appeared immediately'.[87]

The practice of dubbing films threw up more questions. Dubbing seems to remove from film what the talkies 'returned', and for this reason, perhaps, has produced heated discussion over the years. Dubbing certainly alters the relationship between film and the voice, and also the rhythm and tempo of the film's original language.

Some stars have been dubbed by different actors in the same country (Brando had ten different Spanish doubles, Connery a dozen), while others are always dubbed by the same foreign voice-over actors, in order to preserve a sense of 'authenticity' in the voice.[88] Curious situations follow from this practice. An Italian would be shocked to hear Laurence Olivier's 'true' voice – to him it will always be Gino Cervi's that attaches itself to Olivier's face and body. Similarly an Italian woman, hearing Marlon Brando speak for the first time, was disappointed and even saddened by how 'unbeautiful' his voice sounded. 'He would never have been a successful actor in Italy with that voice,' she said.[89] The voices of both Robert De Niro and Al Pacino were dubbed by the same Italian actor for a number of years until the two stars met on the set of *Heat* in 1995, when another actor had to substitute for one of them. This was deeply disturbing to Italian audiences, who felt there was something wrong with Pacino's voice because it wasn't what they were used to hearing.[90] The dubbed voice had become more authentic than the actor's own.

Anyone dubbing Tom Cruise into another language has to sign an agreement that they won't be photographed in a compromising position,[91] as if they were not simply re-voicing the star but also in some way becoming him, or his surrogate, or perhaps just his representative in another country, entrusted with the loan of his charisma, and obliged to keep it safe. Even in the talkies, the voice could be dismembered or appropriated.

## THE SPEAKING SCREEN

The arrival of television seemed like a rebuke to the voice – a declaration that radio, with only the oral at its disposal, was inadequate.[92] The differences between the two media were shown up dramatically by the 1960 Kennedy–Nixon presidential debates, the first ever to be televised. Those who heard the debate on radio pronounced Nixon the winner. The seventy million television viewers, though, saw an underweight, pale Nixon debating with a rested, tanned, and confident Kennedy, and overwhelmingly judged Kennedy winner of the first debate.[93]

Unlike the cinema, live television enabled an audience to see a speaker at the very moment of speaking, seeming to serve as some kind of guarantee of authenticity. But as the medium developed, it was clear that the relationship between the mass-mediated body and voice was far more complex. Television came to be indicted of all kinds of crimes against the voice – of making everyone sound the same,[94] of damaging children's listening skills and attention span, of killing off quality public speaking. The research on all these accusations isn't conclusive, although it's hard to believe that hearing so much speech from an electronic medium that doesn't hear you back is without consequences.

At the same time the sheer number of different speakers and voices that the public has been exposed to through television has made its impact on vocal style. It has certainly played a part in the death of deferential speech, which is also reflected in volume. *Big Brother* participants not only talk at high volume, but also are encouraged to say whatever comes to mind.

TV advertisements, too, have impacted upon the voice, as well as echoing broader social changes.[95] The booming 'announcer's voice' and the 'dark-brown' voice – a deep, smooth voice that makes every product sound like a sexual aid – have given way to a voice of the street. The advertising voice of today is more likely to be laid-back, and belong to a celebrity (no stigma about commercials any more), or a stand-up comedian. 'Why have we ditched the voice of God? It's partly generational. Younger consumers aren't eager to be ordered around by a stern baritone. Rather than obeying an authoritative voice, they look to the voice of a friend for guidance. Thus, all the pros stress how they can do "next-door neighbour" and "real person" and "quirky best friend".'[96]

If television was held responsible for eroding vocal skills, it was soon joined in the dock by the computer. But what began as a medium with more in common with the book than the voice soon came to add acoustic elements and even, in voice telephony, promised to turn into a substitute for the phone itself.[97] Although sceptics doubted that it would ever really transform itself into a proper aural medium, it soon became clear that, shockingly easily, people could be nudged into reacting to computers as if they were humans.[98] Hal, the intelligent talking computer with a personality in Stanley Kubrick's 1968 film, *2001: A Space Odyssey*, is no longer science fiction.

## DISEMPOWERED

At an international conference in Austria the keynote speaker is using PowerPoint. He brings up the first slide which includes, among other things, a joke. While he's speaking the audience is listening but also scanning the slide so that we reach the joke before he arrives at the stage in his presentation where he's actually going to deliver it. A suppressed titter passes round the hall – suppressed because the joke will only formally exist once he's voiced it.

But here's the dilemma. In a few seconds, he'll be saying the joke out loud, and how then should we react? Do we repeat our titter, only mildly, to acknowledge the fact that we've already read the joke ourselves? Or should we simulate exuberant laughter, as if it were fresh to us? In the event, roughly half the hall opts for the mild titter while the other half gives a good impersonation of authentic hilarity. Simply by using a visual aid, the speaker has changed our relationship to his voice, decentring and demoting it. By now the subject of his presentation must be obvious – the human voice.[99]

Though some say that PowerPoint, used an estimated thirty million times a day, helps the mumbling, forgetful, or plain incompetent speaker to get their main points across, others argue that it's designed 'to close down debate, not open it up'.[100] PowerPoint tries to tame the unpredictability and delinquency of the voice, and transform it instead into writing, with all that medium's stability. PowerPoint tries to resolve the conflict between the now unavoidable requirement to make public presentations, and the anxiety about speaking with the body. It expresses a loss of faith in the voice. Today we want machines to do the talking for us.

The voice has travelled a long way since Thomas Watson, Bell's assistant, first heard him say, 'Mr Watson, come here. I want you,' down the wire. Disembodied voices no longer have the same power to disturb. We're not fazed by the preservation of dead people's voices. And awash with radio stations, listeners have become blasé about hearing ordinary people speak over the airwaves, but can be thrown by an ineptly dubbed voice. Today we constantly hear people we could never hope to meet: through the telephone, phonograph (and its successors), radio, cinema, and the television, we hear a greater number and variety of voices in a month than people heard in a lifetime in the past.

By giving us access to such a profusion and diversity of voices, the new technologies have encouraged the belief that everyone is entitled to be heard. Yet, after decades of exposure to radio and television, people being interviewed now speak like other people they've heard being interviewed – we know how we're meant to sound.

Although it gave a good impression of doing so, technology never really severed the body from the voice or later restored it: theirs always has been an on-off partnership, a relationship that has continued to shift throughout the emergence of even newer technologies at the end of the twentieth century.

# 16

# *Voiceprints and Voice Theft*

IN 1907 A Florida woman identified the man who had raped her, even though she hadn't seen him during the attack or met him before. He had, though, spoken two short sentences, on the basis of which she recognised his voice. The trial judge accepted her testimony, arguing that such a terrifying assault had made her extremely alert. 'Who can deny that under these circumstances that voice so indelibly and vividly photographed itself upon the sensitive plate of her memory as that she could forever afterwards promptly and unerringly recognize it on hearing its tones again.'[1]

The case posed a question that has been asked many times since: how long and how accurately can we retain the impression of a voice in our minds? For many American courts the judge's ruling in this case constituted the legal precedent for the admissibility of earwitness identification.[2]

A century later the belief that the individual's voice is as distinctive as their fingerprint has become so unshakable that voice verification has been welcomed by both commerce and government, offering the promise of security in transactions and surveillance. And with the spread of the mobile phone and the development of an extraordinary array of new techniques for storing and remodelling the voice, the process of separating voice and speaker that had begun with Alexander Graham Bell seemed to have come full circle.

## VISIBLE SPEECH

It was Alexander Melville Bell, father of Alexander Graham Bell and an Edinburgh professor of elocution and speech, who began the process that led to voiceprinting, the identification of people by their voice, by trying to represent every single sound made by the human voice in symbols.[3] But voice identification only really ambushed the public imagination during the infamous Lindbergh trial. In 1932 the 20-month-old son of Charles Lindbergh, the first person to fly solo across the Atlantic, was kidnapped and later found dead. A German immigrant, Bruno Hauptmann, was arrested: his 1935 court case was the O.J. Simpson trial of the day. Lindbergh testified that he recognised Hauptmann's voice as that of the kidnapper, even though he'd only heard Hauptmann say four words, 'Hey, doc – over here,' in a cemetery, addressed to the person collecting the ransom while he himself was sitting in a car with the windows closed over half a block away. Lindbergh's testimony was also delivered more than two years and eight months after the evening on which the words were spoken, although in the intervening period he'd insisted to police that he wouldn't be able to remember the voice. Can hearing be this keen, and memory this reliable? Lindbergh's testimony caused a sensation and Hauptmann was executed in 1936.

When a psychologist, Frances McGehee, tried to replicate Lindbergh's identification in a pair of experiments based on similar circumstances she found that, although listeners were able to distinguish a particular voice with some accuracy soon after hearing it, this fell off gradually but steadily over time,[4] a conclusion backed up by more recent experiments.[5]

Examples of misidentification were recorded even in biblical times (Isaac misidentified Jacob's voice). The whole process is bedevilled by the fact that some people are more aurally sensitive (and have plain better hearing) than others, while some speakers have more distinctive voices than others. It's also far easier to identify the voice of someone you already know than that of a stranger.[6] Earwitness parades, in which a suspect's voice is embedded in a recording of up to ten other people, are used today by criminal investigations in the same way as eyewitness ones, even though they aren't comparable, partly because auditory memory is processed differently from visual memory, and voices, unlike faces, have to be paraded sequentially.[7] Also, fear and anger (not to mention dentures and asthma) have a much more distorting effect on the voice than on the face. The accuracy rate of voice parades can be as low as 30 per cent.[8] Indeed the more confident you are about your ability to identify a voice, the less accurately you're able to do so.[9] Hearing the voice of a perpetrator also can have a profoundly traumatising effect on the person making the identification, especially if they're the victim of a violent crime or rape.[10]

Perhaps technology might do better. The invention of the sound spectrograph in the early 1940s produced a visual record of voice patterns based on sound waves. During the Second World War acoustic scientists thought spectrographs might be able to identify enemy radio voices and, after an assassination attempt on Hitler, American phoneticians were asked to compare two broadcasts of Hitler to verify that the speaker in the second wasn't a double. After analysing the recordings and matching the voices they concluded that Hitler was still alive.[11]

But it wasn't until the 1960s that the technology was refined and the word 'voiceprint' coined. As social unrest grew and crime increased, voice identification seemed an ideal method of nailing terrorists, anonymous tipsters, and assorted ransom-seekers.[12] It also raised the issue around which the Soviet writer Alexander Solzhenitsyn built his novel *The First Circle*, set in a Stalinist penal institution where political prisoners are engaged in top-secret research 'to find a way of identifying voices on the telephone and to discover what it is that makes every human voice unique'.[13]

Could such a perpetually elusive aim ever be accomplished?

Voice identification is a romantic idea – that, in an age of homogenisation, we're each the possessor of an inimitable object, our
voice – and a dramatic one: it seems to accord with some theatrical
fantasy of the unmasking of the criminal – betrayed by their voice.
And voiceprinting has had its successes. In 1974, for instance, it
was used to decide if someone purporting to be the eccentric
reclusive tycoon Howard Hughes was Hughes and not an impos-
ter.[14] Yet the very word 'voiceprint', suggesting an analogy with
fingerprints or footprints, is misleading. While fingerprints are an
infallible method of identification, voiceprints in the end always
come down to opinion, even if expert. However consistent the
machines may be, there's inevitably an interpretive element to voice
identification, because it involves a visual task (comparing two sets
of spectrograms), and an aural one (comparing breathing patterns,
inflections, accents, as well as idiosyncratic speech habits in two
different recordings) which, to some extent at least, must be
subjective.[15]

What's more, most people's pitch alone varies hugely even in the
space of a single day. Might Fred, sober and scared, sound more
like Jim than Fred, after three triple whiskies and with an attractive
person pouring him a fourth? If we add in factors like attempted
voice disguise and distorting background noise, can we be sure that
every one of the six billion people in the world will invariably
produce speech more like their own than anyone else's,[16] especially
since no one person says the same word twice in the same way,
even in the same sentence? Voiceprints are actually more comparable in their accuracy to lie detection than fingerprinting.[17] One of
the scientific critics of the procedure has been less equivocal. 'I
believe voiceprinting to be a fraud being perpetuated upon the
American public and the Courts of the United States.'[18]

In fact, despite a series of precedent-making cases, American
courts have not wanted to move at a faster pace than the scientific
community in admitting forensic speaker-identification as evidence, so in some American states it's still not admissible, and
in others it works on a case-by-case basis. The UK law courts seem

a little less tormented over the subject: voice evidence is already being used in English criminal trials, and a series of precedents since Crown against Robb in 1993 has made voice verification admissible as evidence if backed by an expert.[19] Ultimately it's down to British juries as to whether they accept evidence based on speaker recognition or not.[20] Voice verification returned to the headlines after 9/11, with the CIA and the US National Security Agency poring over each new audio tape purporting to come from Osama bin Laden, in order to compare it with confirmed recordings of him.

## THE VOICE OF SECURITY

Hardly has the dark-chocolate voice finished uttering the words, 'The name's Bond. James Bond,' than the high-tech doors open – they've recognised his voice. But what was once the apotheosis of sci-fi is today, increasingly, reality. Speaker recognition is an example of biometrics – establishing identity on the basis of unique physiological and behavioural characteristics.[21]

'Your voiceprint is your key.'[22] 'Unlike tokens, cards, PINS and passwords, your voice cannot be shared, stolen, or forgotten.'[23] 'The future is hear.'[24] Slogans of the brave new vocal world. All sorts of uses are envisaged for this emerging technology. We'll be able to withdraw money from cash tills simply by speaking into them.[25] Even safes and bank boxes will be regulated by the voice. Travel, health services, immigration and border control – all might employ it. At their most euphoric, the biometrics companies envisage a world without the need to carry money, keys, credit card or even identity cards. So ubiquitous will the technology become that the voice will allow us secure access to buildings, and joggers will be able to run unencumbered, paying for a bottle of water en route purely with their voice.[26]

The growth in telephone banking and call centres has created the need for some sort of remote verification and using the voice certainly reduces the demands on customers' memory. To its advocates voice authentication is superior to the other kinds of biometrics, because all you need, no matter where you are in the world, is access to a phone[27] – and, since most Westerners are familiar with the telephone, are now comfortable using it for confidential communication, and own one, the infrastructure is already in place.[28]

Voice verification, according to the prevailing propaganda, is the biometric of choice for most people because they regard it as the least intrusive. Chase Manhattan Bank, the first American domestic bank to introduce it, found that 95 per cent of its customers would accept it, compared with an 80 per cent acceptance of fingerprinting.[29]

Voice verification, say its advocates, also prevents fraud, something about which consumers express considerable anxiety. In reality much of the impetus for voice verification comes not from its capacity to safeguard the funds of individuals but from its potential to save those of commercial institutions. Statistics suggest that help desks spend between 40 and 60 per cent of their time on password problems – resetting forgotten, lost, or shared ones.[30] What costs on average $1.75 a minute when carried out by a live operator can be done by voice recognition for 10–20 cents per minute.[31] And the system is low maintenance: it doesn't require the installation of thousands of scanners, yet it can continuously update and refine its databank each time an individual calls in.

It's also supposedly a way of protecting against identity theft, something the voice-verification lobby is obsessed with, although precise figures of its incidence are hard to come by. Those very same bodies that fan fears of identity theft are, by coincidence, on hand to provide a technological solution to it.

Similarly, much of the research about the acceptability of voice verification has been conducted either by the firms developing the technology or by consulting firms employed by them, so it's hard to get a clear sense of just how readily the public will embrace it. Two-thirds of those questioned in one survey hadn't heard of voice authentication before and yet, in answer to the very next question, over 80 per cent felt that it could help prevent fraud or identity theft. One can't help but wonder what sort of dialogue occurred between the first and second questions.[32]

Above all, the procedure depends on participant compliance: in other words, speakers must be willing to cooperate with the system, and must want to be recognised. Perhaps this is why so much effort is being expended on selling it to consumers – since they have the capacity to sabotage or reject the system, it needs to be depicted as something in their interests, whose accuracy is beyond question, even though this clearly isn't the case.

For the idea of the voice as a kind of audio DNA is misleading. Short utterances, strong background noise, poor quality recording equipment, a speaker who's modified or disguised their voice – all of these, as we've seen, are obstacles to an accurate system. With identical twins there's a false-error rate as high as 50 per cent.[33] The technology simply isn't secure, and academic caution is at odds with commercial euphoria.[34] At best speaker verification can supplement other systems, helping to increase their accuracy, but it isn't a replacement for them yet.[35]

## THE FUTURE IS MOBILE

Public resistance to procedures like voice verification has diminished because of the remarkably rapid spread of another voice technology – the mobile phone. This has brought about major changes in voice use, significantly changed the relationship between parents and their children, and removed the intimate connection between voice and place that the telephone established.

Bell's telephone may have distanced the body from the voice, but it still allowed callers to locate other people's voices with precision: just as in face-to-face contact, you knew exactly where to find and place the voice of the person you called. The phone today follows the voice; the voice doesn't have to go to the phone. It's now almost impossible to pin down an adolescent geographically, because their phone has become as mobile as they are. Children's whispered, midnight, under-the-duvet calls also can elude parental scrutiny.

Parents don't hear the voices of their teenagers' friends so much either, because the youngsters now communicate mobile-to-mobile, without being mediated by adults answering the phone.[36] Mobiles create the possibility, even the need, for teenagers always to be available to friends – although not to parents. With an increasing number of 8–10-year-olds and now a quarter of British under-8s owning one,[37] it seems inevitable that communal voice time in the family will diminish.

From young people's point of view, mobile phones have become a way to demarcate the boundary between themselves and their parents,[38] and overcome the spatial boundaries of the home. The home landline is shared but the mobile is private, an expression of individual personality, customised in appearance and ring tone. (McLuhan called the new electronic media 'extensions of man'.[39] He couldn't have anticipated how true this would be of the mobile, and how much closer it would bring the phone to the body.) The mobile liberates the teenager and pre-teen: texting transcends limits on vocal communication in public places like school, and allows adolescents to target precisely the person to whom they want to speak – no longer do they have to make polite small talk to friends' parents.

A new etiquette of the phone has developed. It's perfectly acceptable for people to answer their mobile even if it means putting their face-to-face conversation on hold.[40] Mobiles have brought about a conflict of registers, so that people now speak to their caller in a style inappropriate for their overhearer. By having access to only one side of a phone conversation and not the other, bystanders find themselves in limbo, 'neither fully admitted nor completely excluded'.[41] Thanks to the mobile, there's also a new exhibitionism in British life: callers speak as if the mobile were conferring privacy, even when they're surrounded by strangers. (Will future generations of mobile users have got so used to its mobility that they'll no longer feel obliged to announce where they're calling from?) The mobile has penetrated remote beaches,[42] and turned even the Finns – famous both for their dislike of small talk and virtuosity in mobile-phone production – into chatter-boxes, of sorts. The very inescapability of mobiles is leading to curbs and controls on their use. In Japan most trains, buses, and many restaurants display 'No mobile phone' signs[43] and other countries are following suit.

Accounts of mobile phones saving lives – of people rescued at sea, or on a mountain, after calls to their family – have become common, while the plangent sound of an unanswered ringing phone after a train crash has also become part of our contemporary soundtrack, an emblem of finality. One of the most distinctive and poignant features of 9/11 was the reporting, in some cases relaying, of last, loving calls made to relatives by people on the doomed planes or in the twin towers. 'This is death being faced in real time,' wrote one writer, 'this space between life and death . . . has been wired up and switched on, electronically illuminated.'[44] Another saw such experiences as 'harbinger of a time to come when no one will die alone, and will make their dying peace with partner, child, parent, friend, even answering-machine or operator,'[45] curiously one of the functions that Edison imagined for the phonograph. When the voice of Herbert Morrison, eyewitness to the crash of the Hindenburg, cracked with emotion, it was unprecedented; now the voices of victims themselves can be heard, and by the time of the London bombings in July 2005 camera phones had become so ubiquitous that pictures too were added to the grim first-person electronic testimony.

The telephone voice isn't the same as the face-to-face voice because the telephone doesn't cover the full range of frequencies of the human voice, favouring certain voices over others (German men, for example, transmit better than Japanese women).[46] On the other hand, mobiles are also reaching remote parts of the developing world that landlines haven't begun to touch. Only 3 per cent of Africans have mobiles, but they represent 53 per cent of all phone subscribers on the continent.[47] The voice will travel everywhere.

## VOCAL TAGGING

The mobile itself, they tell us, might soon be always-connected, a teat for our times. Already the baby monitor allows us to tune in to our infants continuously. Today aural surveillance seems uncontroversial, and the idea that it's legitimate for governments to use the voice to locate and identify its citizens has attracted almost no public criticism. British and American non-custodial government surveillance programmes for serious young offenders now include voice tagging, which requires the young person to call in from a landline at specified times every day. Within a few seconds, the computer can check the voiceprint, phone location, and time.[48]

Voice tagging is a fabulous money-saver. It costs $9,000 to keep a young person in detention for three months, but to track them for the same period, American researchers have found, costs only $300.[49] Judges like the system because not only does it mean that a young offender doesn't have to be housebound – they can go to school or wherever else the terms of their probation allow – but also that a large element of unpredictability can be built into the system, making it almost impossible to cheat it.[50] At the very least such schemes must operate as a kind of aversion therapy, turning the phone into a potential instrument of persecution. Put like this, parents of teenagers might want to sign theirs up for it immediately.

Yet vocal tagging also has a curious historical resonance. In the nineteenth century the philosopher Jeremy Bentham imagined prisoners being confined in brightly lit cells in a Panopticon, a circular cage from where they would be permanently visible to a supervisor in a high central tower. Thus would the inmate become the instrument of their own surveillance. The Panopticon dispensed with the need for physical confrontation or the corporal exercise of power 'by its preventative character, its continuous functioning and its automatic mechanisms'.[51] In an almost exact parallel, the 24/7, 365-day dimension of vocal tagging is what makes it attractive to the police and probation service, who praise the way that it automates the supervision of offenders. Has voice verification given birth to a modern Panauditorium?[52]

A British home secretary proposed another phone solution to youth antisocial behaviour – confiscating mobiles for a few months.[53] So the voice has become a means of social control not only through its presence but also its absence.

## ALL TALK

The Walkman was the first portable device to allow a constant aural presence, one now vastly extended by the iPod and mobile. A remarkable number of other machines, procedures, and software can or soon will be either operated by voice, responsive to it, or possess it. These include a programme that allows us to dub our own voices on to DVDs of animated movies like *Shrek*;[54] Kismet, a robot being developed by MIT's artificial intelligence lab, that not only shows through its voice if it's sad, angry, happy, or calm,[55] but can also tell from its instructor's voice whether it's being praised, scolded, or comforted;[56] and a voice-box transplant, transferring the larynx of a dead person to a living one.[57] This in addition to all the telephone helplines, answering-machines, blogs, karaoke machines, text-to-speech systems[58] already in existence. If the producers of the new technologies are correct, we're going to be using our voices more and more – accessing voice-enabled services on voice portals through our mobiles, for instance. As the chief executive of a company developing speech-synthesis solutions put it chillingly, 'Voice as a brand is becoming more and more important.'[59]

The ability to chop up and reassemble the digital voice is an irrevocable and intriguing one. On the one hand it opens up all sorts of new vocal possibilities, demonstrated in the sonic art now emerging.[60] The liveliest of today's audio artists, far from complaining about the disembodiment of the voice, revel in it and play with it instead, using it to make something new. Similarly the spread of rap, sampling, and ambient sound has produced an exceptionally audio-aware generation, one open to vocal experiment, even if the voices they hear aren't necessarily voice as we've known it. On the other hand, some of the new voice technologies are touching off serious disquiet.

## CRESCENDO OF CONCERN

Over the past ten years warnings about the diminishing role of the voice caused by technology have come from many different quarters. In 2001, for example, it was revealed that 45 per cent of telephone calls have been replaced by email and we now email more than we talk. To encourage staff to talk to each other some large British firms began introducing 'email-free Fridays'.[61] Email, shouted another anxious headline, 'could replace talks with teacher'[62] (as if the culprit were technology rather than teachers' workloads). This was followed by a poll claiming that text messaging was harming students' speaking skills, with three out of 100 company directors believing that email is detrimental to spoken communication.[63] The United States has even passed a Human Voice Contact Act, stipulating that state agencies that use automated telephone-answering equipment must also provide callers with the option of a live operator.

There have also been recurring alarms about young children's poor communicative skills: the British Chief Inspector of Schools warned that many children were unable to speak properly when they started school because parents had left them in front of the television instead of engaging with them through talking and play,[64] while at the other end, large numbers were leaving school unable to express themselves adequately or follow what others say.[65] According to the Director of the Basic Skills Agency, routine communication in families with exhausted parents and too few family meals now amounts to little more than a 'daily grunt' from monosyllabic schoolchildren. (While most of the newspaper columnists pitched in with laments for the lost art of conversation, one demurred, arguing that grunts had a grammar of their own, their meaning varying with length and tone. 'Gaad' on its own means 'There is a depressing but predictable item in the newspaper that I'm reading', but by extending it and lowering the note, the speaker is saying 'You're not really going to watch this television programme, are you?'[66])

So concerned was the British government about children's communicative skills that in 2003 it instigated a national drive to improve 'oracy', setting term-by-term objectives for speaking and listening.[67] It also piloted children taking exams by mobile phone 'because this way they'll use their voices'.[68]

While this renewed focus on speaking is welcome, it often rides on a crude technophobia. *The De-Voicing of Society: Why We Don't Talk to Each Other Anymore*, a recent book by an American neurolinguist, argued that interpersonal conversation was dramatically declining in Western societies, causing a waning of intimacy.[69] To the human, the book claimed, talk was the equivalent of grooming in apes and monkeys, and a critical factor in the creation of intimacy. The author indicted the usual suspects – email, the Internet, home shopping, television, telecommuting – claiming that anomie was their inevitable consequence. Yet mobile phones and answering-machines also stood accused – and neither of these, you might think, were impediments to talk so much as facilitators. On closer observation, the book appeared to be lamenting not so much the loss of talk but of face-to-face talk, and in so doing it appealed to a disappearing (sociobiological) golden age where everyone was busily involved in sustaining conversations. Others claim that technology has amplified the voices of the powerful.[70]

This vanished nirvana isn't convincing. Pre-technological societies were also pretty stratified, with some voices counting for more than others (there were kings and subjects, masters and slaves, even before the birth of the loudspeaker or email). What's more, our knowledge of how voices worked in the pre-technological era only comes from written accounts or transcribed oratory. It seems unlikely that medieval families, say, conducted their everyday business entirely through the rhetorical curlicues preserved in public documents: private life then probably had daily grunts of its own. And far from talking less than we did, we may well be using our voices more. With the arrival and growth of call centres and other service industries where the voice is pivotal, a growing proportion of people now speak for a living. What's more, public speaking, once derided as an outmoded and declining activity in an increasingly informal world, has become a necessary skill in a growing number of spheres. Over thirty years ago a researcher found that almost 50 per cent of the blue-collar people she interviewed had given at least one speech to at least ten people in the past two years, and conjectured that middle-class people had probably given more.[71] Those figures must surely have grown exponentially in the intervening years while, as we've seen, voice-management has become an essential component in politicians' arsenal of spin. Does this sound like a devoiced society?

In addition, new technologies allow us to hear things that in previous generations we would only have read, such as recordings of calls to emergency services. Whether this enlarges human experience, or is just another stage in the banalisation and desensitising of the culture is open to debate, but it certainly doesn't strengthen the devoicing argument.

Of course none of this is speaking in the face-to-face sense that devoicers have in mind, but my interviews suggest that even people who aren't in obviously voice-related jobs, or who lack rhetorical prowess, rely on vocal skills to a remarkable extent, using them to make vital social and personal connections. In a therapeutic culture that lionises talk and pays lip service, at least, to qualities like emotional intelligence[72] that are expressed to a great extent vocally, communicating through the voice is growing rather than diminishing in status.

Indeed, as I've suggested again and again, technological change has made the voice more rather than less important. The irresistible spread of answering-machines, mobile phones, help lines, phone-ins, CD-ROMs, talk shows, and voice-recognition techniques, along with the proliferation of radio stations, is making ours much more of a voice-centred society. Mobile phones, reality TV shows, video diaries, weblogs and webcams have also eroded the boundary between public and private, meaning that we now hear people speak 'privately' in public. (Perhaps the devoicers are looking for vocal intimacy in the wrong place: it's now moved into the public domain.) The critics lament the eruptions of capitals, exclamation marks, and emoticons in email and text messages as a poor expressive substitute for tone of voice. In fact the growth of email and text messaging (with all their stylistic conventions), far from eclipsing the voice, has drawn attention to its irreplaceability. New technologies invariably provoke anxiety: sometimes this turns out to have been warranted, but it usually takes more than a couple of decades to judge what has really been lost, and what has simply morphed into a new shape or been replaced by something similar.

One can't help thinking, too, that some of these pessimists have recruited the voice for their own purposes, as a form of technology-bashing. They seem less interested in the voice per se, and more in using it as an antidote to modernity, a means of somehow pre-venting the future from getting out of hand. Imagining that the voice was free but is now being tamed or enslaved is romantic fantasy – especially since for at least 150 years, as we've seen, the voice has been in a process of almost perpetual transformation.

## WE KNOW WHO YOU ARE

On the other hand not all the anxieties about the effects of technology on the voice can be dismissed. For instance, although speaker authentication is being sold as a guarantor of personal privacy, it might actually invade it. How do we feel about our voices being appropriated by commercial companies and state institutions? The voice-authentication lobby emphasises its potential contribution to workplace safety, especially in the post 9/11 world, yet there are also social, personal, and ethical implications. Will it, for instance, change our relationship with our voice?

Voices are increasingly being used to sell; in the commercial applications being developed they're also requisitioned to buy. They've come to be seen as a potential money- and labour-saver: just as supermarkets have displaced some of the labour of selling from the retailer on to the customer so, by replacing call agents with voice-recognition systems, administrative labour is increasingly being transferred from producers' to consumers' voices. Promoters of the new technology say that the voice is so valuable because it can't be stolen, but aren't they, in some small way, doing precisely that? Or is rueing the fact that our chief tool of communication is now being purloined by the corporate world just anti-modern nostalgia?

Certainly, before the nineteenth century, the voice wasn't seen as a marker of individuality or of an internal disposition, but rather the outward sign of social traits, resulting from the speaker's cultural position and role.[73] Speaker recognition, by contrast, focuses attention on what's distinctive and not shared in the voice, treating the voice as a sign of the self that can be exploited for bureaucratic or corporate needs. Although it may become a boon to the visually impaired and people with other kinds of disabilities, a huge international database of voices is horribly open to abuse. Big Brother no longer needs to watch when he can listen so effectively – truly His Master's Voice.

At the moment the users of voice-recognition software aren't eavesdropping on our conversations, only getting us to say our names along with other, more banal information. The human voice, in any case, has always served multiple functions – the means by which we buy potatoes as well as declare our love. And yet we're undoubtedly living through a major paradigm shift, one in which the voice has travelled even further from the body.

## LOSING YOUR VOICE?

So should we be concerned about the new voice machines? And who, now, does the voice belong to? Because it's not a possession or accessory but an attribute, the voice can't belong to anyone, and yet the question isn't as far-fetched as it might sound.

Although the idea of body and voice existing in perfect harmony in some mythical past is untenable, it's true that for centuries the human voice was (at least to some extent) a guarantor of the body – you couldn't have one without the other. Today voice and body have been prised apart, and the voice is in the throes of becoming just another component – a digitalised bit to be reconstituted and remixed, in the same way as music and song are now broken down into separate segments, sampled and reassembled into a new whole.

Who does the voice belong to? Ask Nancy Cartwright, the voice of Bart Simpson in the TV cartoon series, who is regularly requested by interviewers to 'do Bart'. She always refuses because 'the prohibition has to do with legalities, copyright infringement. It is a funny thing but although I own my own voice, I don't own Bart's voice. It would stand to reason then that I couldn't just go out and in Bart's voice say, "Even though you think I am Bart Simpson, I am actually Nancy Cartwright." These legalities are common in television/film production. The actors don't *own* the characters they are doing. They are owned by the creator and/or the studio/production company.'[74]

Dan Castellanata, the voice of Homer Simpson, faced the same problem when he performed a segment on an American alternative comedian's album in Homer's voice, only for the album's release to be blocked by Fox Television's lawyers, who insisted that Homer's voice was part of their intellectual property.[75]

Who does the voice belong to? Ask Nick Campbell, researcher into speech processing. 'At the moment, the law seems to be undecided about the use of very small samples of a person's voice – it seems that they probably don't have ownership of the "sounds" of their speech . . . currently the sound of a voice is legally similar to the colour of a painting or the words of a book . . . We can't copyright colour or words, only the shapes or ideas that are made up by sequences of them.'[76] Campbell is developing a speech-synthesising system in Japan that could, potentially, enable machines synthetically to produce convincing versions of the voices of real-life people like the American president saying something incriminating. It could bring back old, beloved, long-dead movie stars and get them speaking new scripts (putting live actors out of work in the process).[77]

The new technology introduces the possibility of a new kind of crime – voice theft. If, for example, a synthesised version of Nick Campbell ordered hundreds of pizzas to be delivered to all his friends from the neighbourhood pizza parlour that had recognised and trusted his voice, who would be legally bound to pay for them?

The synthesised-speech programmes being developed today will be used by call centres, text-to-speech systems, and help desks. Eventually they may be indistinguishable from the voice of a living, breathing person, so various security measures to prevent abuse are being entertained. There's even talk of embedding in the digitalised version some almost imperceptible feature (like a watermark on paper) that could be picked up by a special detector but not the human ear, in order to allow an artificially generated voice to be differentiated from a genuine one.[78] Clearly, hearing is no longer believing.

Why are these technological developments so disturbing? One answer might lie in Freud's concept of 'the uncanny'. Certain situations, Freud suggested, cause dread and fear because they make us doubt whether an apparently animate being is really alive or, conversely, whether a lifeless object might not, in fact, be animate.[79] Ghosts and spirits fall into this category (so do wax-works or ventriloquists' dummies[80]).

The dismembered, reconstituted voice is just such an example. Is the voice of some long-deceased star, newly digitalised to sell the latest model limo, a dead or a living thing? Is it still the speaker's own, or has it been so digitally remade that it's now a voice-in-the-machine? Anxieties about synthesised voices express fears about the impact of technology on our very idea of the human, and the way we define alive.

But Freud also believed that uncanny fears express primitive beliefs, often originating in fantasies in the womb, which we think we've surmounted but which reappear with force. The processed, synthesised voice, by this reckoning, may re-stimulate infantile anxieties about the loss, through the act of being born, of the mother's embodied voice. As we've seen, the foetus hears but, even more importantly, also feels the maternal voice: in entering the world, the baby loses the feel of it and now only hears it. Technology continues this process, putting the voice at a further remove from the body – no wonder it excites so many primitive fears.

Freud's concept of 'the uncanny' also helps us understand why the anxieties about the voice kindled by these latest synthesised techniques are almost identical to those evoked by Bell's telephone and Edison's phonograph a century before. New technologies, but old fears. In fact the human voice has proved remarkably resilient and, as we'll see, astonishingly versatile.

# 17

## How People and Corporations are Trying to Change the Voice

THE VOICE IS CHANGING, but governments and corporations aren't the only agents of transformation. Never before have so many individuals tried to alter their voice through so many different methods. If they sometimes underestimate the difficulties involved, this desire for reinventing the voice suggests nevertheless that there's a new aural sensitivity around – we're becoming far more aware of how we sound. A Westerner is now expected to improve their voice just as much as an Ancient Greek was. At the same time, today's multitude of different voice-changing techniques reveals a tension between two common ideas – the voice as a natural phenomenon and as an infinitely malleable personal feature.

### THE HEALING VOICE

The idea that the voice can heal isn't new. Almost every religion uses some form of chanting to achieve physical and spiritual wholeness. In meditation, the mantra – a sacred sound said to embody the deity – supposedly restores the chanter to a state of harmony. The Hindu and Tibetan traditions believe that the throat is one of the seven main chakras, the energy centres of the body, and unblocking it through sound can improve health: the heart rate can be stabilised, blood pressure reduced, circulation improved, left and right sides of the brain synchronised, molecular structure altered, kidney stones dissolved,[1] and even – since chanting is said to cause the release of the hormone melatonin – tumours shrunk.

These Eastern practices have now spread to Westerners, increasing numbers of whom believe that Western society is disenchanted and that the voice is the key to spiritual transformation. They credit it with the ability to reduce stress levels, liberate the inhibited, and create a sense of shared identity.[2]

Behind most of these methods lies the idea that vibrations can help heal the body: toning bells, bowls or forks, it's claimed, can unblock stagnant energy.[3] Rudolph Steiner even predicted, 'There will come a time when a diseased condition will not be described as it is today by physicians and psychologists, but it will be spoken of in musical terms, as one would speak of a piano that was out of tune.'

## FREEING THE VOICE

The idea that we each possess a natural voice is also a popular one today. This voice 'achieves a natural balance of breathing, phonation . . . and resonance',[4] only not so natural that it can emerge without the assistance of workshops, books, and other training. The natural voice, it transpires, has been lost: curbed and constricted by life, it needs to be found and freed.[5]

Already in the 1890s Frederick Matthias Alexander, founder of the Alexander Technique, had suggested that vocal difficulties might be the result of problems of posture, tension, and holding the breath. Most of the therapeutic voice work practised today, though, has grown more from the theories of Wilhelm Reich and Alexander Lowen, both of whom believed that psychological conflicts, when repressed, led to a damned-up voice. For the voice to be freely expressive of the personality, these conflicts needed to be discharged.[6]

## DOCTORING THE VOICE

For a long time ideas like these seemed outlandish – not so in the new culture of the voice. We may have only limited ways at our disposal in which to talk about the voice when it functions well, yet it seems that we're beginning to develop quite a lexicon to describe it when it fails. Vocal problems are coming to be seen less as an annoying but essentially random personal difficulty and more as an industrial health hazard for those working in a noisy environment,[7] or whose jobs are vocally demanding. The voices of teachers and doctors have attracted particular attention, especially since 1994, when a British teacher who'd lost her voice working in an open-plan classroom won a claim for disability benefit.[8] Voice training is increasingly included in teacher training[9], and workshops run by networks of speech and language therapists offering practical help with the voice have proliferated.[10]

Interest in doctors' voices articulates rather different concerns – the importance of doctors' nonverbal skills in creating good therapeutic relationships.[11] It sometimes seems as if doctoring has become entirely a problem of communication: a prominent figure in the medical profession recently claimed, 'The poorly performing doctor doesn't quite know what he is doing scientifically or how to talk to his patients properly.'[12] Patients obviously benefit from being treated by doctors who can communicate effectively and with empathy, but already over-burdened, under-resourced junior medical staff might regard the task of developing better speaking skills as one target too far. Since doctors who are more expressive communicators have been found to reduce the length, and therefore cost, of consultations,[13] and face fewer malpractice claims,[14] the impetus to improve doctors' nonverbal skills may have come at least as much from the need to save money as from the desire to improve patients' lives.

Today's school curriculum also places more emphasis on the development of speaking and listening skills. In some schools pupils now have to give oral presentations of drafts of coursework.

In others there are games designed specifically to draw out shy, self-conscious speakers.[15] After consultation with employers, the teaching of English is being revamped. 'Employers tell us that, whereas two years ago managers spent 20 per cent of their time in discussions, now it is 50 per cent. To succeed in business you have to be good at persuasive talking: selling an idea to your bosses, raising the funding, communicating at the workface.'[16] There's even a campaign called 'Talk to Your Baby', which tries to raise awareness of how important it is for parents to speak and listen to their child.[17]

## YOU SAY TOMAHTO, AND I SAY TOMAHTO

Changing the voice isn't only deliberate and conscious, but also happens instinctively. When we hear someone we know talking on the phone, we can tell almost invariably who they're talking to, purely through the sound of their voice, because we modify our voices from situation to situation, picking up other people's tempo, pause pattern, pitch, volume, and accent – a process known as 'convergence'. 'One partner must "set the pace" in speaking as with walking with another, and the other partner must, in turn, comply.'[18] 'Divergence' is the opposite – where a speaker emphasises the vocal differences between themselves and someone else.[19] Together they form 'speech accommodation theory'.

You can converge upwards, to a more prestigious style of speaking, or downwards, to a less prestigious variety,[20] and people do it in Hungarian, Frisian, Hebrew, Taiwanese, Thai, as well as many other languages.[21] Convergence, like divergence, can be mutual or non-mutual, total or partial.[22]

Why does it happen? Convergence reproduces the feeling of being in harmony with another person that, as we saw, parents and babies can achieve – it's an adult version of attunement. It's also a way of trying to gain another person's approval (the more you crave social approval, the more you'll converge[23]). A converger might be trying to improve communication between themself and someone else: because it reduces the vocal and verbal differences between people and brings them psychologically closer,[24] convergence also expresses the desire to integrate into a group, or identify with another person.[25]

It rests on the fact that we compare other people's behaviour with our own, and the more they resemble us, the more we like them. People in one experiment judged those with a speaking rate similar to their own more competent and socially attractive than those who spoke slower or faster than them.[26] In another experiment the more that the other speakers matched the length of university students' pauses, the warmer the students thought them and the more they wanted to invite them to dinner.[27] On a psychiatric ward, psychotic patients are less likely to be labelled 'difficult' when staff spontaneously match their rhythm and prosody.[28] People were more willing to buy a book after hearing an audio review on the Internet from a voice that sounded like their own.[29] Another experiment got computer users to work with animated screen characters that mimicked, to a greater or lesser extent, their speech patterns – rhythm, intonation, loudness, and pitch. When the users were asked to rate the characters for friendliness, sympathy, comfort, and cooperation, those that had imitated 80 per cent of a user's own vocal qualities got the highest approval.[30] We like other people who are, literally, on our wavelength and share the same beat; if these are very different, communication might even be jeopardised.[31]

Convergence provides a fascinating guide to social status, because lower-status people converge up to higher status. In fact you can almost read off status or dominance from the degree to which one speaker accommodates to another.[32] A major study in Taiwan found that salespeople converged much more towards moneyed customers than the other way round.[33] When interviews between the host and guests on an American talk show, CNN's *Larry King Live*, were analysed, it was found that King shifted his pitch towards that of high-status celebrities or politicians like Bill Clinton or Elizabeth Taylor or Barbara Streisand, but lower-status guests like Henry Kissinger or Dan Quayle accommodated their pitch to his.[34] A study of the televised debates of American presidential candidates between 1960 and 2000 also found that the less dominant candidates accommodated to the more dominant ones: you could predict the winner of the election from the amount and direction of convergence alone.[35] In close elections, it's even been claimed, pitch may exert a strong influence over the popular vote, favouring the candidate with the most commanding presence.

But does this tell us anything other than that confident people impose themselves on the less confident, vocally as well as by other means? Or that candidates accurately assess their chance of winning and communicate it through their voice? And though accommodation may be a form of social glue, a way of speaking in tune, doesn't it also express a fear of difference, a conformism, even narcissism? Doesn't authentic communication take place between people who are distinct rather than merged? There's certainly a limit beyond which accommodation turns into over-accommodation. Japanese businessmen, for instance, don't want their Western counterparts to try to act Japanese because they believe that the Westerners sound false or patronising.[36]

Convergence can also exact a price: a speaker can lose their sense of identity.[37] A 15-year-old girl said, 'I'm conscious of adapting my accent – I do it quite a lot. With the childminder I also have a bit of a cockney accent, with Mum and Dad [who are both Americans] I have a bit of an American accent. I do it to try and build up a connection with people.' Does she ever wonder what she really sounds like? 'Yes, I do.'[38] Not surprisingly women, traditionally, converge far more to men than the other way round.[39]

In divergence we maintain our nonverbal characteristics no matter what the situation, sometimes even accentuating the differences between ourselves and another speaker to establish a unique social identity.[40] Sometimes we deliberately diverge to influence the person we're talking to, slowing down our own speech rate, for instance, in order to slow down theirs.[41]

Convergence is now being exploited commercially. Japanese

researchers are trying to produce robots, computer games, and toys that mimic users' speech patterns in order to build 'rapport' between them.[42] And here's a celebrity voice coach, giving advice on how to use your voice to deliver a sales pitch 'virtually guaranteed to sell'. 'People respond best to people who are like them . . . If [the client has] . . . a high voice, raise the pitch of your voice a little; if they speak quickly, speak more quickly yourself; if they speak slowly, you should slow down too.'[43]

NLP, or Neuro-Linguistic Programming, a New Age type of behavioural training, promises to enhance communication partly by getting adherents to 'model' other people's voices. 'Voice matching is another way that you can gain rapport. You can match tonality, speed, volume and rhythm of speech. This is like joining another person's song or music: you blend in and harmonise.'[44]

Although some people believe that trying to mimic other people always sounds inauthentic,[45] a 52-year-old woman described using convergence when cold-calling companies. 'I hate it, so I made it into a game, because most of the time you get voice-mail. I listen to the voice-mail and hear them speak, whether they sound up or down, and then I just mimic it, sounding exactly like them, hoping they'd feel some kind of affinity with me because I'm sounding more like them. It seems to be rather successful.'[46]

A 16-year-old girl says:

My voice adapts to the people I'm with. When I started school, I consciously changed my voice to fit in with the other children and sound more working-class. My mother used to say that when I said the word 'no', it sounded as if it had every single vowel in it. When I did a drama course last summer with a lot of working-class people, I adapted to their voices, not entirely consciously, but so much that when I phoned up one of my oldest friends she said, 'Who's this?' and when I told her it was me she said, 'You sound so different.' Then a week later I went to a summer camp and my voice went all posh. Naturally it's somewhere in between those two extremes.[47]

The voices of people talking to each other are like the periods of women living together: they automatically synchronise. We become entrained to other people's rhythms: their voices are our tuning-forks. The urge to converge seems irresistible.

## POLYVOCAL

Most of us are vocally versatile and have different voices for different occasions (the performance artist Laurie Anderson once reckoned we each had fifty), but this 24-year-old British solicitor, with his many voices, is positively promiscuous:

> I'm very slow and formal with clients – I try not to get too animated with people at work. I think it's more professional to sound slow and well-enunciated. I want to sound considered and intelligent. I use a sympathetic voice to achieve what I want from a secretary. Imagine that a 35-year-old woman has been at the firm for ages and suddenly this young man comes straight from law school and starts bossing her around – potentially there'd be a tension there. Your tone of voice has a big effect on how successful your relationship with your secretary is. I try to put a lot of appreciation into my voice. Phoning up a woman I don't know for the first time, I try and sound cool and really relaxed, even if I don't feel it. With friends I am really relaxed, and not nearly as well spoken as in front of my parents. Then there's my Valerie voice (she's my grandmother). That's very clear and polite. With my girlfriend, I make much less of an effort to enunciate.[48]

The process is intentional:

> At work it's a completely conscious decision . . . I think the tone of voice is incredibly powerful, persuasive, and important – it's very easy to manipulate people using your voice. I do it a lot. I get my way through using my voice. My friends know that if anyone

is going to blag anything it's going to be me, especially from a woman. In New York there was a huge queue round eight blocks for a concert, and we got VIP tickets at the very front just from me being really nice and appreciative to the people at the door with my voice.

Another time three people in front of me asked and were turned away from photocopying in an office by the woman saying, 'No, we don't do photocopying here.' I smiled a lot and my voice was low and calm and she photocopied the lot for me . . . I don't feel in any way fake when I do it – I enjoy it. People are willing to accommodate you further just because you're being nice with your voice. Maybe it's for the wrong reasons but . . . it can make life so much easier.

I think there's a big difference between trying to charm someone with your voice and having a smarmy voice – a smarmy voice instantly turns people off, it sounds fake and deliberately manipulative, whereas a charming voice makes you sound like a genuinely nice person . . . But guys respond far less to charming voices than women. You have to use a very polite voice with a man – make them realise that they're superior if they are, use your voice to apologise for disturbing them . . .

Other people talk about their voice as a reflection of their feelings; I think I've got used to adapting my voice to the circumstances – I do it subconsciously but also consciously.[49]

In his groundbreaking 1959 book, *The Presentation of Self in Everyday Life*, the sociologist Erving Goffman argued that social life was like a theatrical performance in which the individual 'guides and controls the impression . . . [others] form of him . . . by expressing himself in such a way as to give them the kind of impression that will lead them to act voluntarily in accordance with his own plan.'[50] The young man quoted above is a striking example.

Goffman's theory, radical and in many ways shocking when first published, is today not only unremarkable but might almost stand as a 'to do' list for self-improvers. Indeed, the failure to transform yourself now practically counts as a dereliction of occupational duty. Professional culture no longer has a concept of intrinsic self – it's all appearance and mirage, refraction upon refraction, in which the voice plays a crucial role. The young lawyer, in what perhaps is a sign of changing gendered expectations, is also doing the kind of emotional housekeeping more usually associated with women.

Either way, the voice is now seen as a critical component in professional success. Voice-training companies call themselves the Winning Voice[51] or the Voice of Influence ('Success is just as much about how you sound as about how you look'). Books targeted at business executives bear titles like *The Leader's Voice*,[52] or promise to teach you to use your voice to deliver a sales pitch that's guaranteed to sell.[53] According to *Presenting to Win*,[54] 63 per cent of company directors believe that presentation skills are more important for career success than intelligence or financial aptitude. Staff these days are expected to do whatever's required in order to create a good impression, whether this means accent reduction (big in the United States[55]), or a 'voice lift'. Flab, it seems, affects not just the stomach but the voice too. Once you've been Botoxed, tummy-tucked and had liposuction, you don't want an ageing voice to let you down: $15,000 will buy you rejuvenating vocal implants or a bulking up or tightening of the vocal folds – cosmetic or vanity surgery for the voice.[56]

According to the *New York Times*, a sonorous voice has now been added to the checklist of perfection. Fifteen years ago, one American speech pathologist observed, he rarely saw people whose only problem was dislike of the sound of their voice: today they constitute one-third of his clients. Having a voice coach is now no more remarkable than having a personal trainer.[57]

No longer is there any stigma attached to changing the voice (Prince Diana's voice coach openly advertises the changes he brought to her public-speaking style). Indeed it's a requirement for call-centre staff. 'Remember voice intonation is also very important, as tone, pace and clarity convey your attitude to the customer. You must never sound bored on a call. Your telephone manner should convey the impression that you have been waiting for that individual call all day. To assist in this try putting a smile on your face when receiving a call.'[58] This kind of managerial appropriation of private feelings has been dubbed 'emotional labour'. A study of airline attendants quotes a stewardess describing a colleague who 'put on a fake voice. On the plane she raised her voice about four octaves and put a lot of sugar and spice into it [gives a falsetto imitation of "More coffee for you, sir?"]. I watched the passengers wince. What the passengers want is real people.'[59]

Ironically, all the emphasis on changing the voice to produce the perfect presentation, or sway a judge, or sell more products, has left listeners with a problem in judging what's authentic, so necessitating the creation of yet another set of primers and guides – this lot explaining how to decode other people's voices and distinguish the genuine voice from the phoney one. Books with titles like *Never Be Lied to Again*,[60] *I Know What You're Thinking*,[61] and *Reading People*,[62] that require us to interpret other people's vocal clues and hear between the lines, to become in effect one-person spectrographs, are the direct result of all the vocal makeovers encouraged by their predecessors. It can't be accidental that the field of deception studies has grown so much – there must be more deception around to study.

## I'D KNOW THAT VOICE ANYWHERE

Some people change their voices for a living. And yet impressionists, while they obviously develop terrific control of the muscles of the throat, tongue, and lips, rarely deliver a voice that is really similar to that which they're imitating, only a stereotyped version, a vocal cartoon. They identify the main characteristics of someone's voice, dialect, and phonetic habits, and then exaggerate some aspects while discarding others.[63] When a professor of acoustics recorded the impressionist Rory Bremner impersonating someone, Bremner and his victim sounded almost indistinguishable, but when the acoustics of the two tapes were compared, the higher frequencies were actually quite different. Bremner, though, had been able to make his lower frequencies – site of most of the message-sending function of the voice – convincing enough for the listener to have been seduced into thinking he was someone else.[64]

Bremner himself is aware of the differences.

> You latch on to a distinctive vocal quality and stretch it and exaggerate it until it becomes ridiculous, and then it starts to be funny. To begin with, your impression is quite accurate but then, as your character starts to develop, it isn't so accurate. After a while, people associate a person's voice more with the caricature than with the original. If you make Ken Livingstone sound whiney, after a while all people hear is that whine.[65]

As Bremner remarks, 'inimitable' is usually a misnomer: it's those people with the most distinctive qualities in their voice that are easy to imitate. If a speaker isn't distinctive or well known enough, then it doesn't matter how accurately an impressionist can mimic them: you have to first recognise a voice in order to be able to recognise it being imitated. So being the butt of an impression is flattering: it suggests that you're famous enough, and your voice unique enough, to be recognised.

But mimicry can easily metamorphose from flattery into subversion. Between them, Rory Bremner and cartoonist Steve Bell comprehensively ridiculed British Prime Minister John Major. Bell drew a grey man who tucked his shirt into his underpants (and eventually wore his underpants over his trousers), while Bremner created an image of a lower-middle-class nerd who said 'wunt' instead of 'want' – a man perplexed that, despite his obvious ineffectiveness, he was 'still here'. Once these characterisations had become established, whenever you heard Major speak you heard Rory Bremner's John Major, in the same way that now Tony Blair has become Bremner's Tony Blair, laboured sincerity and all.

Thus, once a satirical impression has taken hold, do politicians undermine themselves simply by speaking.

Indeed a good impressionist ends up virtually obliterating the person on whom their impression is based. There's the true story of the company that wanted to use the voice of TV chat-show host Russell Harty in an advertisement. For the pilot, the advertising agency brought in impressionist Chris Barrie, one of the *Spitting Image* regulars, to do Harty's voice. The company was happy with the result, and the agency went on to produce the actual advertisement, using the real Russell Harty. But when the company heard the finished product, they protested that the voice didn't sound anything like him. They'd got so used to the caricature that the real Harty sounded like an imperfect copy.[66]

## THIS IS HOW I SOUND

Despite the pressures to transform ourselves vocally, our voice, as we have seen in chapter 8, is so intricately bound up with our sense of self that changing it can seem impossible – like changing a limb. As one voice specialist put it, although most people don't know their own voices, they 'must have a very concrete relation to it because patients violently resist any change of their vocal functioning brought about by therapeutic intervention'.[67] A 43-year-old doctor sought treatment because he thought his voice sounded too high, but when a recording of his post-treatment voice was played back to him, 'he became red in the face, and declared that he felt as ashamed of his new (deep and male-sounding) voice as if he had been exposing himself. Therefore he refused to be treated any further.'[68]

People often express their reluctance to change their voice with the words, 'But this voice is not me.' This expression 'implies a basic identification of their personality with their former voice, and they resist any change of voice as they would oppose a change of their personality'.[69]

That was written fifty years ago. In today's flexi-voice culture, might resistance be less? Not necessarily, because the issues being touched upon are so deep. 'The patient's Gestalt of him or herself ... may be inflexible ... the person may be asked to change a basic vocal attribute ... clinicians must address a client's self-concept of what is normal with as much fervour as for the relatively simple task of manipulating vocal pitch.'[70]

Of course vocal change is possible. Techniques to help control nerves, breathe more deeply, loosen the jaw, neck and spine, get the arms swinging, become more sensitive to rhythm, develop more vocal variety, and speak from deeper in the body can be taught.[71] Others work more from the inside, seeing how speaking is connected with speaking out and speaking up, trying to help people 'find a voice that accommodates their expressive needs ... and doesn't only speak in one part of their range ... a monotone ... [which means] you've squashed your expression of yourself but it's also not very interesting to other people to listen to because ... you're deadening yourself.'[72] Still others conduct a kind of psychotherapy through the voice.[73]

Yet some of the most interesting vocal change takes place almost as a by-product of psychological development. A 38-year-old woman described the process:

> I remember in my first job being challenged about always wearing pale-pastel clothes, always pink. Other members of staff said, 'Where's the non-pink part of you?' I think one's voice is part of one's body, appearance, manner, and relating to the world, so I think my voice fitted my pastel image in those days – it was calm and gentle, I hadn't got in touch with my own aggression and strength and sense of self and I think my voice reflected that. The more I've been able to relate to the outer world with a real self, the more my voice has developed and strengthened.[74]

So what of the tension between the 'natural' and 'winning' voice? Both are ideologies, ways of imagining nature and vanquishing it, of triumphing over personal or social limitations. The voice may be a sensitive barometer of feeling and self, but it's always a social self, for we're born and die in the social world (and so the voice is also a subtle gauge of a culture's values at any one time). Even if an optimum pitch exists, most of us exist in a vocal continuum and what counts as a more or less authentic sound varies according to the occasion and company. No amount of voice coaching will change that.

# Conclusion

THE VOICE IS a remarkable instrument, but also a remarkably unacknowledged one. This book was born out a sense of frustration that its brilliance and importance in human society had not been sufficiently recognised. While language and body language have been analysed and extolled, the voice has languished – at least beyond the academic world – largely unhymned. We've lost a sense of its primacy.

Worse, the pessimists, the nostalgics, and the technophobes have together tried to convince us that voice is no longer a central part of human culture, displaced by the text and downgraded by the image. The eye, they insist, has prevailed over the ear.

As this book I hope has demonstrated, they're wrong. Our vocal skills haven't atrophied. On the contrary, they've proved fabulously resilient: every corner of our lives is animated by talk. Babies possess an astonishing prenatal sensitivity to the voice, one which they exercise within hours of birth. The voice acts as a crucial connective tissue between baby and carer: mothers can contain their babies' fears with it, and infants use it to help them acquire language. Through exposure to their parents' voices children learn to speak expressively.

Adults, too, in managing their personal, professional, and social lives, make use of sublime voice-reading talents. Voice is turning into such an important occupational tool, with so much emphasis placed on improving it, that Aristotle and the other classical analysts of rhetoric would feel quite at home.

Inscribed in our voice are both our deepest feelings about ourselves and shifting ideas of what it is to be male or female. The voice is a distinctive human feature, and yet a constantly evolving one. While the invention of the telephone, phonograph, and radio helped change the relationship between body and voice, they also enlarged the voice's orbit, so that its reach was no longer limited by the body's own transportability. We may use our voice in quite different ways from people in oral societies, yet it's held its own, with all sorts of fantastical new uses for it being devised. The voice, I'm certain, has become a more and not less important medium of human communication over the past century, only we lack a shared public language in which to articulate it. Even though the sonic is clearly booming, paradoxically, we've failed to voice the voice's own talents.

I've wanted to celebrate here the many different, creative ways in which people and their cultures make infinitely complex sequences of meanings out of invisible puffs of air, an ability that grows partly from nurture but also from the mere fact of being born human.

Volume, pace and pitch summon whole worlds. Intonation is a language in itself. The American philosopher and wit, Sidney Morgenbesser, was in the audience at a lecture given by the Oxford philosopher J.L. Austin at Columbia University in the 1950s. When Austin explained that many languages employ the double negative to denote a positive ('He is not unlike his sister'), but none employed a double positive to make a negative, Morgenbesser waved his arm dismissively, and retorted: 'Yeah, yeah.'[1]

With the blast and blare of cinemas, restaurants, concerts, computer games and TV commercials, we live in loud times. A recent study by the Royal Society for the Protection of Birds suggested that birds that live near motorways can't hear each other, leading to difficulties in learning songs and communicating with potential mates. Another study found that 5-year-old children who attended nursery developed more voice problems than those who didn't because of the high noise levels and unsympathetic acoustic environment.[2] What's the effect on humans when our voices are submerged by the din?[3] And how can we create an acoustic space in which this suggestive but perpetually elusive instrument, the human voice, can flourish?

To attune properly to the voice we must develop a keener sensitivity, a 'deep listening'.[4] To start a real conversation about this most vital talent, we need to hear with fresh ears.

# Notes

## Preface

1 Ariel Leve, 'They Shoot Movies, Don't They?' (*Sunday Times* Magazine, 8 June 2003).
2 Interview, 31.10.03.
3 Interview, 16.10.03.
4 Interview, 24.04.05.
5 Interview, 10.11.03.
6 Nalini Ambady et al, 'Surgeons' Tone of Voice: a Clue to Malpractice History', *Surgery*, 2002:132.
7 Walter J. Ong, *Orality and Literacy* (London: Methuen, 1982).
8 Richard M. Harris and David Rubinstein, 'Paralanguage, Communication, and Cognition', in Adam Kendon et al, *Organization of Behaviour in Face-to-Face Interaction* (The Hague: Mouton & Co, 1975).
9 Ernst G. Beier, *The Secret Language of Psychotherapy* pp.12, 13 (Chicago: Aldine, 1966).

## 1. What the Voice Can Tell Us

1 John Laver, *The Gift of Speech* (Edinburgh University Press, 1991).
2 T.H. Pear, *Voice and Personality* (London: Chapman and Hall, 1931), a pioneering study in which radio listeners ascribed jobs to voices.
3 Sue Ellen Linville, 'Acoustic Correlates of Perceived versus Actual Sexual Orientation in Men's Speech' (*Folia Phoniatrica et Logopaedica* 50:35–48, 1998).
4 Laver, op cit.
5 Beth Schucker and David R. Jacobs, 'Assessment of Behavioral Risk for Coronary Disease by Voice Characteristics' (*Psychosomatic Medicine*, vol. 39, no. 4, July–August 1977). This study, however, was based on Ray Rosen-

man's research about Type A and Type B behaviour, categories which have been much critiqued. See chapters 8 and 13 for criticism of some features of experimental voice studies.

6  www.courier-journal.com/index.html.

7  D.J. France et al, 'Acoustical Properties of Speech as Indicators of Depression and Suicidal Risk', *Transactions on Biomedical Engineering*, 47(7), July 2000.

8  Grazyna Niedzielska et al, 'Acoustic Evaluation of Voice in Individuals with Alcohol Addiction' (*Folia Phoniatrica et Logopaedica* 46:115–122, 1994).

9  J. Alexander Tanford et al, 'Novel Scientific Evidence of Intoxication: Acoustic Analysis of Voice Recordings from the *Exxon Valdez*' (*Journal of Criminal Law and Criminology*, vol.82, no.3 1991). Such techniques have never been used in a civil or criminal case but the analysts of the *Exxon Valdez* tapes argue that they should be admitted as valid evidence.

10 Jeffrey Whitmore and Stanley Fisher, 'Speech During Sustained Operations' (*Speech Communication*, 20, 55–70, 1996). Alison D. Bagnall et al, 'Voice and Fatigue', paper given at PEVOC 6, Royal Academy of Music, London, September 2005.

11 Kakuichi Shiomi and Shohzo Hirose, 'Fatigue and Drowsiness Predictors for Pilots and Air Traffic Controllers', paper given at the Annual ATCA Conference, Oct 22–26, 2000, Atlantic City, New Jersey.

12 Personal communication, Mark Lawson, 29.07.05.

13 G.P. Nerbonne, cited in Mark L. Knapp, *Non-verbal Communication in Human Interaction* (New York: Holt, Rinehart, and Winston, 1972).

14 R. Murray Schafer, Keynote Address to the Chorus America Conference, Toronto, 8 June 2001.

15 Tom Paulin, 'The Despotism of the Eye', p.43, in *Soundscape: The School of Sound Lectures 1998–2001*, eds. Larry Sider et al (London and New York: Wallflower Press, 2003).

16 Simon Capper, 'Non-verbal Communication and the Second Language Learner: Some Pedagogical Considerations' (www.langue/hyper.chimbu.ac.jp).

17 Personal communication, Bernard Kops, 15.10.02.

18 Samuel Taylor Coleridge, *Biographia Literaria* (London: Routledge, 1983).

19 Paul D. Erickson, *Reagan Speaks: The Making of an American Myth* (New York: New York University Press, 1985). Similarly a 1987 book on modern presidential communication, *The Sound of Leadership* by Roderick Hart (Chicago: University of Chicago Press, 1987), contained nothing at all on the sound of presidents' voices and how they use them.

20 Michel Chion, *The Voice in Cinema* (New York: Columbia University Press, 1999).

21 Stephanie Martin, 'An Exploration of Factors which Have an Impact on the Vocal Performance and Vocal Effectiveness of Newly Qualified Teachers and Lecturers' (paper given at the Pan European Voice Conference, Graz, 31 August 2003) reviews the literature.

22 With the exception of Stephanie Martin and Lyn Darnley, *The Teacher's Voice* (London: Whurr Publishers, 2nd edition, 2004).

23 Interview, 11.11.03. A teacher I interviewed uses her voice to try and control her pupils. 'If the class is fidgeting I go louder and vary the tone. If I get cross with them I go quite high-pitched so it doesn't sound so harsh. They do say that quiet teachers have quiet classes. The voice is such an important tool in teaching.' (Interview, 7.03.04).

24 David Crystal, *The English Tone of Voice* (London: Edward Arnold, 1975).

25 Despite various attempts to isolate and catalogue them, like that by the French

composer of electro-acoustic music Pierre Schaeffer, who tried to create a terminology of sound independent of its origins in the 1960s, by classifying sound objects into families or genres – see Pierre Schaeffer, *Traite des objets musicaux* (Paris: Editions du Seuil, 1966).

26 James Rush, *The Philosophy of the Human Voice* (Philadelphia: Grigg and Elliott, 1833).

27 John Laver, *The Phonetic Description of Voice Quality* (Cambridge: Cambridge University Press, 1980).

28 David Crystal, *A Dictionary of Linguistics and Phonetics* (Oxford: Blackwell, 2003).

29 Walter J. Ong, *Orality and Literacy* (London: Methuen, 1982).

30 R. Murray Schafer, *The Soundscape: Our Sonic Environment and the Tuning of the World* (Toronto: McClelland and Stewart, 1977).

31 Interview with Stephanie Martin, speech and language therapist, 3.11.03.

32 Ong, op cit.

33 Mladen Dolar, 'The Object Voice', p.11, in Renata Salecl and Slavoy Zizekl (eds.), *Gaze and Voice as Love-objects* (Durham: Duke University Press, 1996).

34 Talking is always temporary – the life-cycle eventually silences us all, so there's a poignancy about the voice: I speak therefore I'll die. Orgasm has been called 'une petite mort', a little death. So too is the voice.

35 Richard Thorn, 'The Anthropology of Sound', paper given at 'Hearing is Believing' conference, University of Sunderland, 2 March 1996.

36 Brian Stross, 'Speaking of Speaking: Tenejapa Tzeltal Metalinguistics', in Richard Bauman and Joel Sherzer, eds., *Explorations in the Ethnography of Speaking* (Cambridge: Cambridge University Press, 1989).

37 Laura Martin, ' "Eskimo Words for Snow": A Case Study in the Genesis and Decay of an Anthropological Example' (*American Anthropologist*, vol.88, no.2, 1986).

38 Interview, 30.10.03.

39 Interview, 16.03.04.

40 Interview, 31.10.03.

41 Interview, 14.11.03.

42 Interviews conducted in the UK, October–November 2003.

43 Michael Argyle, *Bodily Communication* (London: Routledge, 1975).

44 W.H. Thorpe, 'Vocal Communication in Birds', in R.A. Hinde, ed., *Non-Verbal Communication* (Cambridge: Cambridge University Press, 1972).

45 Lea Leinonen et al, 'Shared Means and Meanings in Vocal Expression of Man and Macaque' (*Logopedics, Phoniatrics, Vocology*, vol.28: 2, 2003).

46 Argyle, op cit.

47 Uwe Jurgens, 'On the Neurobiology of Vocal Communication', in Hanus Papousek et al, eds., *Nonverbal Vocal Communication: Comparative and Developmental Approaches* (Cambridge: Cambridge University Press, 1992). See chapter 11 for an elaboration.

48 Jan G. Svec et al, 'Vocal Dosimetry: Theoretical and Practical Issues', in G. Schade et al, eds., *Proceeding Papers for the Conference, Advances in Qualitative Laryngology, Voice and Speech Research* (Stuttgart: IRB Verlag, 2003). This research therefore makes an interesting contribution to the heated topic of whether men or women talk more – see chapter 10.

49 We know because researchers have attached neck devices to people that measure the vibrations on their skin. See Jan G. Svec et al, 'Measurement with Vocal Dosimeter: How Many Metres do the Vocal Folds Travel During a Day?', paper given at the Pan European Voice Conference, Graz, Austria, 28.9.2003.

50 Erkki Vilkman, 'The Role of Voice (quality) Therapy in Occupation Safety and Health Context', paper given at the Pan European Voice Conference, Graz, Austria, 29.8.2003; Ingo Titze, 'How Far Can the Vocal Folds Travel', paper given at the American Speech-Language-Hearing Association Annual Conference, Atlanta, Georgia, 2002; Ingo R. Titze et al, 'Populations in the US Workforce Who Rely on Voice as a Primary Tool of Trade: A Preliminary Report', *Journal of Voice*, vol. 11, no. 3, 1997.

51 Bernadette Timmermans et al, 'Vocal Hygiene in Radio Students and in Radio Professionals', *Logopedics, Phoniatrics, Vocology*, vol. 28: 3, 2003.

52 Vilkman, op cit. But see chapter 17 for a discussion of how this is changing.

53 Stephanie Martin interview, op cit.

54 Leon Thurman and Graham Welch, eds, *Bodymind and Voice: Foundations of Voice Education* (Minnesota: The Voicecare Network, 1997).

55 Leslie C. Dunn and Nancy A. Jones, *Embodied Voices* (Cambridge: Cambridge University Press, 1994).

56 George Bernard Shaw, Preface to *Pygmalion*, p.5 (London: Penguin Books, 1951). Shaw did plenty of the despising himself, viz the prejudice pouring out of him later in the Preface – 'An honest slum dialect is more tolerable than the attempts of phonetically untaught persons to imitate the plutocracy.' (ibid., p. 10).

## 2. How the Voice Achieves its Range and Power

1 Manuel Patricio Rodriguez Garcia, 'Transactions of the Section of Laryngology, International Congress of Medicine', London, 1881, in *British Medical Journal*, 17 February 2001.

2 Garcia wasn't the first person to think about how the voice was produced. Aristotle, Galen and Leonardo had all puzzled over the larynx. But his serendipitous idea 150 years ago began literally to illuminate the dark tunnel in which air turns into voice. Garcia was interested in singing and not medicine: his contraption was later refined for clinical use by the addition of artificial light and the redesign of the laryngeal mirror. However much the story of Garcia's invention seems pure Archimedes – the inspired ideation of an idling great man – in reality it arrived at the same moment as a whole array of other attempts to see into the human body, to turn the inside out. In 1851 the first binaural stethoscope was commercially marketed, followed by the ophthalmoscope (making visible the eye), the gastroscope (stomach – after an assistant of Adolf Kussmaul persuaded a sword-swallower to swallow half a metre of pipe plus lamp and lens. What became of him has not been recorded), the otoscope (ear drum), and in 1895 the X-ray.

3 Joseph C. Stemple, 'Voice Research: So What? A Clearer View of Voice Production, 25 Years of Progress: the Speaking Voice', *Journal of Voice*, vol. 7, no. 4, 1993. Some of the research on the voice, it has to be said, is so desiccated as to make one forget that what is under scrutiny is the vital subject of vocal exchange.

4 Alfred Tomatis, *The Ear and Language*, p.59 (Ontario: Moulin Publishing, 1996).

5 ibid., p.60.

6 William J.M. Levelt, *Speaking: from Intention to Articulation* (Cambridge, Mass: The MIT Press, 1989).

7 Robert Thayer Sataloff, 'Vocal Health' (www.voicefoundation.org).

8 Alexander Graham Bell, *Lectures upon the Mechanism of Speech* (New York: Funk and Wagnalls, 1906).

9 Thanks to John Rubin for drawing my attention to this function of the larynx.

10 Aristotle, *De Anima* (London: Routledge, 1993).

11 Harry Hollien, *The Acoustics of Crime* (New York: Plenum Press, 1990).

12 W.A. Aikin, *The Voice* (London: Longmans, Green, and Co, 1951).

13 Steven Pinker, *The Language Instinct*, p.164 (London: Penguin Books, 1994).

14 Hollien, op cit.

15 Peter B. Denes and Elliot N. Pinson, *The Speech Chain* (W.H. Freeman, 1993).

16 Paul Newham, *Therapeutic Voicework* (London: Jessica Kingsley Publishers, 1998).

17 Denes and Pinson, op cit.

18 Leon Thurman and Graham Welch (eds.), 'Bodymind and Voice: Foundations of Voice Education. (Minnesota: The Voicecare Network, 1997). While the vocal tract changes shape and even length during speech, its basic dimensions are physiologically determined, particular to each individual (no two vocal tracts are exactly the same size and shape) and beyond their control.

19 Margaret C.L. Greene, *Disorders of the Voice* (Austin, Texas, PRO-ED, 1986).

20 Thurman and Welch, op cit. The subglottic vocal tract, i.e. that which lies below the folds, also plays a resonant role – think of the chesty sound of a baritone – with the diaphragm providing important support. If we don't get proper support from the chest, we rely excessively on the neck and throat muscles to do a job they weren't designed for, resulting in vocal strain or hoarseness (dysphonia). Clothes, too affect the voice. The high military collars worn by British regimental officers in the late eighteenth and nineteenth centuries helped create the 'stiff upper-lip' voice. Similarly, people anxious about ruining their hair, or broadcasters wearing tight clothes that mustn't betray a wrinkle, suffer similarly from constricted neck muscles that strangulate their voices (Patsy Rodenburg, *The Right to Speak*, London: Methuen, 1992).

21 Aikin, op cit.

22 V.C. Tartter, 'Happy Talk: Perceptual and Acoustic Effects of Smiling on Speech', *Perception and Psychophysics*, vol 27(1), 1980.

23 John Laver and Peter Trudgill, 'Phonetic and Linguistic Markers in Speech' in Klaus R. Scherer and Howard Giles, eds., *Social Markers in Speech* (Cambridge: Cambridge University Press, 1979).

24 Terrence Deacon, *The Symbolic Species* (London: Penguin Books, 1998).

25 Aikin, op cit. Children born without any sound-perception are unable to produce any sound at all.

26 Stemple, op cit.

27 Edward Sapir, *Language: An Introduction to the Study of Speech* (Thomson Learning, 1955).

28 D.B. Fry, *The Physics of Speech* (Cambridge: Cambridge University Press, 1979). These comments, of course, are based on the Western tonal system.

29 Sapir, op cit. We've swapped labial freedom, he suggested, for precision in speaking our mother tongue.

30 Ingo Titze, 'The Human Voice at the Intersection of Art and Science', paper given at PEVOC 2003, 28.8.03, Graz, Austria.

31 Fry, op cit.

32 Alfred A. Tomatis, *The Conscious Ear* (New York: Station Hill Press, 1991).

33 Tomatis 1996, op cit, p.55.

34 Titze, op cit.

35  R. Murray Schafer, *The Soundscape: Our Sonic Environment and the Tuning of the World* (USA: Destiny Books, 1994).

36  Diana Van Lancker, 'Speech Behaviour as a Communication Process', in John K. Darby, *Speech Evaluation in Psychiatry* (New York: Grune & Stratton, 1981).

37  It also helps us locate a sound since, as Darwin pointed out, we've lost the ability (retained by many animals) to move our ears towards a sound. Thankfully we're binaural creatures, with two ears that work together to provide this information. By comparing the relative intensity or loudness heard in each of them, along with the different arrival time of the signal in each ear, we can identify the source and direction of a voice or other sound. And once heard, we can hang on to a particular sound despite the existence of loud and continuous random noise that almost overwhelms it. This is known as the cocktail-party phenomenon: it allows us to listen to several conversations simultaneously but only consciously tune into one of them. It's also what enables us to pick out our own name from a buzz of conversation.

38  Schafer, op cit. Gradual, age-related loss of hearing acuity, called presbycusis, is usually assumed to be an inescapable consequence of ageing – wrinkles of the audio kind. Yet the Mabaan Africans of the Sudan suffer very little hearing loss due to presbycusis: at the age of 60 their hearing is as good if not better than the average 25-year-old North American. The presumed reason is their noise-free environment: the loudest sounds that the Mabaan (at least when they were studied back in 1962) heard were the sounds of their own voices, singing and shouting at tribal dances. Words aside, the unamplified human voice doesn't easily cause injury.

39  Frequency range isn't the same as fundamental frequency.

40  Named after the Italian anatomist, Alfonso Corti. In 1851, three years before Garcia first saw his vocal folds, Corti identified through his microscope a part of the cochlea that now bears his name.

41  Elise Hancock, 'A Primer on Hearing' (*Johns Hopkins Magazine*, September 1996).

42  Thinning hair cells are a major cause of hearing loss, and are permanent, for hair cells, like a woman's eggs, are all present at birth: thereafter they can only be lost, and not gained.

43  Denes and Pinson, op cit.

44  ibid.

45  Murray Sachs, quoted in Hancock, op cit.

46  Pascal Belin et al, 'Voice-selective Areas in Human Auditory Cortex', *Nature*, vol. 403, 20 January 2000.

47  Frans Debruyne et al, 'Speaking Fundamental Frequency in Monozygotic and Dizygotic Twins', *Journal of Voice*, vol. 16, no. 4, 2002.

48  ibid.

49  Robert Thayer Sataloff, 'Genetics of the Voice', *Journal of Voice*, vol. 9, no. 1. See also Steven Gray et al, 'Witnessing a Revolution in Voice Research: Genomics, Tissue Engineering, Biochips and What's Next!', *Logopedics, Phoniatrics, Vocology*, vol. 28: 1, 2003.

50  Tomatis 1996, op cit, p.51.

51  Patricia Liehr, 'Uncovering a Hidden Language: The Effects of Listening and Talking on Blood Pressure and Heart Rate', *Archives of Psychiatric Nursing*, vol. 6, no. 5 (October), 1992.

52  Denes and Pinson, op cit.

53  Tomatis 1991 and 1996, op cit. In response, he developed an Electronic Ear that enhanced certain frequencies, which, he maintained, allowed the ear to sing with

greater facility, and the voice to respond accordingly. When they wore head-phones and had their own voices filtered in this way, all his patients – even those who weren't singers – reported greater well-being and incomparably better singing. By re-educating his patients' ears, Tomatis claimed, his machine could even help people to learn to speak foreign languages more effectively because they now heard them differently. Listening problems, he argued, were also at the root of many learning disabilities. Maverick though he then seemed, today a network of 250 Tomatis Centres around the world treat children (and adults) with communication disorders, attention-deficit problems, motor skills difficul-ties, and even autism. His method attracted recent media attention when the French actor Gerald Depardieu described how Tomatis had helped him learn how to complete his sentences and develop his thoughts (Paul Chutkow, *Depardieu* (New York: Knopf, 1994).

## 3. How We Colour Our Voices with Pitch, Volume, and Tempo

1  David Abercrombie, *Problems and Principles* (London: Longmans, 1956).
2  Kenneth Pike, *The Intonation of American English* (Ann Arbor: University of Michigan Press, 1945); Rulon Wells, 'The Pitch Components of English', *Language*, 21, 1943.
3  George L. Trager, 'Paralanguage: A First Approximation', reprinted in Dell Hymes, *Language in Culture and Society* (New York: Harper-Row, 1964).
4  Morner, Fransesson, and Fant, cited in Harry Hollien, 'On vocal registers', *Journal of Phonetics*, 2, 1974.
5  Philip Lieberman, 'Linguistic and Paralinguistic Interchange', in Adam Kendon et al, eds., *Organisation of Behaviour in Face-to-Face Interaction* (The Hague: Mouton Publishers, 1975). Moroever, as Lieberman points out, what might be paralinguistic in one language may have a linguistic function in another.
6  Bentley's conviction was posthumously overturned by the Court of Appeal in 1998.
7  There are also fashions in the way we stress. The writer Michael Frayn once wrote a comic newspaper column on modern mis-stressing, arguing that flying today wasn't a high-stress occupation but a wrong-stress one, because of the capricious emphases used by cabin crews in announcements. These often seem to suggest a Churchillian resolve. The plane, they tell us, *will* be landing shortly in Cincinnati. We *are* requested to make sure we have all our belongings with us. (Michael Frayn, *Speak After the Beep*, London: Methuen, 1997).
8  David Crystal, *A Dictionary of Linguistics and Phonetics* (Oxford: Blackwell Publishing, 2003). The idea that, unlike English speakers, Mandarin speakers use intonation to distinguish between completely different meanings of words has been challenged as language-specific bias by a Western linguist working in Beijing. From an Aboriginal point of view, it's equally bizarre that changing 'bed' into 'bet' completely alters its meaning (Edward McDonald, Linguist List, 2 July, 2003).
9  See David Crystal, *The English Tone of Voice* (London: Edward Arnold, 1975), and J. Lyons, 'Human Language', in R.A. Hinde, ed., *Non-Verbal Commu-nication* (Cambridge: Cambridge University Press, 1972)
10  Some have even used paralanguage to undermine the power of writing, which they claim is a meagre system compared with the many different meanings that can be communicated by stress, pitch, pause, tempo, etc. (Eleanor J. Gibson and

Harry Levin, *The Psychology of Reading*, Boston: The MIT Press, 1975.) This takes us back to the zero sum idea (and playground game: which would you rather be – deaf or blind?) that we must somehow choose between sound and vision, that by praising one, we inevitably devalue the other. Personally, I value both.

11 Crystal, 1975, op cit.

12 David Crystal, *Prosodic Systems and Intonation in English* (Cambridge: Cambridge University Press, 1969). Here I use the word pitch to refer to both.

13 In 1919 the psychologist Carl Seashore set out to prove that musical ability was inborn, and that we arrive in the world with a good ear or a tin one. He gave listeners fifty pairs of tones, and asked them to say whether the second was higher or lower than the first. Seashore's Measures of Musical Talent discovered that there was a 200-fold difference in listeners' ability to distinguish the minimum frequency between two successive tones (Carl Seashore, *The Psychology of Musical Talent*, Boston: Silver, Burdett & Company, 1919).

14 Maria Schubiger, quoted in Crystal, 1969, op cit.

15 Alan Beck, *Radio Acting* (London: A&C Black, 1997).

16 D.B. Fry, *The Physics of Speech* (Cambridge: Cambridge University Press, 1979).

17 Jacqueline Martin, quote in Beck, op cit.

18 Richard Luchsonger and Godfrey Arnold, *Voice-Speech-Language* (California: Wadsworth, 1965).

19 When we sing, by contrast, we hold pitch steady – or try to – for the length of a note (Fry, op cit).

20 Atkinson, cited in Robert F. Coleman and Ira W. Markham, 'Normal Variations in Habitual Pitch', *Journal of Voice*, vol. 5, no. 2.

21 Coleman and Markham, ibid.

22 Kathryn L. Garrett and E. Charles Healey, 'An Acoustic Analysis of Fluctuations in the Voices of Normal Adult Speakers Across Three Times of Day', *Journal of the Acoustical Society of America*, vol. 82, no. 1, July 1987.

23 See M.C.L. Greene and Lesley Mathieson, *The Voice and Its Disorders* (England: Whurr Publishers, 1989).

24 Interview with Jon Snow, 3.3.04.

25 Interview with Rory Bremner, 30.3.04.

26 Beck, op cit.

27 Daniel Boone, *Is Your Voice Telling on You?* (San Diego: Singular, 1997).

28 Snow, op cit.

29 Interview with Dan Rather, 18.7.03.

30 Demonstrating, perhaps, how news reporting has been redefined as a branch of the leisure industry.

31 *Newsweek* quoted in Mark L. Knapp, *Non-verbal Communication in Human Interaction*, pp.147–8 (New York: Holt, Rinehart and Winston, 1972).

32 E. Sapir, 'Speech as a Personality Trait', p.76, in John Laver and Sandy Hutcheson, eds., *Communication in Face-to-Face Interaction* (England: Penguin Books, 1972).

33 ibid, p.77.

34 John J. Gumperz, *Discourse Strategies* (Cambridge: Cambridge University Press, 1982).

35 Aron W. Siegman, 'The Telltale Voice: Nonverbal Messages of Verbal Communication', p.425, in Aron W. Siegman and Stanley Feldstein, *Nonverbal Behaviour and Communication* (New Jersey: Lawrence Erlbaum Associates, 1987).

36 Crystal, 1969, op cit.
37 Richard M. Harris and David Rubinstein, 'Paralanguage, Communication, and Cognition', in Adam Kendon et al, eds., *Organization of Behaviour in Face-to-Face Interaction* (The Hague: Mouton Publishers, 1975).
38 E.T. Hall, 'A System for the Notation of Proxemic Behaviour' in Laver and Hutcheson, op cit.
39 Interview, 23.10.03.
40 Interview, 31.10.03.
41 Donna Jo Napoli, *Linguistics* (New York: Oxford University Press, 1996). In fact there's no such thing as a decibel: decibels always express a ratio, yet the second sound with which a first is being compared invariably gets erased (Fry, op cit).
42 R. Murray Schafer, *The Soundscape: Our Sonic Environment and the Tuning of the World* (Toronto: McClelland and Stewart, 1977).
43 As the British Conservative leader Iain Duncan-Smith and Labour leader Neil Kinnock both found to their cost.
44 Interview with Stephanie Martin, 3.11.03.
45 Diana Van Lancker and Jody Kreiman, 'Familiar Voice Recognition: Patterns and Parameters. Part 2: Recognition of Rate-altered Voices', *Journal of Phonetics*, 13, 1985.
46 Alan Cruttenden, *Intonation* (Cambridge: Cambridge University Press, 1986).
47 Edward D. Miller, *Emergency Broadcasting and 1930s American Radio* (Philadelphia: Temple University Press, 2003). See chapter 14 for a discussion of how Roosevelt used his voice.
48 Cruttenden, op cit.
49 Peter Roach, 'Some Languages are Spoken More Quickly than Others' in Laurie Bauer and Peter Trudgill, eds, *Language Myths* (London: Penguin Books, 1998).
50 William Apple et al, 'Effects of Pitch and Speech Rate on Personal Attributes', *Journal of Personality and Social Psychology*, vol. 37, no. 5, 1979.
51 Stanley Feldstein et al, 'Gender and Speech Rate in the Perception of Competence and Social Attractiveness', *Journal of Social Psychology*, 141(6), 2001. See also John Laver and Peter Trudgill, 'Phonetic and Linguistic Markers in Speech', in Klaus R. Scherer and Howard Giles, eds., *Social Markers in Speech* (Cambridge: Cambridge University Press, 1979).
52 Norman Miller et al, 'Speed of Speech and Persuasion', *Journal of Personality and Social Psychology*, vol. 34, no. 4, 1978. This study pre-dated Ronald Reagan's presidency. Did his ambling, folksy pace do anything to sever the association of speed with persuasiveness?
53 Feldstein et al, op cit.
54 Robert G. Harper, ed., *Nonverbal Communication: The State of the Art* (New York: John Wiley, 1978).
55 Interview, 31.10.03.
56 Roach, op cit.
57 Rather, op cit.
58 Aaron Barnhart, 'Speeded-up *Frasier* gives KSMO Extra Ad-time', *Kansas City Star*, 7.2.99.
59 Rather, op cit. American TV reporters, argues Jon Snow, so hurtle through their commentary that they omit certain structural elements of the language, getting the pictures to do the work of the verbs (Snow, op cit).
60 Beck, op cit.
61 London: Guinness World Records 2001.

62  Miller, op cit.
63  Rocco Dal Vera, 'The Voice in Heightened Affective States', in Rocco Dal Vera, ed., *The Voice in Violence*, p.64 (Cincinnati: Voice and Speech Trainers Association, 2001).
64  Copy, cited in Siegman and Feldstein, op cit.
65  Klaus R. Scherer, 'Personality Markers in Speech', in Klaus R. Scherer and Howard Giles, eds., *Social Markers in Speech* (Cambridge: Cambridge University Press, 1979).
66  Interview, 17.5.03.
67  Interview, 21.3.04.
68  Interview 16.6.04.
69  Interview, 16.19.04.
70  Interview, 23.11.04.
71  Hillel Halkin, p.vi, *Sholom Aleichem: Tevye the Dairyman and The Railroad Stories* (New York: Schocken Books, 1987).
72  ibid.
73  Quintillian, *Institutio Oratorio*, Book XI, chapter 3 (Loeb Classical Library, 1920).
74  Diana Van Lancker, 'Speech Behaviour as a Communication Process', in John K. Derby, ed., *Speech Evaluation in Psychiatry* (New York: Grune and Stratton, 1981).
75  Crystal, 1975, op cit. Just to confuse matters further, Crystal, 1969, op cit, made a distinction between timbre and voice quality.
76  Greene and Mathieson, op cit.
77  Van Lancker, op cit.
78  Alfred Lang, 'Timbre Constancy in Space: Hearing and Missing Spatially Induced Sound Variations' (Bern: Psychologisches Institut, 1988).

## 4. What Makes the Voice Distinctly Human

1  Erich D. Jarvis et al, 'Behaviourally Driven Gene Expression Reveals Song Nuclei in Hummingbird Brain', *Nature*, 406, 10 August 2000.
2  Michael Argyle, *Bodily Communication* (London: Routledge, 1990).
3  Allison J. Doupe and Patricia K. Kuhl, 'Birdsong and Human Speech: Common Themes and Mechanisms', *Annual Review of Neuroscience*, vol. 22, 1999.
4  W.H. Thorpe, 'Vocal Communication in Birds' in R.A. Hinde, ed., *Non-Verbal Communication* (Cambridge: Cambridge University Press, 1972).
5  ibid. Contrast this with the Misgurnus fish that makes a noise by gulping air bubbles and expelling them forcibly through its anus (R. Murray Schafer, *The Soundscape: Our Sonic Environment and the Tuning of the World* (Toronto: McClelland & Stewart, 1997)).
6  Thorpe, op cit.
7  Gibson, quoted in ibid.
8  This has been demonstrated by cross-fostering experiments, taking birds from the wild as eggs or nestlings of one species and exposing them to the songs of other species, which they then proceed to learn.
9  California Academy of Sciences, www.calacademy.org.
10  Joseph Conrad, *Under Western Eyes*, Part 1, Prologue (London: Penguin Books, 1973).
11  Jarvis, op cit. These claims haven't gone uncontested. Some critics are sceptical about what they see as extravagant extrapolation: they suggest that to treat

songbirds like some kind of surrogate human is misleading – human speech is infinitely more complex than birdsong (*The Osgood File*, CBS Radio, 12.12.03).

12 Jarvis is now considering transgenic experiments to see if the insertion into non-learning species of those genes associated with vocal learning could convert a non-learner into a learner (Dennis Meredith, 'Singing in the Brain', *Duke Magazine*, vol. 88, no. 1, Nov–Dec 2001). Already in *The Descent of Man* Darwin had noted that some species of birds that don't naturally sing 'can without much difficulty be taught to do so – thus a house sparrow has learnt the song of a linnet'. Charles Darwin, *The Descent of Man* (UK: Prometheus Books, 1991).

13 Argyle, op cit. Darwin argued that the human speaking voice originated in music: 'I have been led to infer that the progenitors of man probably uttered musical tones, before they had acquired the power of articulate speech; and that consequently, when the voice is used under any strong emotion, it tends to assume, through the principle of association, a musical character' (Charles Darwin, *The Expression of the Emotions in Man and Animals*, UK: Fontana Press, 1999). From here, he believed, it developed a sexual function: 'Although the sounds emitted by animals of all kinds serve many purposes, a strong case can be made out that the vocal organs were primarily used and perfected in relation to the propagation of the species: the chief and, in some cases, exclusive purpose appears to be either to call or claim the opposite sex' (Darwin, 1991, op cit).

14 Schafer, op cit.

15 Lea Leinonen, 'Shared Means and Meanings in Vocal Expression of Man and Macaque', *Logopedics, Phoniatrics, Vocology*, vol. 28: 2, 2003. This Finnish researcher has found that human adults and children over the age of nine are able correctly to categorise macaque vocalisations into angry, frightened, dominant, pleading, or content, and that human and monkey pleadings come out at a similar pitch and volume.

16 John J. Ohala, 'Cross-Language Use of Pitch: An Ethological View', *Phonetica*, 40, 1983.

17 Dwight Bolinger, 'Intonation Across Languages', in Joseph H. Greenberg, ed., *Universals of Human Language* (California: Stanford University Press, 1978).

18 Interestingly, in Japanese, giving up or admitting defeat is conveyed by the phrase '*ne wo ageru*'. This means 'to raise one's sound or tone' – and they don't mean volume (Norman D. Cook, *Tone of Voice and Mind*, Amsterdam: John Benjamins, 2002).

19 Ohala, op cit.

20 It's found support from music psychologists who contend that ascending pitch contours, both in speech and music, invariably convey a feeling of tension and uncertainty, whereas descending contours suggest resolution (S.Brown cited in Cook, op cit). Cook argues that there are further musical parallels, in that 'a decrease in pitch resolves towards a major mode, and an increase in pitch resolves toward a minor code', p.125. And see chapter 10 for another intriguing possible example of the frequency code.

21 ibid.

22 M.C.L. Greene and Lesley Matheson, *The Voice and Its Disorders* (England: Whurr Publishers, 1989).

23 Edmund S. Crelin, 'The Skulls of Our Ancestors: Implications Regarding Speech, Language, and Conceptual Thought Evolution', *Journal of Voice*, vol. 3, no. 1, 1989.

24 Philip Lieberman, *Eve Spoke* (London: Picador, 1998).

25 ibid.

26 ibid.

27 B. Arensburg et al, 'A Reappraisal of the Anatomical Basis for Speech in Middle Palaeolithic Hominids', *American Journal of Physical Anthropology*, 83, 1980.

28 Baruch Arensburg and Anne-Marie Tillier, 'Speech and the Neanderthals', *Endeavour*, vol. 15, no. 1, 1991. Red and fallow deer, it transpires, also have descended larynxes but can't talk – proof, say some, that a low-lying larynx isn't unique to humans, and isn't necessarily tied to speech production but rather to the fact that it allows a creature to make low-frequency sounds that exaggerate their size. In other words the descended larynx is a useful component in the arsenal of intimidation, through which a deer (or person) can generate the deeper sound of a much larger animal (W. Tecumseh Fitch and David Reby, 'The Descended Larynx Is Not Uniquely Human', *Proceedings of the Royal Society, London*, B, vol. 268, 2001).

29 Richard F. Kay et al, 'The Hypoglossal Canal and the Origin of Human Vocal Behaviour, *Proceedings of the National Academy of Sciences, USA*, vol. 95, issue 9, 28 April 1998.

30 David DeGusta et al, 'Hypoglossal Canal Size and Hominid Speech', *Proceedings of the National Academy of Sciences, USA, Anthropology*, vol. 96, February 1999. The hypoglossal canal, they allege, isn't related to the size of the hypoglossal nerve or tongue function and so isn't a reliable indication of speech and can't be used to date its origins.

31 Terrence Deacon, *The Symbolic Species* (London: Penguin Books, 1998), a thrilling exploration of the evolution of speech.

32 A.G. Clark et al, 'Inferring Nonneural Evolution From Human-chimp-mouse Orthologous Gene Trios', *Science*, 302, 2003. One of them makes a protein that plays a vital role in the membrane in the inner ear. (This, to put it in perspective, out of some 30–35,000 genes in the human genome.)

33 ibid.

34 Nicholas Wade, 'Comparing Genomes Shows Split Between Chimps and People', *New York Times*, 12.12.03.

35 Pierre Paul Broca, 'Loss of Speech, Chronic Softening and Partial Destruction of the Anterior Left Lobe of the Brain', *Bulletin de la Société Anthropologique*, 2, 235–8, 1861, translated by Christopher D. Green, *Classics in the History of Psychology* (www.psychclassics.yorku.ca).

36 Paul Broca, 'Remarks on the Seat of the Faculty of Articulated Language, Following an Observation of Aphemia (Loss of Speech)', *Bulletin de la Société Anatomique*, 6, 1861, translated by Christopher D. Green, 'Classics in the History of Psychology' (www.psychclassics.yorku.ca)

37 Tan's brain is still on display in the Museum of Pathological Anatomy, the Musée Dupuytren, in Paris.

38 Of course it wasn't quite so simple. Just as Manuel Garcia's invention of the laryngoscope rested on many former and simultaneous inventions, so Broca's work didn't take place in a vacuum. In fact he was responding to a challenge made just a fortnight earlier. At a public meeting at the Anthropological Society on 4 April Ernest Aubertin had declared that he'd renounce his belief in cerebral localisation if just one case of speech loss could be produced without a frontal lesion. An inspired Broca provided evidence in support of Aubertin.

39 Robert M. Young, *Mind, Brain and Adaptation in the Nineteenth Century: Cerebral Localization and its Biological Context from Gall to Ferrier* (USA: Oxford University Press, 1990).

40 It doesn't do to mock Victorian gullibility, though: some aspects of phrenology

anticipated our understanding of cerebral hemispheres, and some of the wild modern assertions made about hemispheres, as we'll see, rest on no firmer foundation than did the phrenologists'. See, for instance, John van Wyhe, 'The History of Phrenology on the Web' (http://pages.britishlibrary.net/phrenology/, 2002). And despite the acclaim, Broca's method has not been without its critics. Tan's brain was so decayed that pinpointing the exact origin of spoken language had something of the flavour of blind man's bluff.

41 Until the arrival of functional magnetic resonance imaging – the brain scanner – in the next century any other method of investigation was too invasive.

42 In Wernicke's aphasia, people, after an accident or stroke, speak fluent, grammatical, but meaningless babble, sometimes called a 'word salad'. On one occasion a patient with damage in Wernicke's area who heard the sentence 'It's raining outside?' spoken in the intonation appropriate to 'How are you today?' replied 'Fine' (Rhawm Joseph et al, *Neuropsychiatry, Neuropsychology,* \ *Clinical Neuroscience*, USA: Lippincott, Williams, and Wilkins, 1996).

43 Deacon, op cit.

44 See Sophie K. Scott et al, 'Identification of a Pathway for Intelligible Speech in the Left Temporal Lobe', *Brain*, 2000, 123, and Matthew H. Davis and Ingrid S. Johnsrude, 'Hierarchical Processing in Spoken Language Comprehension', *Journal of Neuroscience*, 15 April 2003. Before speech can become intelligible, we have to process its phonetic and acoustic elements. This takes place in the superior left temporal lobe, the bottom part of the left hemisphere just above the ear, inside the temple. It's here that the auditory signals first arrive in the cerebral cortex after their journey from the cochlear (see chapter 2).

45 Cook, op cit, makes the point that, if one hemisphere is surgically removed at a young age, the remaining hemisphere alone is capable of almost the full range of language tasks. Hemisphere specialisation is a developmental process.

46 Marc Pell, 'Surveying Emotional Prosody in the Brain', in B. Bel and I. Marlien, *Proceedings of Speech Prosody,* 2002 Conference (Aix-en-Provence: Laboratoire Parole et Langage, 2002).

47 Safer and Leventhal, cited by Don M. Tucker, 'Neural Control of Emotional Communication', in Peter David Blanck et al, eds., *Nonverbal Communication in the Clinical Context* (USA: The Pennsylvania State University Press, 1986).

48 Sandra Weintraub et al, 'Disturbances in Prosody: A Right-Hemisphere Contribution to Language', *Archives of Neurology*, vol. 38, December 1981.

49 Jorg J. Schmitt et al, 'Hemispheric Asymmetry in the Recognition of Emotional Attitude Conveyed by Facial Expression, Prosody and Propositional Speech', *Cortex*, 33, 1977. So do psychopaths. When thirty-nine psychopaths incarcerated in a high-security prison were compared with non-psychopaths, they had particular difficulty recognising a fearful voice (R. James Blair, 'Turning a Deaf Ear to Fear: Impaired Recognition of Vocal Affect in Psychopathic Individuals', *Journal of Abnormal Psychology*, vol. 1, 2002).

50 Ellen Winner et al, 'Distinguishing Lies from Jokes: Theory of Mind Deficits and Discourse Interpretation in Right Hemisphere Brain-Damaged Patients', *Brain and Language*, 62, 1998.

51 Joseph, op cit.

52 Diana Roupas Van Lancker et al, 'Phonagnosia: a Dissociation Between Familiar and Unfamiliar Voices', *Cortex*, 24, 1988.

53 Elliott D. Ross, 'How the Brain Integrates Affective and Propositional Language into a Unified Behavioural Function', *Archives of Neurology*, vol. 38, December 1981.

54 Cook, op cit.

55 Guy Vingerhoets et al, 'Cerebral Hemodynamics During Discrimination of Prosodic and Semantic Emotion in Speech Studied by Transcranial Doppler Ultrasonography', *Neuropsychology*, vol. 17(1), January 2003. See also Ralph Adolphs et al, 'Neural Systems for Recognition of Emotional Prosody: A 3-D Lesion Study', *Emotion*, vol. 2, no. 1, 2002.

56 Patrick J. Gannon et al, 'Asymmetry of Chimpanzee Planum Temporale: Humanlike Pattern of Wernicke's Brain Language Area Homolog', *Science*, 279, 9 January, 1998.

57 People with Parkinson's disease, for instance, experience difficulties in understanding other people's tone of voice and using their own expressively because of damage to the basal ganglia, a subcortical part of the brain that doesn't lie in either hemisphere (Pell, op cit). Prosody can also be impaired when the corpus callosum is damaged. You can't simply extrapolate from brain damage to brain function – deterioration of a function isn't just the reverse of its development. Data from schools for simultaneous translators, of the kind hired by the United Nations, suggests that, under pressure, both hemispheres can to some extent become language hemispheres. The problem for these interpreters is to keep the two languages from getting in each other's way. Before training, most students of simultaneous translation begin with a left-hemisphere (right ear) preference for both languages. By the time they've finished, some have developed a separate ear (and therefore hemisphere) for each language. (Deacon, op cit). There's also a theory that stutterers process language differently: brain scans seem to show in them a shift from the left hemisphere to the right. And see also J.S. Morris et al, 'Saying it with Feeling: Neural Responses to Emotional Vocalizations', *Neuropsychologia*, 37, 1999.

58 Tracy L. Luks et al, 'Hemispheric Involvement in the Perception of Syntactic Prosody is Dynamically Dependent on Task Demands', *Brain and Language*, 65, 1998. Yet still we get plied with news stories implying that the right hemisphere goes off duty while the left attends to language, like the recent BBC news report claiming that, while English people only needed to use the left side of their brain to understand English, Mandarin Chinese speakers used both sides of their brain because of the higher intonational demands of the language ('Chinese "takes more brainpower"', *BBC News*, 30.6.03).

59 Daniel Goleman, *Emotional Intelligence* (London: Bloomsbury, 1996).

60 S.K. Scott et al, 'Impaired Auditory Recognition of Fear and Anger after Bilateral Amygdala Lesions', *Nature*, 385, 16 January 1997.

61 This has been challenged: recognition of fear in the voice is impaired not by damage to the amygdala, say some researchers, but by damage to the basal ganglia (A.K. Anderson and E.A. Phelps, 'Intact Recognition of Vocal Expressions of Fear Following Bilateral Lesions of the Human Amygdala', *Neuroreport*, 9 (16), 16 November 1998.

62 R. Adolphs and D. Tranel, 'Intact Recognition of Emotional Prosody Following Amygdala Damage', *Neuropsychologia*, 37, October 1999. The amygdala may be crucial for detecting fear in the voice, but not for perceiving disgust (M.L. Phillips et al, 'Neural Responses to Facial and Vocal Expressions of Fear and Disgust', *Proceedings of the Royal Society, London, Biological Science*, 265, 7 October 1998). More recent work seems to suggest that the amygdala binds together the emotional information that we receive from the face and the voice, so that when we see both a fearful face and a frightened voice, the twin signals of danger combine in the amygdala. This allows us to feel fear instinctively, without the data ever having to pass through the conscious, cortical part of the brain. With a scary (non-facial) picture and a fearful voice, however, the cortex needs

to be activated in order to process the information before fear can be registered (Beatrice de Gelder et al, 'Fear recognition in the voice is modulated by unconsciously recognised facial expressions but not by unconsciously recognized affective pictures', *Proceedings of the National Academy of Sciences of the United States of America*, 99 (6), 19 March, 2002.

63 Just as various parts of the brain confer in the process of voice-reading, so too no one discipline can single-handedly explain how emotion gets encoded in the speaker's voice and decoded by the listener. Over the past decade there's been a fascinating convergence between psychoanalysis and neurobiology. The burgeoning fields of psychobiology and neuropsychology are adding to our understanding of the neurological dimensions of psychological development – some have even argued that the unconscious is located in the right hemisphere (Allan N. Schore, *Affect Regulation and the Repair of the Self*, New York: W.W. Norton, 2003). We no longer need to think of the mind and the brain as providing alternative, mutually incompatible explanations of how human beings feel and behave – or indeed speak. When neuroscience and psychoanalysis treat each other as equal and mutually enriching partners, the resultant insights can be exciting. It's less fertile, in my view, when science is seen as providing 'hard' proof and corroboration for the 'soft' discipline of psychoanalysis. A neuropsychological account of how the human voice is shaped by the emotions would be a thrilling affair.

64 Pamela Davis, 'Emotional Influences in Singing', paper given at the Fourth International Congress of Voice Teachers, London, 1997. Intriguingly, attempts to understand the reasons that tinnitus is so vexatious – why its ringing tones are so much more distressing than, say, the sound of a fridge – have also helped explain the close connection between the auditory system and those parts of the brain that control emotions and alert us to danger. After all we're only aware of a small proportion of incoming sounds – our brain filters out the majority of the unwanted ones. According to the neurophysiological theory of tinnitus, it's less the loudness and pitch associated with the condition that are so troublesome, but rather that when the autonomic nervous system hears the characteristic tinnital whistle, it responds as it does to danger and fear, causing the sufferer to perceive the tinnitus as threatening. By changing the neural pathways and retraining the brain to reclassify tinnitus as less significant, Pawel Jastreboff has helped habituate tinnitus sufferers to their condition, muting its discordant, piercing effects. (See, for example, www.tinnitus-pjj.com. Thanks to John Rubin for drawing my attention to this work.) If the meaning of sounds can be deliberately altered so that the brain responds to them in a new fashion, then enhancing and developing our receptiveness to vocal cues must be a footling task in comparison.

## 5. The Impact of the Mother's Voice (even in the Womb)

1 Denis Vasse, *L'ombilic et la voix: deux enfants en analyse* (Paris: Editions du Seuil, 1974). Even better than the umbilicus, in fact, because when that's ruptured, the voice still remains, and contact with the mother then 'becomes mediated by the voice' (Vasse quoted by Michel Chion, *The Voice in Cinema*, p. 61, New York: Columbia University Press, 1999).

2 Guy Rosolato, 'La voix: entre corps et langage', *Revue francaise de psychanalyse*, 37, no. 1, 1974.

3 Didier Anzieu, 'The Sound Image of the Self', *International Review of Psycho-analysis*, 1979: 6. Kaja Silverman describes the maternal voice as 'the acoustic mirror in which the child first hears itself' (Kaja Silverman, *The Acoustic Mirror*, Bloomington: Indiana University Press, 1988).

4 Paul Newham, *Therapeutic Voicework* (London: Jessica Kingsley, 1998).

5 'One can hear and be heard in the dark, in blindness and through walls' (Anzieu, op cit).

6 90 per cent of the mother's utterances suggesting a change in her engagement with her child were accompanied by a significant change in voice quality, whereas only 57 per cent were reflected in changes in facial expression (Marwick unpublished study, cited in Colwyn Trevarthen, 'Emotions in Infancy: Regula-tors of Contact and Relationships with Persons', in Klaus R. Scherer and Paul Ekman, eds., *Approaches to Emotion*, Hillsdale, New Jersey: Lawrence Erlbaum Associates, 1984).

7 Otto Isakower, 'On the Exceptional Position of the Auditory Sphere', *Interna-tional Journal of Psychoanalysis*, 20, 1939.

8 Irinia Ziabreva et al, 'Separation-Induced Receptor Changes in the Hippocam-pus and Amygdala of Octogon degus: Influence of Maternal Vocalizations', *Journal of Neuroscience*, 23(12), 15 June 2003. Of course a human baby isn't a South American rodent, but it's interesting nevertheless that the limbic region in which maternal vocalizations can alter the activity patterns in the rodent pup are the same area that's activated when a human mother responds to her baby's cries. There's clearly a psychological link for a human between having the experience of being soothed by the mother's voice, and later developing the ability to soothe an infant, but might there be a neurological one too?

9 8.3.96, www.drgreene.com.

10 Hilary Whitney, 'My husband says that Stephen points and smiles when he hears my voice', *Guardian*, 30.5.01. A similar scheme involving fathers is in operation in another British prison.

11 Edward D Murphy, 'A tough time inspires a toy', www.maintoday.com, 30.8.02.

12 'Mother's Voice is Music to a Sick Child's Ears', 29.1.02, www.cosmiverse.com.

13 This position was slowly challenged, although by somewhat dubious scientific methods: in a 1925 experiment a car horn was honked a few feet from the abdomen of a woman in late pregnancy, while a hand on her stomach confirmed – eureka! – that in response 25–30 per cent of the time the foetus moved (B.S. Kisilevsky and J.A. Low, 'Human Foetal Behaviour: 100 Years of Study', *Developmental Review* 18, 1998).

14 It's even been claimed that, by the third trimester, the foetus moves in rhythm to the mother's voice. Bernard Auriol, in 'Les eaux primordiales: La vie sonore du fœtus' (www.cabinet.auriol/free.fr), along with writers on other websites, quotes Dr Henry Truby, a professor from the University of Miami, as making this claim, although I haven't been able to track down any study on which it's based.

15 William P. Fifer and Chris M. Moon, 'The Effects of Fetal Experience with Sound', in Jean-Pierre Lecanuet et al, eds., *Fetal Development : A Psychobio-logical Perspective* (New Jersey : Lawrence Erlbaum Associates, 1995).

16 With a five beat-per-minute increase in heart rate in foetuses exposed to their mother's voice compared to a four beat-per-minute decrease in heart rate in those exposed to a female stranger's voice (Barbara S. Kisilevsky et al, 'Effects of Experience on Fetal Voice Recognition', *Psychological Science*, vol. 14, no. 3, May 2003).

17 ibid. Unborn babies can also discriminate between their mother speaking and a

recording of her voice (P.G. Hepper et al, 'Newborn and Fetal Response to Maternal Voice', *Journal of Reproductive and Infant Psychology*, vol.11, 1993).

18 At 36–39 weeks' gestation their heart rate changes when a woman starts reading the same sentence after a man (or vice versa). (J.P. Lecanuet et al, 'Prenatal Discrimination of a Male and Female Voice Uttering the Same Sentence', *Early Development and Parenting*, vol. 2(4), 1993). We know that foetuses can discriminate between syllables, because one study found them detecting the differences between the speech sounds 'ba' and 'bi' at 27 gestational weeks of age, and at different frequencies at 35 weeks (S. Shahidullah and P.G. Hepper, 'Frequency Discrimination By the Fetus', *Early Human Development*, 36(1), January 1994. See also Jean-Pierre Lecanuet et al, 'Human Fetal Auditory Perception', in Jean-Pierre Lecanuet et al, eds., *Fetal Development : A Psychobiological Perspective*, (New Jersey : Lawrence Erlbaum Associates, 1995). In another study the brains of premature infants detected when a repeated vowel was replaced by another, supporting the view that the human foetus learns to discriminate sounds while still in the womb. This study claims to constitute the 'ontologically earliest discriminative response of the human brain ever recorded' (M.Cheour-Luhtanen et al, 'The Ontogenetically Earliest Discriminative Response of the Human Brain', *Psychophysiology*, 33(4), July 1996.

19 W.P. Fifer and C.M. Moon, 'The Role of the Mother's Voice in the Organization of Brain Function in the Newborn', *Acta Paediatric Supplement* 397, 1994.

20 Sherri L. Smith et al, 'Intelligibility of Sentences Recorded from the Uterus of a Pregnant Ewe and from the Fetal Inner Ear', *Audiology and Neuro-Otology*, 8, 2003. The chief acoustic properties of the mother's voice remain the same inside the womb and outside (Benedicte De Boysson-Bardies, *How Language Comes to Children: From Birth to Two Years* (Massachusetis: MIT Press, 1999).

21 Anthony DeCasper quoted in 'Sounds Inside the Womb Revealed', *Guardian*, 5.2.04. This research is based on a pregnant ewe: other researchers insist that a human foetus isn't the same as a sheep, with the species receptive to different frequencies. Rossella Lorenzi, 'Can unborn babies hear oohs and aahs?', *Discovery News*, Australian Broadcasting Corporation Online, 20.2.04.

22 D Querlu et al, 'Intra-amniotic Transmission of the Human Voice', *Revue Francaise de Gynecologie et Obstetrique*', 83, January 1988.

23 Petitjean, cited in Leon Thurman and Graham Welch, 'Bodymind and Voice: Foundations of Voice Education' (The Voice Care Network, 1997).

24 William G. Niederland, 'Early Auditory Experiences, Beating Fantasies, and Primal Scene', *Psychoanalytic Study of the Child*, 1958.

25 ibid.

26 See Athena Vouloumanos and Janet F. Werker, 'Tuned to the Signal: the Privileged Status of Speech for Young Infants', *Developmental Science*, 7:3, 2004.

27 Querleu and Renard, cited in Melanie J. Spence and Anthony J. DeCasper, 'Prenatal Experience with Low-Frequency Maternal-Voice Sounds Influence Neonatal Perception of Maternal Voice Samples', *Infant Behaviour and Development*, 10, 1987; and D. Querleu et al, 'Reaction of a Newborn Infant Less Than Two Hours After Birth to the Maternal Voice', *Journal de gynecologie, obstetrique, et biologie de la reproduction* 13(2), 1984.

28 Anthony J. DeCasper and William P.Fifer, 'Of Human Bonding: Newborns Prefer Their Mothers' Voices', *Science*, vol. 208, 6 June 1980. Although DeCasper and Fifer are pioneering researchers in the field and have come up with a series of fascinating results, their 1980 study warrants a word of caution. To begin with, it's based on a very small sample – in the first instance 10

neonates, 16 when the experiment was repeated. And then, its very title, 'Of Human Bonding', and premise both rest on assertions about mother-infant bonding that were formulated by two paediatricians whose research has since been dismissed by most of the scientific community as nothing more than 'scientific fiction' (see Diane E. Eyer, *Mother–Infant Bonding: A Scientific Fiction* (New Haven: Yale University Press, 1992). DeCasper and Fifer considered their finding that newborns prefer their mother's voice as confirmation of the (now discredited) theory that the period shortly after birth is uniquely important for mother–infant bonding. The newborns' preference, they suggested, may even be an integral aspect of that bonding. Like many other sociologists I believe that you can find neonates' vocal preferences tantalising and touching without accepting the whole bonding hypothesis.

A 1988 study found that, while the heart rate of infants under 24 hours old decelerated when they heard their mother's voice, that of older infants speeded up. Elizabeth M. Ockleford et al ('Responses of Neonates to Parents' and Others' Voices', *Early Human Development*, 18, 1988) hypothesised that, beyond a few hours after birth, babies respond defensively to the sudden termination of their mother's voice.

29 To conduct a study like this you need an awake baby who's calm and slightly hungry. (How else do you get it to suck of its own accord?) A flexible arm or a research assistant holds a dummy in the baby's mouth. Attached to it is a pressure transducer: this detects and measures the rate and intensity of the baby's sucking, in response to a stimulus (until the baby has got used to it and so the effect has worn off), and sends the information to a computer. When a baby resumes sucking or sucks harder after a change of stimulus, this shows that the baby has noticed the difference. (Laboratoire de Sciences Cognitives et Psycholinguistiques Infant Lab, Paris, www.ehess.fr/centres/lscp.babylab). (There are other ways of measuring a baby's auditory capabilities, but they tend to treat the infant more like one of Pavlov's dogs.)

30 John Bowlby, *Attachment* (London: Penguin Books, 1984). In one of the earliest experiments, conducted in 1927, a three-week-old baby, when he heard a human voice, began to suck and gurgle with pleasure, and wore an expression that suggested pleasure. When the voice stopped the baby started to cry and show other signs of displeasure. Was this because he had already come to associate the female voice with food, as the researchers thought, or (as the psychoanalyst John Bowlby argued) because he wanted to hear his mother? The sounds arising from the preparation of a bottle of milk didn't evoke a similarly pleasurable response. (Hetzer and Tudor-Hart study, cited in ibid).

31 Wolff, cited in Anne Fernald, 'Meaningful Melodies in Mothers' Speech to Infants', in Hanus Papousek et al, *Nonverbal Vocal Communication: Comparative and Developmental Approaches* (Cambridge: Cambridge University Press, 1992). Hearing its mother immediately after birth (along with recognising her smell) causes the baby to look at her, which then teaches it to recognise her face. The mother's voice, therefore, helps launch the sighted infant's visual abilities (Colwyn Trevarthen, 'Intrinsive Motives for Companionship in Understanding: their Origin, Development, and Significance for Infant Mental Health', *Infant Mental Health Journal*, vol. 22(1–2), 2001).

32 Wolff, cited in Bowlby op cit.

33 Bowlby put it this way: 'On the one hand, her infant's interest in her voice is likely to lead a mother to talk to him more; on the other, the very fact that his attention to her has the effect of increasing the mother's vocalisations . . . is likely to lead the baby to pay even more attention to the sounds she makes. In

this mutually reinforcing way the vocal and auditory interaction between the pair increases.' (Bowlby in ibid, p.274).

34 Spence and DeCasper, op cit; Fifer and Moon 1995, op cit.

35 Piontelli cited in Suzanne Maiello, 'The Sound-object: A Hypothesis about Prenatal Auditory Experience and Memory', *Journal of Child Psychotherapy*, vol. 21, no. 1, 1995.

36 Caroline Floccia et al, 'Unfamiliar Voice Discrimination for Short Stimuli in Newborns', *Developmental Science*, 3:3, 2000.

37 See review of research in ibid.

38 Jacques Mehler and Josiane Bertoncini, 'Infants' Perception of Speech and other Acoustic Stimuli', in John Morton and John C. Marshall, eds., *Psycholinguistics 2: Structures and Processes* (London: Paul Elek, 1979). If, on the other hand, the word order and therefore cadences are reversed, the baby shows no marked preference for the mother's voice over that of a stranger.

39 Caron et al, cited in John Wilding et al, 'Sound Familiar?', *Psychologist*, vol. 13, no.11, November 2000.

40 Anzieu, op cit. In fact they have a pretty sophisticated understanding of how sight and sound cohere. In one experiment 6–12-week-old infants became distressed when watching a video of their mothers in which the speech and visual content were discrepant, suggesting that they analyse voice characteristics from a very young age, and can skilfully combine voice and face (Murray and Trevarthen, cited in Wilding et al, op cit). At 2 months they look longer at filmed images of faces where the vowel shape being made by someone's lips matches the vowel sound coming from their mouth than images where they don't synchronise (Michelle L. Patterson and Janet F. Werker, 'Two-month-old Infants Match Phonetic Information in Lips and Voice', *Developmental Science* 6:2, 2003).

41 Christopher W. Robinson and Vladimir M. Sloutsky, 'Auditory Dominance and Its Change in the Course of Development', *Child Development*, vol.75, no.5, September/October 2004.

42 Vladimir M. Sloutsky and Amanda C. Napolitano, 'Is a Picture Worth a Thousand Words? Preferences for Auditory Modality in Young Children', *Child Development*, vol. 74, no. 3, May/June 2003. It's also been argued, though, that four-year-olds attend more to the content of speech than to its paralinguistic features, where the two are discrepant, while adults focus on paralanguage (J. Bruce Morton and Sandra E. Trehub, 'Children's Understanding of Emotion in Speech', *Child Development*, vol. 72, no. 3, May/June 2001). Younger children have also been found to be less skilled at interpreting the emotional meaning of prosody than older ones and adults (Christiane Baltaxe, 'Vocal Communication of Affect and Its Perception in Three to Four-Year-Old Children', *Perceptual and Motor Skills*, 72, 1991), though in this study three- and four-year-olds were able to perceive emotions expressed through intonation. No psychological study that I've come across, however, seems to build in the modifying factor of the children's home environment, which presumably is hugely influential in shaping small children's interpretive abilities. Children raised in homes where emotions are more easily expressed and decoding skills encouraged and valued must surely develop these abilities younger and more fully.

43 Diane Mastropieri and Gerald Turkewitz, 'Prenatal Experience and Neonatal Responsiveness to Vocal Expressions of Emotion', *Developmental Psychology*, 35(3), November 1999.

44 ibid.

45 Charles Darwin, 'A Biographical Sketch of an Infant', *Mind*, 2, 1877.

46 Baltaxe, op cit.

47  Anzieu, op cit.

48  J.T. Manning et al, 'Ear Asymmetry and Left-Side Cradling', *Evolution and Human Behaviour*, 18, 1997.

49  We know this because, in an important experiment, babies didn't suck extra hard to activate the tape recording of their father reading Dr Seuss's *And to Think That I Saw It on Mulberry Street* – and this shouldn't be read as some precocious judgement on Dr Seuss (Anthony J. DeCasper and Phyllis A. Prescott, 'Human Newborns' Perception of Male Voices: Preference, Discrimination, and Reinforcing Value', *Developmental Psychobiology*, 17(5), 1984).

50  Cynthia D. Ward and Robin Panneton Cooper, 'A Lack of Evidence in 4-Month-Old Human Infants for Paternal Voice Preference', *Developmental Psychobiology*, 35, 1999 – a paper whose very title seems to betray a desire to find support for the father. In fact the results of the experiments it describes demonstrate the opposite: that four-month-old babies don't prefer their father's voice.

51  DeCasper and Prescott, op cit. Perhaps this is also because the father's voice, usually a lower frequency than the mother's, is unlikely to have been as audible in the uterus as a woman's voice. Other theories are that babies prefer the harmonic tones in their mother's voice (Masashi Kamo and Yoh Iwasa, 'Evolution of Preference for Consonances as a By-product', *Evolutionary Ecology Research*, 2, 2000), and that postnatally too they remain more sensitive to women's higher frequency ranges (Jacqueline Sachs, 'The Adaptive Significance of Linguistic Input to Prelinguistic Infants', in Catherine E. Snow and Charles A. Ferguson, *Talking to Children: Language Input and Acquisition*, Cambridge: Cambridge University Press, 1977).

52  Cited in Paul Fraisse, 'Rhythm and Tempo', in Diana Deutsch, ed., *The Psychology of Music* (London: Academic Press, 1982). Peter Auer et al, quoting the French poet and linguist Henri Meschonnic: *'qu'on ait pu être si longtemps indifferent a la voix, à ne voir que des structures, des schemas, des arbres, toute une spatialisation muette du langage'* ('that one could have been indifferent to the voice for so long, seeing only structures, schemas, trees, a whole mute spatialisation of language'), even regard our long indifference to the voice as a consequence of the neglect of rhythm, in *Language in Time: the Rhythm and Tempo of Spoken Interaction* (New York: Oxford University Press, 1999).

53  Alvarez, cited in Maiello, 2003.

54  Suzanne Maiello, 'The Sound-object: a Hypothesis about Prenatal Auditory Experience and Memory', *Journal of Child Psychotherapy*, vol. 21, no. 1, 1995.

55  Salk, cited in Suzanne Maiello, 'The Rhythmical Dimension of the Mother–infant – Transcultural Considerations', *Journal of Child and Adolescent Mental Health*, 15(2), 2003.

56  Papousek, cited in ibid.

57  Fraisse, op cit, and Michael P. Lynch et al, 'Phrasing in Prelinguistic Vocalizations', *Developmental Psychobiology*, 28(1), 1995.

58  Lenneberg, cited in Joseph Jaffe et al, 'Rhythms of Dialogue in Infancy' *Monographs of the Society for Research in Child Development*, vol. 66, no.2, serial no. 265, April 2001. Elsewhere Joseph Jaffe and Stanley Feldstein suggest that the average, uninterrupted vocalisation in conversation lasts 1.64 seconds, spanning about five words, forming a syntactic and rhythmic unit before a juncture, with a slight slowing of stress and intonation change (*Rhythms of Dialogue*, New York: Academic Press, 1970).

59  Fraisse, op cit; Lynch et al, op cit. Newborns suck rhythmically, at intervals from 600 to 1,200 milliseconds.

60  Frederick Erickson, 'Timing and Context in Everyday Discourse: Implications

for the Study of Referential and Social Meaning', *Sociolinguistic Working Paper no.67*, (Austin, Texas: Southwest Educational Development Laboratory, 1980).

61 Lewcowicz, cited in Daniel N. Stern, 'Putting Time Back Into Our Considerations of Infant Experience: A Microdiachronic View', *Infant Mental Health Journal*, vol. 21(1–2), 2000.

62 Colwyn Trevarthen and Stephen Malloch, 'Musicality and Music Before Three: Human Vitality and Invention Shared with Pride', *Zero to Three*, vol. 23, no. 1, September 2002. Music theory is throwing up interesting deliberations on this theme – see, for example, Carolyn Drake and Daisy Bertrand, 'The Quest for Universals in Temporal Processing in Music', *Annals of the New York Academy of Sciences*, part 930, 2001.

63 Gro E. Hallan Tonsberg and Tonhild Strand Hauge, 'The Musical Nature of Human Interaction', Voice: A World Forum for Music Therapy, 2003, www.voices/no/mainissues.

64 Keiko Ejiri, 'Relationship between Rhythmic Behaviour and Canonical Babbling in Infant Vocal Development', *Phonetica*, vol. 55, no. 4, 1998. See also K. Ejiri and N. Masataka, 'Co-occurrences of Preverbal Vocal Behaviour and Motor Activity in Early Infancy', *Developmental Science*, vol.4, no.1, March 2001.

65 Stephen N. Malloch et al, 'Measuring the Human Voice: Analysing Pitch, Timing, Loudness and Voice Quality in Mother–Infant Communication', paper presented at the International Syposium of Musical Acoustics, Edinburgh, August 1997 (reprinted in Proceedings of the Institute of Acoustics, vol. 19, part 5).

66 Bullowa, cited in Jaffe et al, op cit.

67 William S. Condon and Louis W. Sander, 'Neonate Movement Is Synchronised with Adult Speech: Interactional Participation and Language Acquisition', *Science, New Series*, vol. 183, no. 4120, 11 Jan 1974. Since the infant wasn't looking at the adult, the role of eye contact could be ruled out, and since the synchronisation occurred whether the adult speaker was present or their voice came out of a tape recorder, it couldn't be the case that the speaker was coordinating with the baby rather than the other way round.

68 Daniel Stern, *The Interpersonal World of the Infant* (London: Karnac Books, 1998).

69 Colwyn Trevarthen and Kenneth J. Aitken, 'Infant Intersubjectivity: Research, Theory, and Clinical Applications', *Journal of Child Psychology and Psychiatry*, vol. 42: no. 1, 2001.

70 William Condon, quoted in Carole Douglas, 'The Beat Goes On: Social Rhythms Underlie All our Speech and Actions', *Psychology Today*, November 1987, p.40.

71 Cynthia L. Crown et al, 'The Cross-Modal Coordination of Interpersonal Timing: Six-Week-Old-Infants' Gaze with Adults' Vocal Behaviour', *Journal of Psycholinguistic Research*, vol. 31, no. 1, January 2002.

72 Miriam K. Rosenthal, 'Vocal Dialogues in the Neonatal Period', *Developmental Psychology*, vol. 18, no. 1, 1982.

73 Sachiyo Kajikawa et al, 'Rhythms of Turn-taking in Japanese Mother–child Dialogue', paper given at Linguistics and Phonetics 2002 conference, Meikai University, Japan.

74 Mechthild Papousek, 'Early Ontogeny of Vocal Communication in Parent–infant Interactions', in Hanus Papousek et al, eds., *Nonverbal Vocal Communication: Comparative and Developmental Approaches* (Cambridge: Cambridge University Press, 1992).

75 Beatrice Beebe et al, 'Systems Models in Development and Psychoanalysis: The

Case of Vocal Rhythm, Coordination, and Attachment', *Infant Mental Health Journal*, vol. 21 1–2, 2000, pp.105–6.

76 ibid.

77 Mary Catherine Bateson, 'Mother–Infant Exchanges: the Epigenesis of Conversation Interaction, *Annals of the New York Academy of Sciences*, 263, 1975.

78 Philippe Rochat, 'Dialogical Nature of Cognition', in Jaffe et al, op cit.

79 Daniel N. Stern, *The First Relationship: Infant and Mother* (Cambridge, Mass: Harvard University Press, 2002).

80 Jaffe et al, op cit. Interestingly, there was more vocal coordination between strangers and infants in a lab than between confident mothers and their secure babies at home, suggesting that you need more predictability in timing with someone you don't know, in an unfamiliar setting.

81 Beebe et al, op cit.

82 Maya Gratier, 'Expressive Timing and Interactional Synchrony Between Mothers and Infants: Cultural Similarities, Cultural Differences, and the Immigration Experience', *Cognitive Development*, 18, 2003. It's even been argued that mothers able to synchronise with their 3-month-old babies, and whose babies at 9 months are also able to synchronise with them, are more likely to produce children who, at 2 years of age, are able to exercise self-control and self-regulation (Ruth Feldman et al, 'Mother-Infant Affect Synchrony as an Antecedent of the Emergence of Self-Control', *Developmental Psychology*, vol. 35, no. 5, 1999).

83 Jaffe et al, op cit.

84 Crown et al, op cit.

85 Erickson and Shultz, quoted in Auer et al, p.21. Auer et al argue that both parties to an interaction collaborate to produce the rhythm of their conversation, but language and culture also help shape them. They compare the contrasting rhythms of ending a phone conversation in German and Italian.

86 Stern et al, 1985, cited in Christine Kitamura and Denis Burnham, 'Pitch and Communicative Intent in Mother's Speech: Adjustments for Age and Sex in the First Year', *Infancy*, 4(1), 2003.

87 Rosenthal, op cit.

88 Hanus Papousek and Marc H. Bornstein, 'Didactic Interactions: Intuitive Parental Support of Vocal and Verbal Development in Human Infants', in Hanous Papousek et al, *Nonverbal Vocal Communication: Comparative and Developmental Approaches*' (Cambridge: Cambridge University Press, 1992).

89 Daniel N. Stern, *The Interpersonal World of the Infant*, 1998, op cit.

90 Stern 2002, op cit, p.106.

91 Stern 2002, op cit, p.14.

92 Maiello 2003, op cit, p.85.

93 Gratier, op cit.

94 Suzanne Maiello, 'Prenatal Trauma and Autism', *Journal of Child Psychotherapy*, vol. 27, no. 2, 2001. Maiello speculates that autistic babies are unable to filter out the sounds of a depressed or mentally disturbed mother. At some psycho-auditory level they cut out, thereby damaging the prenatal 'sound object' through which the foetus introjects the maternal voice, and which is the precursor of the postnatal maternal object.

95 Feldman et al, op cit.

96 Adena J. Zlochower and Jeffrey F. Cohn, 'Vocal Timing in Face-to-Face Interaction of Clinically Depressed and Non-depressed Mothers and Their 4-Month-Old Infants', *Infant Behaviour and Development*, 19, 1996.

97 Trevarthen and Malloch, op cit.

98  Peter S. Kaplan et al, 'Child-Directed Speech Produced By Mothers with Symptoms of Depression Fails to Promote Associative Learning in 4-Month-Old Infants', *Child Development*, vol. 70, no. 3, May/June 1999.

99  Nadja Reissland et al, 'The Pitch of Maternal Voice: a Comparison of Mothers Suffering from Depressed Mood and Non-depressed Mothers Reading Books to Their Infants', *Journal of Child Psychology and Psychiatry*, 44:2, 2003.

100  Betty Hart and Todd Risley, *Meaningful Differences in the Everyday Experience of Young American Children*, p.77 (Baltimore, Maryland: Paul H. Brookes Publishing, 1995).

101  ibid. Hart and Risley are careful not to stigmatise welfare parents, acknowledging their resilience and persistence in the face of repeated defeats and humiliations. They argue that professional families prepare their children to participate in a culture concerned with symbols and analytic problem-solving, while welfare families school their children in obedience, politeness, and conformity as the keys to survival. Their findings have a class gradient, and seem to be suggesting that class-mediated social disadvantage penetrates even into the voice, through which a sense of competence and confidence can be transmitted down the generations. Though Hart and Riseley's study was painstaking and sensitive, research like this can end up simply endorsing middle-class parenting styles and indicting working-class ones, a criticism some levelled at Basil Bernstein's work.

102  ibid, pp. 102–3.

103  Francis Spufford, 'Pillow Talk', *Guardian*, 13.3.02.

104  Interview, 31.10.03.

105  See Amanda Craig, 'Listen with Mother', *Guardian*, 9.6.04, on how losing her voice temporarily through thyroid cancer affected her. In another case a 17-year-old was hit by a train while riding his bike, sending him into a minimally conscious state. He was played a tape of his mother reading a story, followed, after a brief pause, by a tape of an age-matched voice reading the same story. Magnetic resonance imaging showed that the mother's voice strongly activated the amygdala, one of the organs responsible for emotional processing (T. Bekinschtein et al, 'Emotion Processing in the Minimally Conscious State', *Journal of Neurology, Neurosurgery and Psychiatry*, 75, 2004).

106  Interview, 17.11.03.

107  Interview 31.10.03.

108  Martin H. Teicher, 'Wounds That Time Won't Heal: The Neurobiology of Child Abuse', *Cerebrum*, vol. 2, no. 4, Fall 2000.

109  Anne Karpf, 'Loud But Not Proud', *Guardian*, 21.03.01.

110  Robert Fulghum, *All I Really Need to Know I Learned in Kindergarten* (London: HarperCollins, 1990).

111  Title of a new report in *The Times*, by Glen Owen, 9.1.03.

112  This advice, quoted in Amelia Hill, 'Science Tells How to Bring up Baby', *Observer*, 7.11.04, purports to come from 'The Definitive 21st Century Child-rearing Book' by Professor Margot Sunderland, 'a leading expert in the development of children's brains', and 'Director of Education and Training at the Centre for Child Mental Health in London', a private organisation that she herself founded. I can find no details of Sunderland's qualifications, nor of her book, and yet one of her claims in this article – that ignoring a crying child can cause serious damage to its brain – was widely disseminated in its wake.

113  Anzieu, op cit, p.32.

114  Susan Milmoe et al, 'The Mother's Voice: Postdictor of Aspects of Her Baby's

Behaviour', Proceedings of 76th Conference of the American Psychological Association, 1968.

115 For a stimulating discussion of maternal ambivalence see Rozsika Parker: *Torn in Two: the Experience of Maternal Ambivalence* (London: Virago, 1995).

116 Chion, op cit, p.62. 'In the beginning, in the uterine darkness', laments the French writer Michel Chion, 'was the Voice, the Mother's voice. For the child once born, the mother is more an olfactory and vocal continuum than an image. Her voice originates in all points of space, while her form enters and leaves the visual field. We can imagine the voice of the Mother weaving around the child a network of connections it's tempting to call the umbilical web. A rather horrifying expression, to be sure, in its evocation of spiders.' (ibid, p.61).

117 This entrapping maternal voice resembles the Freudian concept of the vagina dentata (vagina with teeth), ever ready to castrate the approaching man.

118 Kaja Silverman, op cit.

119 Maiello 1995, op cit, p.27. 'If so, the child's listening ear would no longer be completely fused in the primary sonic one-ness. There might already be some distance and differentiation between the voice and the ear, the germ of distinction between a listening "me" and a speaking "not-me".' Maiello goes on to make a delightful, but perhaps far-fetched, connection between the fact that, from the fifth month of interuterine life, the foetus is able to put its thumb in its mouth and suck it, and the fact that this ability appears at the time when its hearing capacity is fully developed. She conjectures that, feeling an emptiness following the silence of the mother's voice, the foetus attempts literally to fill the gap by putting its thumb into its mouth. Of course the foetus doesn't know that voices come from mouths – perhaps it plugs the only gap available for it to fill by itself. I'm grateful to Gianna Williams for drawing my attention to Suzanne Maiello's work.

120 Interview 25.5.03.

## 6. Mothertalk: the Melody of Intimacy

1 In fact it's often called 'musical speech' because the exaggerated prosody gives it a sing-song quality. Laurel J. Trainor et al, 'Is Infant-Directed Speech Prosody A Result of the Vocal Expansion of Emotion?', *Psychological Science*, vol. 11, no. 3, May 2000.

2 Charles A. Ferguson, 'Baby Talk in Six Languages', *American Anthropologist, New Series*, vol. 66, no. 6, part 2: 'The Ethnography of Communication', December 1964. Ferguson focused largely on the lexical and syntactic aspects of baby talk, which lie beyond the scope of this book.

3 Sally-Anne Ogle and J.A. Maidment, 'Laryngographic Analysis of Child-directed Speech', *European Journal of Disorders of Communication*, 28, 1993.

4 Olga K. Garnica, 'Some Prosodic and Paralinguistic Features of Speech to Young Children', in Catherine E. Snow and Charles A. Ferguson, *Talking to Children: Language Input and Acquisition* (Cambridge: Cambridge University Press, 1977). Baby talk or motherese is now considered a 'register' – a special code or style of speaking – in the linguistic repertoire of adult speakers of any language (Bella M. DePaulo and Lerita M. Coleman, 'Evidence for the Specialness of the "Baby Talk" Register', *Language and Speech*, vol. 24, part 3, 1981). The British may be famously reluctant to learn other tongues, yet even the most determined monoglot successfully speaks many different variants of their own language, as the situation demands, because we're all skilled

code-switchers. Code-switching affects vocabulary and grammar, but can also be heard in tone of voice and inflection. Vocal registers or codes are a way of signalling our understanding of the social context, and acknowledging the person to whom we're speaking. Some of the features of baby talk, like elevated pitch, for instance, can be found in infant-directed singing too (Tali Shenfield and Sandra E. Trehub, 'Infants' Reponse to Maternal Singing', paper given at the scientific conference of ISIS, the International Study on Infant Studies, Brighton, 17.7.00).

5 See A.S. Holzrichter, 'Motherese in American Sign Language', MA Thesis, University of Texas at Austin, 1995.

6 Indeed non-parents with experience of infants exaggerate their pitch even more than parents or inexperienced non-parents. Those who see themselves as particularly 'good' with babies might be trying to prove it by doing everything they can to attract the child's attention: the flamboyant use of motherese is probably designed to display their own skill (Joseph L. Jacobson et al, 'Paralinguistic Features of Adult Speech to Infants and Small Children', *Child Development*, 54, 1983).

7 Catherine E. Snow, 'Mothers' Speech Research: from Input to Interaction', in Snow and Ferguson, op cit.

8 Jacqueline Sachs, 'The Adaptive Significance of Linguistic Input to Prelinguistic Infants', in Snow and Ferguson, op cit.

9 K. Niwano and K. Sugai, 'Maternal Accommodation in Infant-directed Speech During Mothers' and Twin-infants' Vocal Interactions', *Psychological Reports*, 92(2), April 2003.

10 Patricia K. Kuhl et al, 'Cross-Language Analysis of Phonetic Units in Language Addressed to Infants', *Science, New Series*, vol. 277, no. 5326, 1.8.97.

11 Anne Fernald, 'Meaningful Melodies in Mothers' Speech to Infants', in Hanus Papousek et al, eds., *Nonverbal Vocal Communication: Comparative and Developmental Approaches* (Cambridge: Cambridge University Press, 1992). See also Anne Fernald and Patricia Kuhl, 'Acoustic Determinants of Infant Preference for Motherese Speech', *Infant Behaviour and Development*, 10, 1978.

12 Stratton and Connolly, cited in ibid; Sachs, op cit. A high-pitched voice is much more effective in eliciting a smile from a baby than a low-pitched one (Wolff, cited by Fernald 1992, op cit).

13 Fernald 1992.

14 Roger Brown, Introduction, in Snow and Ferguson, op cit.

15 ibid.

16 Joanne Siu-Yiu Tang and John A. Maidment, 'Prosodic Apects of Child-Directed Speech in Cantonese', in Valerie Hazan et al, eds., 'Speech, Hearing, and Language: Work in Progress 1996, vol. 9,' University College, London: Department of Phonetics and Linguistics.

17 Schafer, cited by Sachs, op cit.

18 Kuhl et al, 1997, op cit.

19 Tang and Maidment, op cit.

20 Naomi S. Baron, 'The Uses of Baby Talk', in *Pigeon-Birds and Rhyming Words: The Role of Parents in Language Learning* (Washington, DC: ERIC Clearinghouse on Language and Linguistics, 1989).

21 Kuhl et al, 1997, op cit.

22 C. Kitamura et al, 'Universality and Specificity in Infant-directed Speech: Pitch Modifications as a Function of Infant Age and Sex in a Tonal and Non-tonal Language', *Infant Behaviour and Development*, 24, 2002.

23 Christine Kitamura and Denis Burnham, 'Pitch and Communicative Intent in Mothers' Speech: Adjustments for Age and Sex in the First Year', *Infancy*, 4:1, 2003. At this age babies can tell the difference between approving and comforting infant-directed speech, but a 4-month-old can't – see Melanie J. Spence and David S. Moore, 'Categorization of Infant-Directed Speech: Development from Four to Six Months', *Developmental Psychobiology*, 42, 2003.

24 Lacerda et al, cited by Kitamura and Burnham.

25 Kitamura and Burnham, op cit. By now fascinating differences in the way mothers speak to boys and girls begin to emerge – see chapter 10.

26 Garnica, op cit.

27 Patrick Craven, 'Motherese, Affect, and the Mother-Infant Dyad: Shedding the Chomskian Notion of Early Development', MSc dissertation, University of Edinburgh Centre for Cognitive Science, 1998. It's even been argued that, since (put crudely) listening to Mozart makes you cleverer, couldn't the musical qualities of motherese do the same? See Craven for a discussion.

28 Steven Pinker, *The Language Instinct*, p.40 (London: Penguin Books, 1995).

29 ibid, p.279.

30 David Miall, 'The Poetics of Baby Talk', *Human Nature*, vol. 14, no. 4, 2003.

31 Dean Falk, 'Prelinguistic Evolution in Early Hominins: Whence Motherese?', *Behavioural and Brain Science*, 27:6, 2004.

32 Fernald, cited by Kitamura and Burnham, op cit.

33 Trainor et al, op cit.

34 Leher Singh et al, 'Infants' Listening Preferences: Baby Talk or Happy Talk', *Infancy*, vol. 3, no. 3, 2002.

35 Ben G. Blount and Elise J. Padgug, 'Prosodic, Paralinguistic, and Interactional Features in Parent–Child Speech: English and Spanish', *Journal of Child Language*, 4, 1976; Anne Fernald et al, 'A Cross-Language Study of Prosodic Modifications in Mothers' and Fathers' Speech to Preverbal Infants', *Journal of Child Language*, 16, 1989.

36 Linnda (sic) R. Caporael, 'The Paralanguage of Care-giving: Baby Talk to the Institutionally Aged', *Journal of Personality and Social Psychology*, vol. 40, no. 5, 1981.

37 Brown, op cit.

38 Di Anne L. Grieser and Patricia Kuhl, 'Maternal Speech to Infants in a Tonal Language: Support for Universal Prosodic features in Motherese', *Developmental Psychology*, vol. 24, no. 1, 1988.

39 Mechthild Papousek, 'Early Ontogeny of Vocal Communication in Parent–Infant Interactions', in Hanus Papousek et al, eds., *Nonverbal Vocal Communication: Comparative and Developmental Approaches* (Cambridge: Cambridge University Press, 1992).

40 Clifton Pye, 'Quiché Mayan Speech to children', *Journal of Child Language*, 13, 1986. In Papua New Guinea, for instance, little or no speech is addressed to infants until they begin to speak. Samoan rules also restrict parents' speech to young children, who are supposed to be seen and not heard (Elinor Ochs, 'Talking to Children in Western Samoa', *Language in Society*, vol. 11, 1982). Kaluli parents don't think that their children are talking to them until they can say the words 'mother' and 'breast' (Schieffelin, cited in Pye, op cit). Mohave parents believe that newborns and even foetuses can understand and respond to rational verbal admonitions (George Devereux, 'Mohave Voice and Speech Mannerisms', in Dell Hymes, *Language in Culture and Society*, New York: Harper and Row, 1967.

41 Paul Ekman, cited in Fernald 1992, op cit; C.Kitamura et al, 2002, op cit.

42 Robert W. Mitchell, 'Americans' Talk to Dogs: Similarities and Differences with Talk to Infants', *Research on Language and Social Interaction*, vol. 34, no. 2, 2001.

43 Levin and Hunter, cited in Bella M. DePaulo and Lerita M. Coleman, 'Verbal and Nonverbal Communication of Warmth to Children, Foreigners, and Retarded Adults', *Journal of Nonverbal Behaviour*, 11(2), 1987.

44 K. Hirsh-Pasek and R. Treiman, 'Doggerel: motherese in a new context', *Journal of Child Language*, 9(1), 1982.

45 Robert W. Mitchell, 'Controlling the Dog, Pretending to Have a Conversation, or Just Being Friendly: Influences of Sex and Familiarity on Americans' Talk to Dogs During Play', *Interaction Studies*, 5:1, 2004.

46 Charles A. Ferguson, 'Baby Talk as a Simplified Register', in Snow and Ferguson, op cit.

47 Denis Burnham et al, 'What's New, Pussycat? On Talking to Babies and Animals', *Science*, vol. 296, issue 5572, May 2002.

48 Baron, op cit.

49 Elaine Slosberg Andersen, *Speaking with Style: The Sociolinguistic Skills of Children*, (London: Routledge, 1992). Although foreigner talk gets us speaking louder rather than softer (as if being more audible somehow equalled being more comprehensible).

50 DePaulo and Coleman, 1981, op cit.

51 Caporael, op cit.

52 Brown, op cit.

53 Ferguson, 1977, op cit.

54 Caporael, op cit. See also DePaulo and Coleman, 1987, op cit, and Ellen Bouchard Ryan et al, 'Evaluative Perceptions of Patronizing Speech Addressed to Elders', *Psychology and Ageing*, vol. 6, no. 3, 1991. This last study, however, exemplifies many of the research flaws enumerated in chapters 8 and 13, in that paralinguistic speech traits were 'inferred' from written dialogue sequences.

55 See, for example, Jacobson et al, op cit, who did find both higher pitch and wider pitch range, but adjusted the differences by dividing men and women's average variability by their mean fundamental frequency, and Mechthild Papouselk et al, 'Infant Responses to Prototypical Melodic Contours in Parental Speech', *Infant Behaviour and Development*, 13, 1990.

56 Fernald et al 1989, op cit. Perhaps mothers and fathers do different things with their babies – the father more inclined to horse-play, the mother more involved in containing and rhythmic care-taking. Had the fathers been recorded doing what they more usually do, would they have sounded different?

57 Malcolm Slaney and Gerald McRoberts, 'BabyEars: A Recognition System for Affective Vocalizations', *Speech Communication*, vol. 39, issues 3–4, February 2003.

58 Alison Gopnik et al, *How Babies Think*, p.128 (London: Weidenfeld and Nicholson, 1999).

59 Andrew Marlatt, 'Is there a "wooka-boo" lurking in every man?', www.babycenter.com.

60 Amye Warren-Leubecker and John Neil Bohannon 111, 'Intonation Patterns in Child-directed Speech: Mother–Father Differences', *Child Development*, 55, 1984.

61 ibid.

# 7. The Emergence of the Baby's Voice

1 It requires the coordination of respiratory, laryngeal, and supralaryngeal (throat, mouth and nose) movements (Barry M. Lester and C.F. Zachariah Boukydis, 'No Language But a Cry', in Hanus Papousek et al, eds., *Nonverbal Vocal Communication: Comparative and Developmental Approaches* (Cambridge: Cambridge University Press, 1992).

2 A subsequent international search uncovered reports of 131 cases between 1546 and 1941, almost all associated with obstetrical procedures (David B. Chamberlain, 'Babies Remember Pain', *Journal of Perinatal and Prenatal Psychology*, 3(4), 1989).

3 Although they can't actually be heard by others until released from the watery amniotic environment, their sounds penetrate the air (Pamela Davis, 'Emotional Influences on Singing', paper given at the First International Conference on the Physiology and Acoustics of Singing, Groningen, The Netherlands, October 305,2002).

4 Paul J. Moses, *The Voice of Neurosis*, p.16 (New York: Grune and Stratton).

5 ibid.

6 Richard Luchsinger and Godfrey E. Arnold, *Voice-Speech-Language* (Belmont, California: Wadsworth, 1965).

7 Ostwald et al, cited in Joanne Loewy, 'Integrating Music, Language, and Voice in Music Therapy', in 'Voices: A World Forum for Music Therapy', vol. 4, no. 1, 1 March, 2004, www.voices.no.mainissues.

8 P.H. Wolff, 'The Natural History of Crying and Other Vocalizations in Early Infancy', in B.M. Foss, ed., *Determinants of Infant Behaviour*, vol. 4 (London: Methuen, 1969).

9 These conditions may get translated into different sounds via the vagus nerve, which produces changes both to the heart rate and vocal intonation that are associated with different emotional states. Stress causes a decrease in vagal tone, which can affect the sound of their cry (Rebecca M. Wood and Gwen E. Gustafson, 'Infant Crying and Adults' Anticipated Caregiving Responses: Acoustic and Contextual Influences', *Child Development*, vol. 72, no. 5, September/October 2001).

10 K. Michelsson et al, 'Cry Score – an Aid in Infant Diagnosis', *Folia Phoniatrica*, 36, 1984; Dror Lederman, 'Automatic Classification of Infants' Cry', M.Sc thesis, Department of Electrical and Computer Engineering, Ben-Gurion University of the Negev, October 2002. One genetic disorder, *cri du chat* syndrome, has been named after the sound made by infants and children suffering from it.

11 ibid.

12 Joseph Soltis, 'The Signal Functions of Early Infant Crying', in *Behavioural Brain and Sciences* (Cambridge University Press, 2006).

13 Corwin et al, cited in Lederman, op cit.

14 Lester and Boukydis, op cit.

15 Katarina Michelsson et al, 'Cry Characteristics of 172 Healthy 1 – 7-Day-Old Infants', *Folia Phoniatrica et Logopaedica*, 54, 2002.

16 ibid.

17 Janice Chapman, 'What Is Primal Scream?', paper given at the First International Conference of the Physiology and Acoustics of Singing, Groningen, October 305, 2002.

18 More upset than when they're exposed to equally loud non-human sounds (Marco Dondi et al, 'Can Newborns Discriminate Between their Own Cry and

the Cry of another Newborn Infant?' *Developmental Psychology*, vol. 35(2), March 1999). A 1971 study found that newborns cried more when they heard another newborn crying than they did when they heard white noise, a 5-month-old crying, or a synthetic cry (M.L.Simner, 'Newborn's Response to the Cry of Another Infant', *Developmental Psychology*, 5, 1971. See also Abraham Sagi and Martin L. Hoffman, 'Empathetic Distress in the Newborn', *Developmental Psychology*, 12: 2, 1976).

19 G.B. Martin and R.D. Clarke, 'Distress Crying in Neonates: Species and Peer Specificity', *Developmental Psychology*, 18, 1982.

20 Patsy Rodenburg, *The Right to Speak*, p.28 (London: Methuen, 1992).

21 Harry Hollien, *Forensic Voice Identification* (San Diego: Academic Press, 2002).

22 Lester and Boukydis, op cit.

23 See for instance Gwene E. Gustafson et al, 'Acoustic Correlates of Individuality in the Cries of Infants', *Developmental Psychobiology*, vol. 17:3.

24 Hollien, op cit.

25 Wood and Gustafson, op cit.

26 Ostwald, cited in Paul Newham, *Therapeutic Voicework* (London: Jessica Kingsley Publishers, 1998).

27 Rodenburg in 'I'd Know that Voice Anywhere', BBC Radio 4, 14.6.03.

28 Natasha Dobie, 'Communicating Voice', Newsletter of the British Voice Association, vol.3, issue 3, March 2003.

29 Dobie, op cit.

30 Rodenburg 1992, op cit.

31 Luchsinger and Arnold, op cit.

32 Juliet Miller, 'The crashed Voice – a Potential for Change: a Psychotherapeutic view', *Logopedics, Phoniatrics, Vocology*,28, 2003.

33 See also Katarina Michelsson et al, 2002, op cit.

34 Reuters, 29 April 2001.

35 C. Hunger, 'Search Your Voice and Find Your Life Stories . . .', *Onktruid* (The Netherlands), 2000.

36 Charles Darwin, 'A Biological Sketch of an Infant', p.294, *Mind*, 1877.

37 Moses, op cit, p.17.

38 Philip Sanford Zeskind and Victoria Collins, 'Pitch of Infant Crying and Caregiver Responses in a Natural Setting', *Infant Behaviour and Development*, 10, 1987.

39 Annette Karmiloff-Smith, cited in Naomi Stadlen, *What Mothers Do* (London: Judy Piatkus, 2004).

40 Lester and Boukydis, op cit.

41 Benedicte de Boysson-Bardies et al, *How Language Comes to Children: From Birth to Two Years* (Boston: The MIT Press, 1999).

42 Lester and Boukydis, op cit.

43 Patricia K. Kuhl and Andrew N. Meltzoff, 'Evolution, Nativism and Learning in the Development of Language and Speech', in M. Gopnik, ed., *The Inheritance and Innateness of Grammars* (New York: Oxford University Press, 1997).

44 Colwyn Trevarthen, 'Learning about Ourselves from Children: Why a Growing Human Brain Needs Interesting Companions', Research and Clinical Centre for Child Development, Annual Report 2002–2003 (No. 26), Graduate School of Education, Hokkaido University. See also M.C. Bateson, 'Mother–Infant Exchanges: The Epigenesis of Conversational Interaction', in D.R. Rieber, ed., *Developmental Psycholinguistics and Communication Disorders: Annals of the New York Academy of Sciences*, vol. 263 (New York: New York Academy of Science, 1975).

45 Lenneberg et al, cited in Danny Steinberg, *Psycholinguistics: Language, Mind, and World* (Harlow, Essex: Longman, 1982).

46 Kuhl and Meltzoff 1997, op cit.

47 Trevarthen, op cit.

48 Wolff, op cit.

49 J.A.M. Martin, *Voice, Speech, and Language in the Child: Development and Disorder* (Vienna: Springer-Verlag, 1981).

50 M.A.K. Halliday, *Learning How to Mean: Explorations in the Development of Language* (London: Edward Arnold, 1975).

51 Allison J. Doupe and Patricia K. Kuhl, 'Birdsong and Human Speech: Common Themes and Mechanisms', *Annual Review of Neuroscience*, 22, 1999.

52 In order to acquire the remarkable skill of learning to speak, hearing ourselves is as important as hearing other people (ibid). See also J.L. Locke, *The Child's Path to Spoken Language* (Cambridge, Mass: Harvard University Press, 1993).

53 Terrence Deacon, *The Symbolic Species* (London: Penguin Books, 1997).

54 Kuhl and Meltzoff, 1997, op cit.

55 Koopmans van Beinum and Van der Stelt, cited in de Boysson-Bardies et al, op cit.

56 ibid.

57 Judee K. Burgoon et al, *Nonverbal Communication: The Unspoken Dialogue* (New York: McGraw-Hill, 1996).

58 Rodenburg, 1992, op cit.

59 John Bowlby, *Attachment and Loss*, vol. 1 (London: Pelican, 1984).

60 Webster et al, cited in Jacqueline Sachs, 'The Adaptive Significance of Linguistic Output to Prelinguistic Infants', in Catherine E. Snow and Charles A. Ferguson, *Talking to Children: Language Input and Acquisition* (Cambridge: Cambridge University Press, 1977).

61 Lieberman, cited in ibid.

62 Weisberg, cited in Bowlby, op cit.

63 Kuhl and Meltzoff, op cit.

64 Roman Jakobson, *Child Language, Aphasia, and Phonological Universals* (The Hague: Mouton de Gruyter, 1972; originally published 1941).

65 D. Kimbrough Oller and Rebecca E. Eilers, 'Development of Vocal Signalling in Human Infants: Towards a Methodology for Cross-species Vocalization Comparisons', in Hanus Papousek et al, op cit.

66 Siobhan Holowka and Laura Ann Petitto, 'Left Hemisphere Specialization for Babies While Babbling', *Science*, vol. 297, 30.08.02.

67 Kuhl and Meltzoff, op cit.

68 Laura Ann Petitto and Paula F. Marentette, 'Babbling in the Manual Mode: Evidence for the Ontogeny of Language', *Science*, vol. 251, 22.03.91.

69 Although young songbirds also 'try out' elements and phrases from their later songs – see Doupe and Kuhl, op cit.

70 Roman Jakobson, 'Why "Mama" and "Papa"?' in Bernard Kaplan and Seymour Wapner, eds, *Perspectives in Psychological Theory* (New York: International Universities Press, 1960, reprinted in 1962). Roman Jakobson, *Selected Writings, Vol. 1: Phonological Studies*, pp.542–3 (The Hague: Mouton, 1962).

71 Kuhl and Meltzoff, op cit.

72 Delack, cited by Jacqueline Sachs, op cit.

73 ibid.

74 Patricia K. Kuhl and Andrew N. Meltzoff, 'Infant Vocalizations in Response to Speech: Vocal Imitation and Developmental Change', *Journal of the Acoustical Society of America*, 100 (4), Pt 1, October 1996. See also Vincent J. van Heuven

and Ludmilla Menert, 'Why Stress Position Bias?', *Journal of the Acoustical Society of America*, 100(4), Pt.1, October 1996.

75 Xin Chen et al, 'Auditory-oral Matching Behaviour in Newborns', *Developmental Science*, 7:1, 2004.

76 Peter D. Eimas et al, 'Speech Perception in Infants', *Science, New Series*, vol. 171, no. 3968, 22.01.71.

77 Kuhl and Meltzoff 1997, op cit.

78 Derek M. Houston and Peter W. Jusczyk, 'The Role of Talker-Specific Information in Word Segmentation by Infants', *Journal of Experimental Psychology*, vol. 26, no. 5, 2000. This experiment found that 32-week-old babies could recognise familiar words spoken by different speakers of the same sex, but not of the opposite sex. The acoustic differences in the voices of men and women were too great, it seems, to be ignored by the infants, and stopped them recognising that the same word was being spoken. When they were three months older, however, they could recognise a familiar word no matter what the sex of the speaker.

79 Jusczyk et al, cited in Derek M. Houston and Peter W. Jusczyk, 'Infants' Long-Term Memory for the Sound Patterns of Words and Voices', *Journal of Experimental Psychology: Human Perception and Performance*', vol. 29(6), December 2003. Another experiment found that when $7\frac{1}{2}$ -month-olds heard words repeated a day later by the same voice, they were more likely to recognise the word.

80 Paul Iverson et al, 'A Perceptual Interference Account of Acquisition Difficulties for Non-native Phonemes', *Cognition*, 87, 2003. Naoyuki Takagi found that even with training, adult monolingual Japanese listeners sometimes misidentified them. ('The Limits of Training Japanese Listeners to Identify English /r/ and /l/: Eight Case Studies', *Journal of the Acoustical Society of America*, 111(6) June 2002.

81 Janet F. Werker and Renée N. Desjardins, 'Listening to Speech in the First Year of Life: Experimental Influences on Phoneme Perception', *Current Directions in Psychological Science*, vol.4, no.3, June 1995. See also Janet F. Werker and Chris E. Lalonde, 'Cross-Language Speech Perception: Initial Capabilities and Developmental Change', *Developmental Psychology*, vol. 24, no. 5, 1988 and Janet F. Werker and Richard C. Tees, 'Speech Perception as a Window for Understanding Plasticity and Commitment in Language Systems of the Brain', *Developmental Psychobiology*, 46, 2005.

82 Patricia K. Kuhl et al, 'Linguistic Experience Alters Phonetic Perception in Infants by 6 Months of Age', *Science, New Series*, vol. 255, no. 5044, 31.01.92. This 'perceptual magnet', which shrinks the distance between some sounds and enlarges that between others, isn't the result of language-learning but, on the contrary, prepares us for it.

83 William Kessen et al, 'The Imitation of Pitch in Infants', *Infant Behaviour and Development*, vol. 2, no. 1, January 1979.

84 Jenny R. Saffran and Gregory J. Griepentrog, 'Absolute Pitch in Auditory Learning: Evidence for Developmental Reorganization', *Developmental Psychology*, vol. 37(1), January 2001; Diane Deutsch et al, 'Tone Language Speakers Possess Absolute Pitch', paper presented at the 138th Meeting Lay Language Papers, Acoustical Society of America, Columbus, Ohio, 4.11.99.

85 William Kessen et al, 'The Imitation of Pitch in Infants', *Infant Behaviour and Development*, vol. 2, no. 1, Jan. 1979, p.99.

86 Erik D. Thiessen and Jenny R. Saffran, 'When Cues Collide: Use of Stress and Statistical Cues to Word Boundaries by 7–9-month-old Infants', *Developmental Psychology*, vol. 39(4) July 2003.

87 Alessandra Sansavini, 'Neonatal Perception of the Rhythmical Nature of Speech: The Role of Stress Patterns', *Early Development and Parenting*, vol. 6(1) 1997.

88 Diane Mastropieri and Gerald Turkewitz, 'Prenatal Experience and Neonatal Responsiveness to Vocal Expressions of Emotion', *Developmental Psychobiology*, 35, 1999.

89 Thierry Nazzi et al, 'Language Discrimination by Newborns: Towards an Understanding of the Role of Rhythm', *Journal of Experimental Psychology: Human Perception and Performance*, vol. 24, no. 3, 1998. (See also Thierry Nazzi et al, 'Language Discrimination by English-Learning 5-Month-olds: Effects of Rhythm and Familiarity', *Journal of Memory and Language*, 43, 2000; Franck Ramus, 'Perception of Linguistic Rhythm by Newborn Infants', 2000, www.cogprints.org; and Alessandra Sansavini, 'Neonatal Perception of the Rhythmical Nature of Speech: The Role of Stress Patterns', *Early Development and Parenting*, vol. 6(1) 1997. We know this partly because of changes in their sucking-rates but also because, when you filter out the prosody, they can no longer accomplish this feat, so it was obviously the prosody of the language that guided them (G. Dehaerne-Lambertz and D. Houston, 'Language Discrimination Response Latencies in 2-month-old Infants', *Language and Speech*, 41, 1, 1998).

90 Christine Moon et al, 'Two-day-olds Prefer their Native Language', *Infant Behaviour and Development*, 16, 1993.

91 Kuhl and Meltzoff, 1997, op cit, p.2.

92 Patricia K. Kuhl, 'A New View of Language Acquisition', *Proceedings of the National Academy of Sciences of the United States of America*, vol. 97, no. 22, 24.10.00.

93 Kuhl and Meltzoff, 1997, op cit, p.3.

94 ibid, p.1.

95 Janet F. Werker and Richard C. Tees, 'Influences on Infant Speech Processing: Towards A New Synthesis', *Annual Review of Psychology*, 50, 1999 .

96 As the laryngologist Paul Moses put it, in 1954, 'Voice production is autoerotic activity before it is communication. The infant derives pleasure from phonation as an oral activity, and secondarily he learns to derive pleasure from hearing himself.' (Paul Moses, *The Voice of Neurosis*, p.18, p.109, New York: Grune & Stratton, 1954).

97 J.A.M. Martin, *Voice, Speech, and Language in the Child: Development and Disorder* (Vienna: Springer-Verlag, 1981).

# 8. Do I Really Sound Like That?

1 Deso A. Weiss, 'The Psychological Relations to One's Own Voice', *Folia Phoniatrica*, vol. 7, no. 4, 1955, p.209.

2 Susan Lee Bady, 'The Voice as a Curative Factor in Psychotherapy', *Psychoanalytic Review*, 72(3), Fall 1985, p.481.

3 ibid, p.486.

4 Interview, 30.5.03.

5 Although Dieter Maurer and Theodor Landis challenge this in 'Role of Bone Conduction in the Self-Perception of Speech', *Folia Phoniatrica et Logopaedica*, 42, 1990.

6 Patsy Rodenburg, *The Right to Speak* (London: Methuen, 1992).

7 Philip S. Holzman and Clyde Rousey, 'The Voice as Precept', *Journal of Personality and Social Psychology*, vol. 4, no. 1, 1966, pp. 84–85. A German

voice teacher told me that she gets students to listen to recordings of other voices alongside her own: the context makes the experience less agonising, and they don't compare what they sound like to what they thought they did, only to how others do. In this way they become slightly distanced from their own voice and develop ways of listening to and describing voices in general (personal communication, Evamarie Haupt, 1.09.05). A speech and language therapist suggested, 'If you hate your voice, you can spend six years in therapy, or you can realise that it's just a muscle setting and do exercises every morning so that eventually you get used to using it in a different way' (personal communication, Christine Shewell, 1.09.05).

8 Interview, 21.11.03.
9 Morris W. Brody, 'Neurotic Manifestations of the Voice', *Psychoanalytic Quarterly*, vol. 12, 1943.
10 Interview, 7.12.03.
11 Interview, 5.11.03.
12 Interview, 1.6.04.
13 Interview, 11.11.03.
14 Interview, 11.11.03.
15 Interview, 10.6.03.
16 Interview, 25.05.03.
17 Paul J. Moses, *The Voice of Neurosis*, p.60 (New York: Grune & Stratton, 1954).
18 Interview, 16.07.03.
19 ibid.
20 Interview, 23.10.03.
21 Interview, 21.3.04.
22 Interview, 6.11.03.
23 Interview, 5.11.03.
24 Interview, 7.05.05.
25 Interview, 10.10.03.
26 Erika Friedmann et al, 'The Effects of Normal and Rapid Speech on Blood Pressure', *Psychosomatic Medicine*, vol. 44, no. 6, December 1982. This paper is misnamed: it studied reading rather than speech, which confirms my point that it's the sound of one's own voice rather than purely the interactive aspects of speech that excite us physiologically.
27 Interview, 25.5.03.
28 ibid.
29 Sir James Frazer, *The Golden Bough*, p.246 (Hertfordshire: Wordsworth Editions, 1993).
30 Michael Frayn, *Speak After the Beep*, p.142–3 (London: Methuen, 1997).
31 Moses, op cit.
32 Sigmund Freud, 'Fragment of an Analysis of a Case of Hysteria', p.28, in *The Standard Edition of the Complete Psychological Works of Sigmund Freud*, vol. 7 (London: The Hogarth Press, 1953).
33 ibid, p.30.
34 ibid.
35 Modern critics have contended that it was Freud's own voice that was hysterical (albeit metaphorically), that he was blind to his own identification with Herr K. (Claire Kahane, 'Introduction: Part 2' in Charles Bernheimer and Claire Kahane, *In Dora's Case: Freud, Hysteria, Feminism* (London: Virago, 1985). Freud, it's been argued, appropriated Dora's story, becoming the third male complicit in the implicit deal through which, Dora felt, her father had bartered her to Herr K.

in return for his wife. The mute Dora became, at least for a while, an emblem of the silenced female, victim of male predatory sexual power – woman gagged. See, for example, Toril Moi, 'Representation of Patriarchy: Sexuality and Episte- mology in Freud's Dora', in ibid; Claire Kahane, 'Freud and the Passions of the Voice' in John O'Neil (ed.), *Freud and the Passions* (The Pennsylvania State University, 1996); and Claire Kahane, *Hysteria, Narrative, and the Figure of the Speaking Woman, 1850–1915* (Maryland: The Johns Hopkins University Press, 1996).

36 'A dispute, in the course of which she suppressed a rejoinder, caused a spasm of the glottis, and this was repeated on every similar occasion.' (Josef Breuer and Sigmund Freud, 'Studies on Hysteria', p.40, in *The Standard Edition of the Complete Psychological Works of Sigmund Freud*, vol. 2 (London: The Hogarth Press, 1955).

37 ibid, p.35.

38 ibid, p.30.

39 Of course hysteria has a different meaning as a psychological term to the colloquial one of exaggerated emotions. As a diagnostic category it refers to physical symptoms without an organic cause – what today we might call 'somatic'. And yet from its very route – the Greek for uterus – hysteria had an ideological meaning: thought to derive from disturbance of the uterus, it made femininity itself seem inherently pathological.

40 Margaret C.L. Greene, *Disorders of Voice* (Texas: Pro-Ed, 1986).

41 J. Nemec, cited in Robert J. Toohill, 'The Psychosomatic Aspects of Children with Vocal Nodules', *Archives of Otolaryngology*, vol. 101, October 1975. Another study describes children with vocal nodules as hyperactive, nervous, tense, or emotionally disturbed (Toohill, op cit). See also Aronson, cited in Caitriona McHugh-Munier et al, 'Coping Strategies, Personality, and Voice Quality in Patients with Vocal Fold Nodules and Polyps', *Journal of Voice*, vol. 11, no. 4, 1997.

42 'Common forms of psychogenic voice disorder arise mostly in women who have personalities that are no more than mildly neurotic, who are suffering from protracted anxiety associated with particular life stresses or events, who tend to have taken on above-average responsibilities, who are frequently caught up in family and interpersonal relationship difficulties, who are commonly having difficulties in assertiveness and the expression of emotions or negative feelings and who, consequently, feel powerless about making personal change. (Peter Butcher, 'Psychological Processes in Psychogenic Voice Disorder', *European Journal of Disorders of Communication*, 30, 1995).

43 Interview, 5.11.03.

44 Valerie Sinason, 'When children are dumbstruck', *Guardian*, 29.5.93.

45 Richard Barber, 'Full power', *Saga*, June 2003. A woman who'd been raped remembered 'episodes, spread over several years, when I couldn't, for the life of me, speak intelligibly . . .) For about a year after the assault, I rarely, if ever, spoke in smoothly flowing sentences' (Susan Brison, 'After I Was Raped', *Guardian*, 6.2.02).

46 'It's as if the psyche relaxes (and here we could replace psyche with voice): the voice relaxes once it is felt that the crucial psychological issues are finally being dealt with.' (Juliet Miller, 'The Crashed Voice – a Potential for Change: a Psychotherapeutic View', *Logopedics, Phoniatrics, Vocology*, 28, 2003). The voice teacher Patsy Rodenburg suggests that psychic trauma that has shut down the voice can be released in voice workshops through what she calls delayed 'vocal shock'. Working with the deepest breath, students of the voice find

themselves unblocking painful memories of grief, rage, and even sexual abuse (Patsy Rodenburg, *The Right To Speak*, London: Methuen, 1992).

47 John Diamond, *The Times*, 24.10.98.

48 John Diamond, *The Times*, 15.5.99.

49 John Diamond, *The Times*, 3.10.98. Touchingly his small daughter alone was able to negotiate the only conversations possible with his post-surgical voice. 'She accepts without second thought the fact that once I spoke like that and now I speak like this, and when she can't understand me . . . she'll say, "What?" and manage to maintain the appropriate remorse or fear or truculence all the way through her incomprehension and my repetition, which is some trick.' (John Diamond, *The Times*, 18.10.97).

50 John Diamond, *The Times*, 20.9.97.

51 ibid. The playwright Peter Tinniswood also articulated his feelings after he, too, was left voiceless by treatment for oral cancer. Rescued eventually by an electronic voicebox, which forced him to slow down his naturally rapid speech, he found that his brain now worked faster than his voice. Tinniswood put the experience to dramatic use by writing a play, *The Voice Boxer*, about the grief, sense of loss, and rage of a man who wakes up after a laryngectomy only to find that his voice has developed a life of its own and is now travelling the world. (Peter Tinniswood, 'Speaking as a Dalek', *Guardian*, 8.2.00).

52 Sheila Johnston, 'What a piece of work', *Independent*, 18.4.91.

53 Interview, 31.5.04.

54 Paul Newham, *Therapeutic Voicework*, p.81 (London: Jessica Kingsley, 1998).

55 Rodenburg, op cit, p.226.

56 ibid.

57 Interview with Andrea Haring, 28.5.03.

58 Interview, 12.4.04.

59 Rodenburg, op cit, pp.168–9.

60 Interview with Andrea Haring, and Elena McGee, 28.05.03.

61 Interview, 7.12.03.

62 Interview, 30.05.03.

63 Interview, 30.05.03.

## 9. How Our Emotions Shape the Sounds We Make (and Other People Hear Them)

1 John Carvel, 'Minister of Sound', *Guardian*, 25.5.00.

2 Of the same kind that set the first ever British radio play, Richard Hughes's '*Danger*' (1924), down a coal mine.

3 Blunkett has described his reliance on his voice-reading skills. '[I] try to imagine the pleasantness of the visual aspects of the person and their voice . . . Not seeing voices makes you a fairer judge . . . Voices that are sharp and harsh reflect – but not exclusively – harsh and sharp personalities; voices that are pleasant and easy often reflect that personality . . . But you do get it wrong and I've certainly got it wrong in a big way once in my life' (*Daily Mail*, 06.09.05).

4 As the neurologist Henry Head called it in his treatise on aphasia, *Aphasia and Kindred Disorders of Speech* (Cambridge University Press, 1926).

5 Oliver Sacks, *The Man Who Mistook his Wife for a Hat*, pp.76–7 (London: Picador, 1986).

6 ibid. In 2000 a quartet of researchers tried to test out Sacks's hypothesis to

discover if aphasics were significantly better at detecting lies than people without language impairment. While ordinary people are usually no better than chance at detecting deceit from a person's voice, the researchers found that aphasics were more accurate in sniffing out lies. But the study also found that the aphasics only did better when they got clues from the liars' facial behaviour or from face and voice combined, and not when the cues were from the voice alone (Nancy L. Etcoff et al, 'Lie Detection and Language Comprehension', *Nature*, vol. 405, May 2000). This suggests that the relationship between face and voice is a complex and mutually reinforcing one and, as I'll be arguing, that statements about the voice's superior ability to reveal all are often wildly exaggerated.

7 Antoinette L. Bouhuys and Wilhemina E.H. Mulder-Hajonides Van Der Meulen, 'Speech Timing Measures of the Severity, Psychomotor Retardation, and Agitation in Endogenously Depressed Patients', *Journal of Communication Disorders*, 17, 1984.

8 Tom Johnstone and Klaus R. Scherer, 'The Effects of Emotions on Voice Quality', unpublished research report, Geneva Studies in Emotion and Communication, 13(3), 1999 www.unige.ch/fapse/emotions.

9 Caitriona McHugh-Munier et al, 'Coping Strategies, Personality, and Voice Quality in Patients with Vocal Fold Nodules and Polyps', *Journal of Voice*, vol.11, no.4, 1997.

10 Paul J. Moses, *The Voice of Neurosis* (New York: Grune & Stratton, 1954).

11 Lindsley, cited in Carl E. Williams and Kenneth N. Stevens, 'Emotions and Speech: Some Acoustical Correlates', *Journal of the Acoustical Society of America*, vol. 52, no. 4, Part 2, 1972.

12 Williams and Stevens, ibid.

13 Moses, op cit, on William Faulkner's research.

14 Moses, op cit.

15 Moses, op cit. Microtremors in our vocal muscles when we speak might also be an expression of our psychological state. Voice specialists have been arguing for some time about the effects that stress has on these microtremors. It's been variously suggested that they get induced, modified, or even suppressed by emotional tension – a subject of particular interest to those developing lie detectors (Elvira Mendoza and Gloria Carballo, 'Vocal Tremor and Psychological Stress', *Journal of Voice*, vol. 13.no. 1).

16 Klaus R. Scherer, 'Expression of Emotion in Voice and Music', *Journal of Voice*, vol. 9, no. 3, 1995. In dangerous situations the autonomous nervous system, by influencing the secretion of mucus and salivation, gives us a dry mouth and, by changing subglottal pressure, affects our breathing pattern, leading to a higher-pitched voice (Klaus R. Scherer, 'Vocal Affect Expression: a Review and a Model for Future Research', *Psychological Bulletin*, vol. 99, no. 2, 1986).

17 John Rubin, 'I'd Know That Voice Anywhere', BBC Radio 4, 14.06.03.

18 Ross Buck, 'The Neuropsychology of Communication: Spontaneous and Symbolic Aspects', *Journal of Pragmatics*, 22, 1994.

19 This kind of research gained impetus from the popularity of radio, then at its most salient culturally.

20 Klaus R. Scherer, 'Personality Inference from Voice Quality: the Loud Voice of Extroversion', *European Journal of Social Psychology*, vol. 8, 1978. Originally coined by Jung, the concepts of extroversion and introversion gained currency in the 1960s through the work of Hans Eysenck. Though Eysenck posited a continuum of these traits, extroversion was essentially seen as the opposite of introversion. Yet, reading the definition of these terms (see, for example, H.J. Eysenck and S.B.G. Eysenck, *Manual of the Eysenck Personality Inventory*

(London: University of London Press, 1964), I can't be alone in finding myself located simultaneously at opposite ends of the spectrum.

21 Moses, op cit.

22 ibid, pp. 83,1.

23 ibid.

24 Bruce L. Brown and Jeffrey M. Bradshaw, 'Towards a Social Psychology of Voice Variations', in Howard Giles and Robert N. St Clair, *Recent Advances in Language, Communication and Social Psychology* (London: Lawrence Erlbaum Associates, 1985).

25 As recently as 2000 one pair of researchers was arguing that people who suffered from dysphonia (difficulty producing sounds) were neurotic introverts, and that 'personality may act as a persistent risk factor for voice pathology' (Nelson Roy et al, 'Personality and Voice Disorders: A Superfactor Trait Analysis', *Journal of Speech, Language, and Hearing Research*, vol. 43, June 2000, p.764). Of course this begs the whole question of what personality is, or whether indeed there is any such thing – see Aron W. Siegman, 'The Telltale Voice: Nonverbal Messages of Verbal Communication', in Aron W. Siegman and Stanley Feldstein, eds., *Nonverbal Behaviour and Communication* (New Jersey: Lawrence Erlbaum Associates, 1987). For a discussion of some problems surrounding personality testing, see Brown and Bradshaw, op cit.

26 Before him the musicologist Friedrich Marpurg had tried to link specific acoustic patterns with particular emotional states. Sorrow, for example, was characterised by a 'slow melody and dissonance', and envy – he suggested – by 'growling and annoying tones'. (L.M. Lovett and B. Richardson, 'Talk: Rate, Tone and Loudness – How They Change in Depression', paper given to the Autumn Quarterly Meeting, Royal College of Psychiatrists, 1992.)

27 ibid. Difficulties with speech, argued the American psychiatrist Adolph Meyer in 1904, were one of the most striking features of severe depression (John F. Greden and Bernard J. Carroll, 'Decrease in Speech Pause Times with Treatment of Endogenous Depression', *Biological Psychiatry*, vol. 15, no. 4, 1980).

28 Kraepelin, quoted in John F. Greden et al, 'Speech Pause Time: A Marker of Psychomotor Retardation Among Endogenous Depressives', *Biological Psychiatry*, vol. 16, no. 9, 1981, p.852.

29 Murray Alpert, 'Encoding of Feelings in Voice', in P.I. Clayton and J.E. Barrett, *Treatment of Depression: Old Controversies and New Approaches* (New York: Raven Press, 1983).

30 William Styron, *Darkness Visible: A Memoir of Madness* (London: Vintage, 2001). The voice can also help distinguish between types of depression, since not all produce the same kind of acoustic change – schizophrenia, depression, and mania all sound different. (Julian Leff and Evelyn Abberton, 'Voice Pitch Measurements in Schizophrenia and Depression', *Psychological Medicine*, 11, 1981.) See also Newman and Mather cited in John K. Darby, 'Speech and Voice Studies in Psychiatric Populations' in *Speech Evaluation in Psychiatry*, ed. John K Darby (New York: Grune and Stratton, 1981). Schizophrenics' voices have been studied almost obsessively, for a very long time. Almost a century ago 'flatness of affect', i.e. speaking in a monotone, was identified as perhaps the most crucial symptom of schizophrenia (Kraepelin, cited in Nancy C. Andreasen et al, 'Acoustic Analysis: An Objective Measure of Affective Flattening', *Archives of General Psychiatry*, vol. 38, March 1981). The schizophrenic's voice becomes as immobile as their face. They also repeat senseless rhythmic sentences over and over again for months and even years, perhaps in a regression to infantile

patterns 'in which words, language, communication have lost meaning, but rhythm has not' (Moses, op cit, p56).

31 John K. Darby and Alice Sherk, 'Speech Studies in Psychiatric Populations', in Harry and Patricia Hollien, eds., *Current Issues in the Phonetic Sciences*, vol. 9, pt 2 (Amsterdam: John Benjamins B.V., 1979).

32 Kraepelin's idea that a depressed person speaks with elongated pauses has been confirmed in research many times over. (E. Szabadi and C.M. Bradshaw, 'Speech Pause Time: Behavioural Correlate of Mood', *American Journal of Psychiatry*, 140:2, February 1983). Describing the change in his voice when he became depressed, William Styron said, 'My speech, emulating my way of walking, had slowed to the vocal equivalent of a shuffle' (William Styron, 'A Journey Through Madness', *Independent on Sunday*, 24 February 1991).

33 Klos et al cited in Lovett and Richardson, op cit. Forensic psychiatrists also use Speech-pause time to help to evaluate the psychological state of an accused person and decide if they're competent to stand trial (Bernard L. Diamond, 'The Relevance of Voice in Forensic Psychiatric Evaluations', in John K. Darby, 1981, op cit). What makes it a particularly useful method is that the person being tested doesn't even need to be present (audio recordings of them can be used), let alone attached to electrodes, nor do they need to be discussing something personal or highly charged – any banal subject will do (Alpert, op cit.)

34 Greden and Carroll, op cit. See also Greden op cit; H.H. Stassen et al, 'The speech Analysis Approach to Determining Onset of Improvement under Anti-depressants', *European Neuropsychopharmacology*, 8, 1998; and G.M.A. Hoffmann et al, 'Speech Pause Time as a Method for the Evaluation of Psychomotor Retardation in Depressive Illness', *British Journal of Psychiatry*, 146, 1985. There are examples of a patient's pause time decreasing by more than 50 per cent within several days of starting treatment. See also Ostwald cited in John K. Darby and Harry Hollien, 'Vocal and Speech Patterns of Depressed Patients', *Folia Phoniatrica et Logopaedica*, 29, 1977. Antidepressants also affect the voice – not just the speed of speaking, its volume, and variety, but also its pitch (Stassen, op cit).

35 David F. Salisbury, 'Researchers Measure Distinct Characteristics in Speech of Individuals at High Risk of Suicide', 'Exploration' (online research journal of Vanderbilt University), 23.10.00.

36 D.J. France et al, 'Acoustical Properties of Speech as Indications of Depression and Suicidal Risk, *Transactions on Biomedical Engineering*, 47 (7) July 2000.

37 Although it's also characterised by a larger pitch range (Johnstone and Scherer, 1999, op cit) as well as slow tempo and regular rhythm.

38 Tom Johnstone and Klaus R. Scherer, 'Vocal Communication of Emotion', in Michael Lewis and Jeannette M. Haviland-Jones, *Handbook of Emotions*, 2nd edition (New York: The Guilford Press, 2000).

39 Klaus R. Scherer and Harvey London, 'The Voice of Confidence', *Journal of Research in Personality*, 7, 1973.

40 Rainer Banse and Klaus R. Scherer, 'Acoustic Profiles in Vocal Emotion Ex-pression', *Journal of Personality and Social Psychology*, vol. 70, no. 3, 1996.

41 For the various studies that produced these findings, see Iain R. Murray and John L. Arnott, 'Toward the Simulation of Emotion in Synthetic Speech: A Review of the Literature on Human Vocal Emotion', *Journal of the Acoustic Society of America*, 93 (2), February 1993.

42 Johnstone and Scherer, 2000, op cit.

43 Allan and Barbara Pease, *Why Men Lie and Women Cry* (London: Orion, 2003).

44 Carl E. Williams and Kenneth N. Stevens, 'Vocal Correlates of Emotional States', in John K. Darby, ed., *Speech Evaluation in Psychiatry* (New York: Grune and Stratton, 1981).

45 William Hargreaves et al, 'Voice Quality in Depression', *Journal of Abnormal Psychology*, vol. 70, no. 3, 1965. See also Newman and Mather in Lovett and Richardson, op cit.

46 Judith Whelan, 'Voice Key to Suicide Intention, Study Finds', *Sydney Morning Herald*, 18.8.00.

47 Younger depressed patients speak faster after treatment, but older people slow down (Ostwald, cited in Darby and Hollien, op cit).

48 Mark L. Knapp, *Non-verbal Communication in Human Interaction* (New York: Rinehart & Winston, 1972).

49 Kenton L. Burns and Ernst G. Beier, 'Significance of Vocal and Visual Channels In the Decoding of Emotional Meaning', *Journal of Communication*, vol. 23, March 1973. This particular study ended with a sentence – 'How do acted mood expressions compare with genuine mood expressions?' – that one can't help feeling the researchers should have asked before they ever started. Klaus R. Scherer, the leading researcher in this field, has been frank about some of the limitations of this work. See, for instance, his 'Expression of Emotion in Voice and Music', *Journal of Voice*, vol. 9, no. 3, and 'Vocal Affect Expression: a Review and a Model for Future Research', *Psychological Bulletin*, vol. 99, no. 2, 1986.

50 Williams and Stevens, 1981, op cit.

51 Klaus R. Scherer, 'Speech and Emotional States', in Darby, op cit. The presence of several simultaneous emotions in the voice, not to mention background noise, is regarded as 'interference' (see for instance Williams and Steven, 1991, op cit).

52 Scherer, 1986, op cit.

53 Except for accommodation theory, examined in chapter 17.

54 Johnstone and Scherer 2000, op cit, p.228. Perhaps what this kind of research on the voice, emotions and personality reveals more than anything is the stereotypes that attach to different kinds of voices, rather than anything significant about real emotional states (Paul Newham, *Therapeutic Voicework*, London: Jessica Kingsley, 1998). Although the voice affects how *we* feel about people, it may not invariably be an accurate reflection of how *they* feel or who they are.

55 Pollack, cited in Murray and Arnott, op cit.

56 Interview, 21.11.03.

57 Interview, 7.12.03.

58 ibid.

59 Interview 26.5.03.

60 Interview 7.5.04.

61 Interview 5.6.04.

62 Interview 31.3.03.

63 Interview 12.7.03.

64 Bugenthal, quoted in Robert G. Harper et al, eds., *Nonverbal Communication: The State of the Art*, p.30 (New York: John Wiley, 1978).

65 Alexia Demertzis Rothman and Stephen Nowicki Jr., 'A Measure of the Ability to Identify Emotions in Children's Tone of Voice', *Journal of Nonverbal Behaviour*, 28(2), Summer 2004.

66 Erin B. McClure and Stephen Nowicki, Jr., 'Associations Between Social Anxiety and Nonverbal Processing Skill in Preadolescent Boys and Girls', *Journal of Nonverbal Behaviour*, 25(1), Spring 2001. Of course this is a chicken-and-egg situation: does good voice-decoding decrease social anxiety, or does less social

anxiety make for better voice-reading, or perhaps both? And these studies, once again, were based upon staged emotions – children told to sound happy, sad, angry, or fearful – rather than spontaneous expressions of feeling.

67 ibid.

68 Cadesky, cited in Rothman and Nowicki, op cit.

69 Mitchell, cited in ibid. Again this begs questions about the causal nature of the relationship between the offence and the misreading – which caused what? And yet most of us have seen aggressive people misread mild comments as overt hostility often enough to appreciate how easy it is to project one's own emotion into the voice of someone else.

70 See, for instance, Stephen Nowicki and Marshall P. Duke, *Helping the Child Who Doesn't Fit In* (Peachtree Publishers, 1992), and Marshall P. Duke et al, *Teaching Your Child the Language of Social Success* (Bay Back Books, 1996). Some of these authors have certainly done significant work on children and adolescents with serious difficulties in voice-reading. They've suggested that such children's low self-esteem, high anxiety and depression, along with unpopularity, may be related to non-verbal processing difficulties (Elizabeth B. Love et al, 'The Emory Dyssemia Index: a Brief Screening for the Identification of Non-verbal Language Deficits in Elementary-School Children', *Journal of Psychology*, no. 1, 1994).

71 Stephen Nowicki, Jr., and Marshall Duke, *Will I Ever Fit In? The Breakthrough Program for Conquering Adult Dyssemia* (Free Press, 2002).

72 Rocco Dal Vera, 'The Voice in Heightened Affective States', p.58, in Rocco Dal Vera, ed., *The Voice in Violence* (2001, Voice and Speech Trainers Association).

73 www.marcsalem.com.

74 His claims have been challenged, as has his contention that he uses psychology rather than clever conjurors' tricks. See, for example, Simon Singh, 'Spectacular Nonsense or Silly Psycho-Babble', *Daily Telegraph*, 5.6.03.

75 Aldert Vrij et al, 'People's Insight into Their Own Behaviour and Speech Content When Lying', *British Journal of Psychology*, 92, 2001.

76 Luigi Anolli and Rita Ciceri, 'The Voice of Deception: Vocal Strategies of Naïve and Able Liars', *Journal of Nonverbal Behaviour*, 21 (4), Winter 1997.

77 Aldert Vrij, 'Detecting the Liars', *Psychologist*, vol. 14, no. 11, November 2001.

78 See Knapp, op cit and Judee K. Burgoon et al, *Nonverbal Communication* (New York: McGraw-Hill, 1996). Burgoon et al suggest that these two may be linked: that a person who finds it difficult to identify fear in others may respond by expressing it more carefully themselves.

79 Michael Argyle, *Bodily Communication* (London: Routledge, 1975). Our ability to detect the emotional nuances in other people's voices becomes more accurate with age, and is connected with our general social competence (Rothman and Nowicki, op cit), but not our gender or IQ (Eric Benjamin Cadesky et al, 'Beyond Words: How Do Children with ADHD and/or Conduct Problems Process Nonverbal Information About Affect?', *Journal of the American Academy of Child and Adolescent Psychiatry*, vol. 39 (9), September 2000).

80 Vrij, 2001, op cit, p.597.

81 Samantha Mann et al, 'Detecting True Lies: Police Officers' Ability to Detect Suspects' Lies', *Journal of Applied Psychology*, vol. 89, no. 1, 2004. As Mann et al point out, many of these studies contain a fundamental flaw because they take place in laboratory settings where subjects have been asked to lie and therefore won't be feeling the kind of anxiety that attends lying in real life. One piece of research evaluating the ability of 509 people – including police officers, judges and psychiatrists, as well as those from the CIA, FBI, and drug-enforcement

agencies – to detect liars found that only the secret service performed better than chance (Paul Ekman and Maureen O'Sullivan, 'Who Can Catch A Liar', *American Psychologist*, vol. 46, no. 9, September 1991).

82  Moses, op cit.

83  T.H. Pear, *Voice and Personality*, pp.34, 75 (London: Chapman and Hall, 1931).

84  Interview 26.10.03.

85  Arthur Miller, *Timebends*, p.380 (London: Methuen, 1988). Thanks to Michael Ball for reminding me of this passage.

86  Interview 25.5.03.

87  Interview 7.10.03.

88  Interview 20.05.03.

89  Paul Newham, *Therapeutic Voicework* (London: Jessica Kingsley, 1998).

90  Susan Lee Bady, 'The Voice as a Curative Factor in Psychoanalysis', *Psychoanalytic Review*, 72(3), Fall 1985.

91  Morris W. Brody, 'Neurotic Manifestations of the Voice', *Psychoanalytic Quarterly*, vol. 12, 1943, p.377. The fascinating, complex relationship between inner and outer voices is beyond the scope of this book.

92  'Much attention may profitably be paid to the telltale aspects of intonation, rate of speech, difficulty in enunciation, and so on . . . Tonal variations in the voice . . . are frequently wonderfully dependable clues to shifts in the communication situation' (Harry Stack Sullivan, *The Psychiatric Interview*, p.5, New York: Norton, 1954). Note that Sullivan was speaking about individual patients, as heard by their experienced practitioner. Attempts to identify the emotions in patients' voices objectively and quantify their intensity have produced some pretty crude results – for example, Robert Roessler and Jerry W. Lester, 'Voice Predicts Affect During Psychotherapy', *Journal of Nervous and Mental Disease*, vol. 163, no. 3, 1976.

93  C. G. Jung, *The Psychogenesis of Mental Disease*, pp.70–71 (London: Routledge and Kegan Paul, 1960).

94  Interview, 30.5.03. One psychotherapist noted in 1943, 'Subtle changes in the timbre, the inflections or monotony, the rate of speech, pitch and intensity or deviations in the use or grouping of words, may all be expressions of underlying emotional conflict' (Brody, op cit, p.371).

95  Laura N. Rice and Conrad J. Koke, 'Vocal Style and the Process of Psychotherapy', in John K. Darby, ed., *Speech Evaluation in Psychotherapy* (New York: Grune & Stratton, 1981). How discrete and clearly identifiable these categories are is another question. The Chicago work challenges the classic study of silence in psychotherapy, which treated disturbances, hesitations, and silences in the patient's speech purely as expressions of anxiety and defences (George Mahl, 'Disturbances and Silences in the Patient's Speech in Psychotherapy', *Journal of Abnormal and Social Psychology*, vol. 53, 1956). The meaning of pauses and hesitations, of course, depends on the individual patient, or indeed session: what might be a defence in one person at one time may herald a glorious insightful breakthrough in another.

96  ibid, p.162.

97  Brody, op cit.

98  Jungian analysts sit beside the couch, within the patient's peripheral vision.

99  Interview, 30.5.03.

100  Bady, op cit, p.483.

101  James Gooch in Panel Report: 'Bion's Perspective on Psychoanalytic Method', *International Journal of Psychoanalysis*, 2002. Of course there's no single

therapeutic voice, whatever the caricatures of psychotherapy would have us believe. The British psychoanalyst Wilfred Bion, for example, could be famously and facetiously sarcastic, according to his patients (James Gooch, 'Bion's Perspective on Psychoanalytic Technique', paper given at the 42nd Congress of the International Psychoanalytic Association, Nice, 26.07.01, www.psychoanalysis.org.uk), and different schools of therapy use different techniques and vocal styles.

102 Interview, 5.11.03.

103 Sharon Zalusky, 'Telephone Analysis: Out of Sight, But Not Out of Mind', www.psychomedia.it. Zalusky discusses the way that the telephone can intensify the transference and, as a place between fantasy and reality, can become a transitional object in itself.

104 Interview, 11.04.05. She also tapes the sessions, which has allowed her to listen to the sound of her own voice 'and hear myself in a way I never have before'.

105 Sigmund Freud, 'Recommendations to Physicians Practising Psychoanalysis', *The Standard Edition of the Complete Works of Sigmund Freud*, vol. 12, p.116 (London: Hogarth, 1958).

106 A 44-year-old woman remarked, 'Someone who didn't understand a word of English could come into one of my therapy sessions and understand what was going on purely on the basis of hearing my voice and my therapist's. I often go off on some tangent – some subject that it's easy for me speak about and doesn't raise any strong feelings, my therapist calls it my "headtalk", all from the top of my head, and she stops me when I get like that. When I touch some painful place I usually talk very little, as if I don't want to distract myself by having to say something, and my voice goes small, there's almost no gap between what I'm feeling and saying, the saying is the feeling. And when that happens, my therapist's voice gets very gentle: it's like she's acknowledging in sound that I'm facing up to something difficult, like she's trying to hold my hand with her voice, not to crash into this moment with myself when I'm crying or shocked or lead it off into some other direction or require me to have to react to her, to let me stay where I am, but also to remind me, just by using her voice, that she's still there, and knows it's painful, but also that it's good that I've connected with it. And that's all there, in the sound of our voices.'

107 Freud, op cit, p.115.

108 Theodor Reik, *Listening with the Third Ear*, p.145 (New York: Farrar-Strauss and Company, 1949).

109 Peter Lewis, unpublished paper, 1987.

# 10. Male and Female Voices: Stereotyped or Different?

1 Although these change with age (Alan Cruttenden, *Intonation*, Cambridge: Cambridge University Press, 1995).

2 Judee K. Burgoon et al, *Nonverbal Communication* (New York: McGraw-Hill, 1996).

3 The larynx increases by 62 per cent between the ages of 10 to 16 for boys, and 34 per cent between the ages of 12 to 16 for girls (Kahane, cited in Ingo R. Titze, 'Physiologic and Acoustic Differences Between Male and Female Voices', *Journal of the Acoustical Society of America*, 85(4), April 1989).

4 Paul J. Moses, *The Voice of Neurosis* (New York: Grune & Stratton, 1954).

5 Changes to the vocal folds (and the resulting vocal differences) are secondary sex characteristics, physical features that develop during puberty, influenced by hormones. These characteristics distinguish the sexes from each other but are not directly concerned with sexual reproduction, the province of the primary sex organs.

6 In Western culture: among the Mohave Indians, the breaking of the male adolescent's voice isn't considered a sign of puberty (George Devereux, 'Mohave Voice and Speech Mannerisms', in Dell Hymes, Language in Culture and Society, New York: Harper and Row, 1964).

7 Deso A. Weiss, 'The Pubertal Change of the Human Voice', Folia Phoniatrica, vol. 2, 1950.

8 Richard Luchsinger and Godfrey E. Arnold, Voice – Speech – Language (California: Wadsworth, 1965).

9 Margaret Mead, cited in Cheris Kramarae, 'Women's Speech: Separate But Unequal?', in Barrie Thorne and Nancy Henley, Language and Sex: Difference and Dominance (Massachusetts: Newbury House, 1975).

10 Birdwhistell, cited in Nancy M. Henley, 'Power, Sex, and Nonverbal Communication', in Thorne and Henley, op cit.

11 Henley, in ibid.

12 Joan Swann, Girls, Boys and Language, p.21 (Oxford: Blackwell, 1992).

13 David Graddol and Joan Swann, Gender Voices, p.22 (Oxford: Blackwell, 1989).

14 'Suite Stuff', March 2000, www.tgni.com. But this is often exaggerated by some transsexuals who, to pass as a woman, fall back on a falsetto (Paul J. Daniel, 'Voice Change Surgery in the Transsexual', Head and Neck Surgery, May–June 1982). The falsetto is a common male way of caricaturing women: the actress Meryl Streep has questioned why Dustin Hoffman chose such a squeaky high voice for the character of Dorothy in Tootsie (Hollywood Greats, BBC1, 8.3.04). It's harder for male-to-female transsexuals to acquire a feminine-sounding voice than the other way round because the injection of male hormones thickens the vocal folds, so automatically lowering the pitch, but female hormones, interestingly, don't raise the pitch. Male-to-female transsexuals sometimes resort to surgery to stretch the vocal folds, but if over-stretched, these produce a Minnie Mouse voice. Over the past few years, however, speech therapists have begun using a non-surgical approach known as voice-feminisation therapy. While some of the techniques are based on altering language and on outrageous caricatures – like getting male-to-female transsexuals to use words like 'cute', seeking confirmation from the listener ('It's a nice day, isn't it?'), or using hyperbole ('That was absolutely the most fantastic, awe-inspiring, and beautiful movie I have ever seen') – others are mainly paralinguistic, like varying their rates of speech (Moya L. Andrews and Charles P. Schmidt, 'Gender Presentations: Perceptual and Acoustical Analyses of Voice', Journal of Voice, vol. 11, no. 3, 1997).

15 Delack and Lowlow, cited in Carole T. Ferrand and Ronald L. Bloom, 'Gender Differences in Children's Intonational Patterns', Journal of Voice, vol. 10, no. 3.

16 Weinberg and Bennett, op cit. Already, by age 8 too, children have begun to adopt the gendered pronunciation of their same-sex parent – see Elizabeth A. Strand, 'Uncovering the Role of Gender Stereotypes in Speech Perception', Journal of Language and Social Psychology, vol. 18, no. 1, March 1999.

17 Jacqueline Sachs et al, 'Anatomical and Cultural Determinants of Male and Female Speech', in Roger W. Shuy and Ralph W. Fasold, eds., Language Attitudes: Current Trends and Prospects (Washington, DC: Georgetown University Press, 1973).

18  Jean Berko Gleason and Esther Blank Greif, 'Men's Speech to Young Children', in Barrie Thorne et al, eds., *Language, Gender, and Society* (Boston: Heinle and Heinle, 1983).

19  Lieberman, cited in Philip M. Smith, 'Sex Markers in Speech', in Klaus R. Scherer and Howard Giles, eds., *Social Markers in Speech* (Cambridge: Cambridge University Press, 1979).

20  Elaine Slosberg Andersen, *Speaking with Style: the Sociolinguistic Skills of Children* (London: Routledge, 1990).

21  Ferrand and Bloom, op cit.

22  Crystal, cited in ibid.

23  See Gill Branston, '. . . Viewer, I Listened to Him . . . Voices, Masculinity, in the Line of Fire', in Pat Kirkham and Janet Thumim, eds., *Me Jane: Masculinity, Movies, and Women* (London: Lawrence and Wishart, 1995) for a discussion of Eastwood's voice. And yet some of the supposed difference between boys' and girls' voices is illusory. When asked to identify the sex of singers in a cathedral choir, listeners were able to identify the gender of choristers correctly only 53 per cent of the time (David Howard et al, 'Can Listeners Tell the Difference Between Boys and Girls Singing the Top Line in Cathedral Music?', in C. Stevens et al, eds., *Proceedings of the 7ᵗʰ International Conference on Music Perception and Cognition*, Adelaide: Casual Publications, 2002). Part of the difficulty of coolly appraising the similarities and differences between men and women's voices is because most of what we know about speech production comes from studies of male speakers: what's taken as normal, the vocal template, is based on male voices (Ingo R. Titze, 'Physiologic and Acoustic Differences Between Male and Female Voices', *Journal of the Acoustical Society of America*, April 85(4)1989), and female voices, as a result, have been seen as deviant or abnormal – at least until the explosion of interest in the subject on the part of women in the 1970s. See Caroline Henton, in 'The Abnormality of Male Speech', in George Wolf, ed, *New Departures in Linguistics*, New York: Garland Publishing, 1992 for a robust challenge to the dominant view.

24  Sachs et al, op cit.

25  ibid, p.75. See also Bernd Weinberg and Suzanne Bennett, 'Speaker Sex Recognition of 5- and 6-year-old Children's Voices', *Journal of the Acoustical Society of America*, vol. 50, no. 4 (pt 2), 1971.

26  Berachot 24a.

27  Subsequent orthodox commentaries have conducted heated debates over whether this also applied to a woman's speaking voice and not just her singing one, if it covered a man's wife as well as strange women, and whether listening to women singing on radio and television was proscribed too. Despite the insistence of some modern female commentators that men's voices might possess a similar power to induce sinful thoughts in women, the Talmud's dictates that women shouldn't sing in the presence of men have prevented women from participating in orthodox synagogue choirs for centuries.

28  I Corinthians 14: 34–35. Like the rabbis' proscriptions, this too had the effect of excluding women from participating in ecclesiastical services. So both Judaism and Christianity grafted ideas of shame and indecency on to women's voices, articulating beliefs about the connection between the voice and sexuality.

29  Aristotle, *Politics*, part XIII (Oxford: OUP, 1998).

30  Sophocles, *Ajax*, p. 293 (Oxford: OUP, 1999).

31  T. Wilson, 'The Arte of Rhetorique', 1553, quoted in Caroline Henton, 'The Abnormality of Male Speech', p.29, in George Wolf, ed., *New Departures in Linguistics* (New York: Garland Publishing, 1992), although it's as well to

remember, as one linguist has pointed out, that 'the fact that social science has neglected women makes women of the past and other cultures *seem* silent, when in fact the silence is that of current Western scholarship' (Susan Gal, quoted in Deborah Cameron, ed., *The Feminist Critique of Language: A Reader*, p.4, London: Routledge, 1998).

32 Described by Ovid in gruesome detail: 'The severed tongue along the ground/lay quivering . . . jerking and twitching' (*Metamorphosis*, Book 3, London: Penguin, 1970).

33 ibid.

34 Hans Christian Andersen, *The Little Mermaid* (London: Faber and Faber, 1981).

35 Kathleen Hall Jamieson, *Eloquence in an Electronic Age*, (New York: Oxford University Press, 1988).

36 George Webbe, quoted in ibid, p.151. In seventeenth-century colonial America, the ducking stool was used to submerge women characterised as a 'scold', 'nag', or just plain 'unquiet': they then had to choose between silence and drowning.

37 ibid.

38 Quoted in Henton, 1992, op cit, p.30.

39 ibid.

40 John C. Steinberg, 1927, quoted in Anne McKay, 'Speaking Up: Voice Amplification and Women's Struggle for Public Expression', in Cheris Kramarae, *Technology and Women's Voices*, p.203 (London: Routledge and Kegan Paul, 1988), reprinted in Caroline Mitchell, ed., *Women and Radio: Airing Differences* (London: Routledge, 2000).

41 'Filing Steel', *Daily Express*, 19.09.28, quoted in Cheris Kramarae, 'Resistance to Women's Public Speaking', in Senta Tromel-Plotz, ed., *Gewalt durch Sprache* (Frankfurt: Fisher Taschenbuch Verlag, 1984). The concept of 'listener' had clearly yet to take root – here they're still 'listeners-in', with that phrase's palpable sense of eavesdropping. Add to that the sound-image of 'groans', and yet again we encounter the female voice's apparent sexual suggestiveness, and the almost shameful impact of transposing into the public realm a voice that obviously belongs in private (in the bedroom).

42 *Weekly Dispatch*, 19.09.26, quoted in ibid.

43 *Evening Standard*, 2.10.03, quoted in ibid.

44 *Sunday Dispatch*, 22.07.45, quoted in ibid.

45 *Southern Daily Echo*, 9.11.28, quoted in ibid.

46 *Radio Broadcast* magazine, 1924, quoted in McKay, p.200, op cit.

47 John Wallace, *Radio Broadcast* magazine, 1926, quoted in ibid, McKay, p.202.

48 *News Chronicle*, 29.07.33, quoted in Kramarae, 1988, op cit.

49 ibid. While we can chuckle with enlightened hindsight at the absurdity of the reporting, there's a more serious aspect to the newspaper coverage. Situating Borrett in a domestic context, in relation to her husband and son, freed her from accusations of having a 'sexy voice', which had so impeded aspiring female announcers before her.

50 *Daily Express*, 17.10.33, quoted in Kramarae, 1988, op cit. When the BBC did eventually account for Borrett's dismissal, it blamed – yet again – not its own prejudices but those of other women, 'tens of thousands' of whom apparently 'sent in complaints, based mainly on the fact that she is a married woman, wife of a pensioned naval officer'. These presumably counted for more than the complaints of children that men's voices were 'too deep' or 'too rough', or the information that 7–9-year-old listeners preferred women's voices because

they were clearer, brighter, and enunciated better than men (*Daily Express*, 7.03.34, quoted in ibid).

51 'Radio Announcer: The "sweetheart of the AEF" joining NBC', *Newsweek*, 12.01.35, quoted in McKay, op cit.

52 The *Evening News* reported that 'the old prejudice against women announcers has disappeared since the war, and listeners, as well as the BBC, have decided that announcing – as distinct from news reading – is a job that women can handle with ability and charm' (*Evening News*, 24.01.42, quoted in Kramarae, 1988, op cit).

53 *Sunday Dispatch*, 22.07.45, in ibid.

54 'Women's Employment at the BBC', October 1945, quoted in ibid.

55 *Daily Telegraph*, 11.11.58, quoted in ibid.

56 Mileva Ross, 'Radio', in Josephine King and Mary Stott, *Is This Your Life?: Images of Women in the Media*, p.18 (London: Virago, 1977).

57 Anne Karpf, 'Foreword' to Caroline Mitchell, ed., *Women and Radio: Airing Differences* (London: Routledge, 2000).

58 Rosalind Gill, 'Justifying Injustice: Broadcasters' Accounts of Inequality in Radio', in Mitchell, op cit, p.141.

59 ibid, p.146.

60 Marylou Pausewang Gelfer and Shannon Ryan Young, 'Comparisons of Intensity Measures and their Stability in Male and Female Speakers', *Journal of Voice*, vol. 11, no. 2, 1997, although this study asked men and women to read at a level 'as if' they were speaking to someone only a few feet away, rather than measuring actual conversational levels.

61 Philip M. Smith, op cit.

62 As the American women's rights leader, Susan B. Anthony, put it in 1848, 'Taught that a low [soft] voice is an excellent thing in woman, she has been trained to a subjugation of the vocal organs, and thus lost the benefit of loud tones and their well-known invigoration of the system.' Quoted in Nancy M. Henley, *Body Politics*, p.76 (New York: Simon and Schuster, 1977).

63 Kirsty Neumann, 'Surgical Techniques in Male-to-female Transsexuals', paper given at PEVOC5, Graz, Austria, August 2003.

64 Michael Frayn, 'The Long and the Short of It', in *Speak After the Beep*, p.35 (London: Methuen, 1997). For a discussion of the trivialisation of women's telephone talk as gossip or chatter, see Lana F. Rakow, *Gender on the Line: Women, the Telephone and Community Life* (Urbana: University of Illinois Press, 1992).

65 Only two out of fifty-six studies appearing between 1951 and 1991 found that women talked more than men (Deborah James and Janice Drakich, 'Understanding Gender Differences in Amount of Talk: A Critical Review of Research', in Deborah Tannen, ed., *Gender and Conversational Interaction* (New York: Oxford University Press, 1993). See also Joan Swann, 'Talk Control: an Illustration from the Classroom of Problems in Analysing Male Dominance of Conversation', in Jennifer Coates and Deborah Cameron, eds., *Women in Their Speech Communities* (London: Longman, 1988), and Aron W. Siegman, 'The Telltale Voice: Nonverbal Messages of Verbal Communication', in Aron W. Siegman and Stanley Feldstein, eds., *Nonverbal Behaviour and Communication* (New Jersey: Lawrence Erlbaum Associates, 1987).

66 And where equal numbers of men and women were present, men asked two-thirds of the questions (Janet Holmes, 'Women Talk Too Much', in Laurie Bauer and Peter Trudgill, *Language Myths*, London: Penguin Books, 1998).

67 Sadker and Sadker, cited in Swann, op cit.

68 Marjorie Swacker, 'The Sex of the Speaker as a Sociolinguistic Variable', in Thorne and Henley, op cit.

69 See Don H. Zimmerman and Candace West, 'Sex Roles, Interruptions and Silences in Conversations', in Thorne and Henley, op cit. Fathers, in two studies, interrupted their children more than mothers did, and both fathers and mothers interrupted their daughters more than their sons (Greif, cited in Jean Berko Gleason and Esther Blank Greif, 'Men's Speech to Young Children', and West and Zimmerman cited in Candace West and Don H. Zimmerman, 'Small Insults: A Study of Interruptions in Cross-Sex Conversations Between Unacquainted Persons', both in Thorne et al, op cit).

70 Carol W. Kennedy and Carl Camden, 'Interruptions and Nonverbal Gender Differences', *Journal of Nonverbal Behaviour*, Winter, 8(2) 1983, speculate that women smiled more 'when they obtained the [speaking] turn' because they were pleased at getting the chance to speak, because they were uneasy with the conversational spotlight being beamed at them or, in their view most likely, to soften the blow of turn-taking – 'as an expression of apology, or as an act of submission' (p.105). This would seem to suggest that for women the very act of speaking is somehow equated with aggression, and needs to be actively counter-vailed – sometimes, as we'll see in the next chapter, by a little-girl voice. Yet despite this, interrupting doesn't seem to be the prerogative of either gender, and its meaning, in any case, differs according to context. 'Simultaneous talk' (as it's been called) might be a way of supporting a speaker, of signalling rapport and solidarity, rather than an expression of disrespect. Deborah James and Sandra Clarke, 'Women, Men, and Interruptions: A Critical Review', in Tannen 1993, op cit. Neither Geoffrey W. Beattie, 'Interruption in Conversational Interaction, and its Relation to the Sex and Status of the Interactants', *Linguistics*, 19, 1981, nor Kristin J. Anderson, 'Meta-analyses of Gender Effects on Conversational Interruption: Who, What, When, Where, and How', *Sex Roles*, August 1998, found evidence of gender differences either. On the other hand measuring interruptions doesn't take account of the amount of time a speaker may be holding the floor. To illuminate this phenomenon, you need to study not just interruptions but also uninterruptibility.

71 Ann Cutler, cited in Jay Ingram, *Talk, Talk, Talk* (Toronto: Penguin Books, 1992).

72 Dale Spender, *Man Made Language*, p.42 (London: Routledge and Kegan Paul, 1980).

73 ibid, p.24. But there are also other possible reasons. Perhaps women talk more than men in certain settings – in 'private' as opposed to 'public' situations? (Deborah Tannen, *You Just Don't Understand*, London: Virago, 1991). And indeed when, out of the forty years of research the sixteen studies dealing with informal talk were isolated, it was found that women were talking as much or more than the men (James and Drakich, op cit).

74 Sally McConnell-Ginet, 'Intonation in a Man's World', in Thorne et al, op cit. Women's greater expressiveness was seen as an expression of vocal versatility and simultaneously of relative powerlessness, which required them to use it as a device to hold the listener's attention. Some early feminist writing had a tendency to treat men's voices as socially constructed, but women's as an expression of superior skill – redirecting the prejudice against women's voices into prejudice against men's.

75 Ruth M. Brend, 'Male–Female Intonation Patterns in American English', in Thorne and Henley, *Language and Sex: Difference and Dominance*, op cit.

76 Robin Lakoff, 'Language and Woman's Place', in Deborah Cameron, *The*

*Feminist Critique of Language: A Reader* (London: Routledge, 1998); Carole Edelsky, 'Question Intonation and Sex Roles', *Language in Society*, 8, 1979.

77 Brend, op cit, p.86.

78 Philip M. Smith, op cit.

79 Except when responding to a female interviewer, and even here, there were many different meanings that could be attributed to such intonation besides politeness. (Edelsky, op cit).

80 On the contrary, because of the way it was calculated, one linguist argued, this had been over-emphasised (Caroline G. Henton, 'Fact and Fiction in the Description of Female and Male Pitch', *Language and Communication*, vol. 9, no. 4, 1989).

81 Nicola Daly and Paul Warren, 'Pitching it Differently in New Zealand English: Speaker Sex and Intonation Patterns', *Journal of Sociolinguistics*, 5/1, 2001.

82 Patsy Rodenburg, 'Powerspeak: Women and Their Voices in the Workplace', in Frankie Armstrong and Jenny Pearson, eds., *Well-Tuned Women* (London: The Women's Press, 2000).

83 Kristin Linklater, 'Overtones, Undertones and the Fundamental Pitch of the Female Voice', in Armstrong and Pearson, op cit.

84 Carol Gilligan, 'Remembering Iphigenia: Voice, Resonance, and the Talking Cure', in Edward R. Shapiro, ed., *The Inner World in the Outer World* (New Haven: Yale University Press, 1997).

85 Lyn Mikel Brown and Carol Gilligan, *Meeting at the Crossroads: Women's Psychology and Girls' Development* (Massachusetts: Harvard University Press, 1992).

86 Henton, 1992, op cit, p.56.

87 The description, she suggested, should have read instead 'the Bengali initial "n" is sometimes pronounced as "l" in pretentious speech, particularly that of status-conscious men' (Ann Bodine on Chatterji in 'Sex Differentiation in Language' in Thorne and Henley, op cit, p.141).

88 ibid, p.302.

89 Tannen, op cit.

90 John Gray, *Men Are from Mars, Women Are from Venus* (New York: HarperCollins, 1992).

91 Aki Uchida, 'When "Difference" Is "Dominance": A Critique of the "Anti-Power-Based" Cultural Approach to Sex Differences', in Cameron, 1998, op cit.

92 In Yana, a now extinct language of Northern California studied by the linguist Edward Sapir, men talking to men 'speak fully and deliberately', whereas when women are involved, either as speakers or listeners, 'a clipped style of utterance in used' (Sapir, quoted in D. Crystal, 'Prosodic and Paralinguistic Correlates of Social Categories', p.189, in Edwin Ardener, ed., *Social Anthropology and Language*, London: Tavistock,, 1971). Interestingly one of my British interviewers volunteered the opposite observation of the men she heard, claiming that they used a wider pitch range in talking to women, but reverted to more of a monotone in conversation with men.

93 Joann M. Montepare and Cynthia Vega, 'Women's Vocal Reactions to Intimate and Casual Male Friends', *Personality and Social Psychology Bulletin*, vol. 14, no. 1, March 1988.

94 Markel et al, cited in Siegman, op cit.

95 Hall and Braunwald, cited in Montepare and Vega, op cit.

96 Woods, cited in Jo Verhoeven, 'The Communicative Setting and Markers in Speech', in Jo Verhoeven, ed., 'Phonetic Work in Progress', University of Antwerp, 2002.

97  Sarah Gracie, 'Body of Evidence', *Guardian*, 12.03.98.

98  Lieberman cited in Graddol and Swann, *Gender Voices*, op cit.

99  See studies cited by Christine Kitamura and Denis Burnham, 'Pitch and Communicative Intent in Mother's Speech: Adjustments for Age and Sex in the First Year', *Infancy*, 4(1), 2003.

100  ibid. If stereotypes are the reason, then why does it take until their baby is 3 months old for these to appear? Perhaps only at this stage, when they're no longer quite so exhausted, do most mothers move beyond seeing their infant as a baby, whose vulnerability is more important than its sex, and begin to treat it as a miniature version of its gender.

101  Barrie Thorne et al, 'Language, Gender and Society: Opening a Second Decade of Research', in Thorne et al, op cit.

102  Gall et al, cited in Thorne et al, op cit. It's even been argued that men stutter more than women because fluency is more highly valued in men than in women, putting too much pressure on men to speak without hesitating.

103  Philip M. Smith, op cit, on Edwards. We don't only hear other people differently, depending on their gender, but also ourselves. The majority of women in a famous Norwich study described themselves as using Received Pronunciation (RP – the upper-middle-class accent, aka Oxford accent or BBC English) when in fact they didn't, while with men it was the other way round: as many as half of them said that they spoke in a more lower-class accent than they actually did. For the men there seemed to be positive connotations to working-class speech (perhaps there's something feminising in the very idea of 'refined' speech, and an entrenched belief that masculinity should sound rougher), while the women aspired to more middle-class ways of speaking that bring overt prestige (Peter Trudgill, *Sociolinguistics*, London: Penguin Books, 2000). There's been much debate about whether women are conservative forces in linguistic change, or innovators. See W. Labov, *The Social Stratification of English in New York City*, (Washington DC: Center for Applied Linguistics', 1966) and *Sociolinguistic Patterns* (Oxford: Basil Blackwell, 1978).

104  David W. Addington, 'The Relationship of Selected Vocal Characteristics to Personality Perception', *Speech Monographs*, vol. 35, 1968. Equally compelling evidence came two years later in a landmark study which, although it wasn't specifically about speech but about health, came to an even more shocking conclusion. Seventy-nine clinically trained psychologists, psychiatrists, and social workers were asked to rate the characteristics of a healthy man, a healthy woman, and a healthy adult, irrespective of sex. These mental-health professionals suggested that healthy women differed from healthy men by being more submissive, less adventurous, less objective, more emotional, more excitable in minor crises, and disliking maths and science (just like the Barbies!). 'This constellation seems a most unusual way of describing any mature, healthy individual' (Inge K. Broverman et al, 'Sex-role Stereotypes and Clinical Judgments of Mental Health', *Journal of Consulting and Clinical Psychology*, vol. 34, no. 1, p.5, 1970). The study also found that the clinicians' idea of a healthy, mature man didn't differ from that of a healthy adult – both were assertive, decisive, and relatively independent – but that of a healthy, mature woman did. In other words, it was impossible for a woman to be considered both a healthy female and a healthy adult. A healthy adult would be a deviant female, a healthy female a deviant adult – women had to choose. The stereotyped characteristics attributed to healthy women match pretty closely those associated with women's voices.

105  Carol Ann Valentine and Banisa Saint Damian, 'Gender and Culture as De-

terminants of the "Ideal Voice'", *Semiotica*, 71–3/4, 1988. In Mexico, however, it's the female voice that's closer to the cultural ideal.

106 Dr Clifford Nass quoted by Anne Eisenberg, 'Mars and Venus, On the Net: Gender Stereotypes Prevail', *New York Times*, 12.10.00.

107 Dilray S. Sokhi et al, 'Male and Female Voices Activate Distinct Regions in the Male Brain', *NeuroImage*, 27, 2005.

108 Stephen McGinty, 'Can't Hear you, Love . . . Blame My Brain', *Scotsman*, 6.8.05.

109 www.netscape.com.

110 Adorno in 1928, quoted in Barbara Engh, 'Adorno and the Sirens: Tele-phono-graphic Bodies', in Leslie C. Dunn and Nancy A. Jones, eds., *Embodied Voices*, p.129 (Cambridge: Cambridge University Press, 1994).

111 To the extent that they've become identified with bodily fluids (Janet Beizer, 'Rewriting Ophelia: Fluidity, Madness and Voice in Louise Colet's "La servante"', in Dunn and Jones, op cit).

112 Michel Chion, *The Voice in Cinema* (New York: Columbia University Press, 1999).

113 Kaja Silverman, *The Acoustic Mirror* (Bloomington: Indiana University Press, 1988).

114 One of the DJs Rosalind Gill, op cit, interviewed indicted women DJs of this very crime. See Nicki Thorogood, 'Mouthrules and the Construction of Sexual Indentities', *Sexualities*, vol. 3(2), 2000 for an interesting discussion of the conventions and practices about mouths that we follow in creating our sexual and gender identities. And for one on telephone sex lines, see Kira Hall, 'Lip Service on the Fantasy Lines', in Cameron, 1998, op cit. There's an extraordinary similarity – almost shocking when you perceive it for the first time – between human vocal anatomy and the female genitalia. The right and left vocal folds, when viewed from above, form a V with a striking resemblance to the vagina and cervix. As one voice teacher memorably put it, 'Both men and women have tiny vaginas in their throats' (Kristin Linklater, 'Vox Eroticus', American Theatre, April 2003).

115 The headscarf, by contrast, although it also signals their religious identity, leaves the mouth uncovered.

116 Laver and Trudgill, cited in Verhoeven, op cit.

117 Seppo K. Tuomi and James E. Fisher, 'Characteristics of Simulated Sexy Voice', *Folia Phoniatrica*, 31, 1979.

118 Branston, op cit.

119 C.G. Henton and R.A.W. Bladon, 'Breathiness in Normal Female Speech: Inefficiency Versus Desirability', *Language and Communication*, vol. 5, no. 3, 1985. See also Dennis H. Klatt and Laura C. Klatt, 'Analysis, Synthesis, and Perception of Voice Quality Variations among Female and Male Talkers', *Journal of the Acoustical Society of America*, February 87(2), 1990.

120 Pamela Jean Trittin and Andres de Santos y Lleo, 'Voice Quality Analysis of Male and Female Spanish Speakers', *Speech Communication*, vol. 16(4), 1995.

121 Moya A. Pattie, 'Voice Changes in Women Treated for Endometriosis and Related Conditions: The Need for Comprehensive Vocal Assessment', *Journal of Voice*, vol. 12, no. 3, 1998.

122 Maureen B. Higgins and John H. Saxman, 'Variations in Vocal Frequency Perturbation Across the Menstrual Cycle', *Journal of Voice*, vol. 3, no. 3, 1989.

123 Jean Abitbol et al, 'Sex Hormones and the Female Voice', *Journal of Voice*, vol. 13, no. 3, 1999. These same doctors have posed the question, 'Does voice have a gender?', and answered it in the affirmative, by referring not to gender but sex.

124 Allen Hirson and Sam Roe, 'Stability of Voice and Periodic Fluctuations in Voice Quality Through the Menstrual Cycle', *Voice*, 2, 1993.

125 Sung Won Chae et al, 'Clinical Analysis of Voice Change as a Parameter of Premenstrual Syndrome', *Journal of Voice*, vol. 15, no. 2, 2001. See also the leading exponent of this theory, Jean Abitbol et al, 'Does a Hormonal Vocal Cycle Exist in Women? Study of Vocal Premenstrual Syndrome in Voice Performers by Videostroboscopy-Glottography and Cytology on 38 Women', *Journal of Voice*, vol. 3, no. 2, 1989.

126 Lacina, cited in ibid.

127 Abitbol et al, 1999, op cit.

128 Graddol and Swann op cit, p.18.

129 Irma M. Verdonck-de-Leuw and Hans F. Mahieu, 'Vocal Ageing and the Impact on Daily Life: a Longitudinal Study', *Journal of Voice*, vol. 18, no. 2, 2004.

130 William Shakespeare, *As You Like It*, Act 2, scene 7.

131 Alison Russell et al, 'Speaking Fundamental Frequency Changes Over Time in Women: A Longitudinal Study', *Journal of Speech and Hearing Research*, vol. 38, February 1995.

132 Henton, 1992, op cit.

133 Abitbol et al, 1999, op cit, p.440. For another example of value judgements posing as objectivity, see Leo Van Gelder, 'Psychosomatic Aspects of Endocrine Disorders of the Voice', *Journal of Communication Disorders*, 7, p.260, 1974, where he talks of voice virilisation symptoms, brought about by anabolic steroids and the use of the pill, in women 'characterized by "nervous" psycho-motor habits and energetic drive, sometimes found in business women, actresses, journalists and women in other emancipated professions'.

134 Ofer Amir et al, 'The Effect of Oral Contraceptives on Voice: Preliminary Observations', *Journal of Voice*, vol. 16, no. 2, 2002.

## 11. How Men and Women's Voices Are Changing (and Why)

1 137.6 Hz, as compared with 118.9 Hz (Majewski et al, cited in David Graddol and Joan Swann, *Gender Voices*, Oxford: Blackwell, 1998).

2 128 Hz, as compared with 161 Hz (Klaus R. Scherer, 'Personality Markers in Speech', in Klaus R. Scherer and Howard Giles, eds., *Social Markers in Speech* (Cambridge: Cambridge University Press, 1979).

3 Carol Ann Valentine and Banisa Saint Damian, 'Gender and Culture as Determinants of the 'Ideal Voice', *Semiotica*, 71–3/4, 1988.

4 Maria DiBattista, *Fast-Talking Dames* (New Haven: Yale University Press, 2003), argues that they drew on the tradition of talky, uppity women that can be traced back to Shakespearean heroines like Beatrice and Rosalind, as well as those in Restoration drama and George Bernard Shaw's plays. Zadie Smith has remarked that Hepburn's Bryn Mawr tang, with its peculiar, long English vowels, gave 'the sense that one is being spoken to from a pinnacle of high-Yankee condescension' (Zadie Smith, 'The Divine Ms H', *Guardian*, 1.7.03).

5 The dialogue in Howard Hawks's *His Girl Friday* ran at 240 words per minute, compared with an average of 100–150 (McCarthy, cited in DiBattista).

6 Gill Branston, '. . . Viewer, I Listened to Him . . . Voices, Masculinity, In the Line of Fire', in Pat Kirkham and Janet Thumim, eds., *Me Jane: Masculinity, Movies, and Women* (London: Lawrence and Wishart, 1995).

7 DiBattista, op cit.

8 ibid.

9 Interview with Stephanie Martin, 3.11.03.

10 American women's voices averaged 214 Hz, Swedish women's 196 Hz, Dutch women's 191 Hz (Renee van Bezooijen, 'Sociocultural Aspects of Pitch Differences between Japanese and Dutch Women', *Language and Speech*, 38(3, 1995)). The pitch levels refer to women between the ages of 20 and 30.

11 In comparison, the pitch of English or American men and women is much less differentiated (Leo Loveday, 'Pitch, Politeness and Sexual Role: An Exploratory Investigation into the Pitch Correlates of English and Japanese Politeness Formulae', *Language and Speech*, vol. 24, part 1, 1981).

12 Van Bezooijen, op cit, has argued that too much emphasis has been placed on femininity as expressed in the Japanese woman's voice, and not enough on how the Japanese man's voice has to establish his masculinity.

13 Hideko Yamazawa and Harry Hollien, 'Speaking Fundamental Frequency Patterns of Japanese Women', *Phonetica*, 49, 1992.

14 Ohara, cited in van Bezooijen, op cit. For a parallel description of voice and gender in China, see Marjorie K.M. Chan, 'Gender Difference in the Chinese Language: A preliminary Report', *Proceedings of the 9th North American Conference on Chinese Linguistics* (Los Angeles: GSIL Publications, University of Southern California, vol.2, 1992).

15 Van Bezooijen, op cit.

16 Smith, cited in ibid.

17 Fairchild, cited in Graddol and Swann, *Gender Voices*, op cit.

18 J.J. Ohala, 'Cross-language Use of Pitch: an Ethological View', *Phonetica*, 40, 1983.

19 Matthew Gordon and Jeffrey Heath, 'Sex, Sound Symbolism, and Sociolinguistics', *Current Anthropology*, vol. 39, no. 4, August-October, 1998.

20 Caroline Henton, 'The Abnormality of Male Speech', in George Wolf, ed., *New Departures in Linguistics* (New York: Garland Publishing, 1992).

21 The average pitch of the 1945 group was 229 Hz, while that of the 1993 cohort was 206 Hz (Cecilia Pemberton et al, 'Have Women's Voices Lowered Across Time? A Cross Sectional Study of Australian Women's Voices', *Journal of Voice*, vol. 12, no. 2, 1998).

22 ibid.

23 C.E. Linke, 'A Study of Pitch Characteristics of Female Voices and their Relationship to Vocal Effectiveness', *Folia Phoniatricia*, 25, p.184, 1973. The transformation of Margaret Thatcher's voice is discussed in chapter 14.

24 *Aera* magazine, 15.6.98.

25 'Women Newscasters Lowering Pitch', *Daily Yomiuri*, 13.6.96.

26 The average pitch of Japanese announcers in 1995 was 216 Hz, compared with 230 Hz in 1991. Professor Kasuya professed himself worried that some announcers 'misunderstand that the lower the voice, the better, and end up speaking in an unnaturally lower voice'. Concerned that 'too low is not good,' he said, 'I feel responsible if I have led the TV industry to worship lower voices' (ibid).

27 'Women Newscasters Lowering Pitch', *Daily Yomiuri*, 13.6.96.

28 Ryann Connell, 'Are Deep-throated Women More Likely to Get Ahead?', *Mainichi Daily News*, 28.6.02.

29 Interview with Jon Snow, 3.03.04.

30 'Best and Worst Voices in America', *Center for Voice Disorders*, Wake University, Winston-Salem, North Carolina, 2001.

31 Valentine and Damian, op cit.

32 Interview, 1.6.03.

33 Interview with Stephanie Martin, 3.11.03.

34 Hershey and Werner, cited in Philip M. Smith, 'Sex Markers in Speech', in Scherer and Giles, op cit.

35 Interview, 31.10.03.

36 Interview, 19.11.03.

37 Interview, 27.11.03.

38 Interview, 26.04.03.

39 Jack W. Sattel, 'Men, Inexpressiveness, and Power', p. 122, in Barrie Thorne et al, eds., *Language, Gender, and Society* (Boston: Heinle and Heinle, 1983).

40 72 per cent of British men own a mobile and spend over an hour a day talking on it, compared with 67 per cent of women, who use it for 55 minutes a day (Alan Travis, 'Men Lead Mobile Phone Revolution', *Guardian*, 11.11.02). American men also talk on their mobiles 35 per cent more than women ('Men Talk More on Cellular Phones, Survey Shows', International Communications Research press release, 14.06.01; Jay Wrolstad, 'Is Wireless a Guy Thing?', www.wirelessnesfactor.com).

41 Although whether women really do invariably possess superior communication skills to men needs questioning, as Deborah Cameron has done, in *Good to Talk?* (London: Sage, 2000).

42 Barrie Thorne and Nancy Henley, 'Difference and Dominance: An Overview of Language, Gender, and Society', in Barrie Thorne and Nancy Henley, *Language and Sex: Difference and Dominance* (Massachusetts: Newbury House, 1975).

43 A singer told me, 'In '60s popular music, so much of the black Motown sound was falsetto, and I marvelled at their willingness to do that. I'm a conservative white guy – I'd never sing that way.' 31.5.03.

44 Austin cited in Thorne and Henley, op cit. It's the same with clothes. As the comedian Eddie Izzard has pointed out, modern dress codes favour women, who are free to wear trousers, while men are mocked if they wear a dress.

45 Interview, 13.08.05. Men have to disidentify with their mother and identify with their father (see Nancy Chodorow, *The Reproduction of Mothering: Psychoanalysis and the Sociology of Gender*, University of California Press, 1979), and this happens on the vocal level too (Kaja Silverman, *The Acoustic Mirror*, Bloomington: Indiana University Press, 1988).

46 Interview, 16.5.03.

47 Interview with Charles Michel, 29.5.03. Of course there are cultural issues at work here too: black culture allows for far more variety of pitch in its males.

48 Amelia Hudson and Anthony Holbrook, 'A Study of the Reading Fundamental Vocal Frequency of Young Black Adults', *Journal of Speaking and Hearing Research*, 24, June 1981.

## 12. Cultural Differences in the Voice

1 Dell Hymes, 'Models of the Interaction of Language and Social Life', in John J. Gumperz and Dell Hymes, *Directions in Sociolinguistics* (New York: Holt, Rinehart and Winston, 1972).

2 The Bella Coola in British Columbia value fluent speech and demand a flow of witty comments (ibid).

3 The Paliyans of south India communicate very little and have turned almost

silent by the age of 40. To them verbal people are regarded as offensive and even abnormal (Gardner, cited in Dell Hymes, *Foundations in Sociolinguistics*, London: Tavistock, 1977).

4 Bronislaw Malinowski, 'Coral Gardens and Their Magic', vol. 2, *The Language of Magic and Gardening* (London: Routledge, 2001).

5 E.D. Lewis, 'A Quest for the Source: the Ontogenesis of the Creation Myth of the Ata Tana Ai', in James J. Fox, ed., *To Speak In Paris* (Cambridge: Cambridge University Press, 1988).

6 Steven Feld, 'From Ethnomusicology to Echo-Muse-Ecology: Reading R. Murray Schafer in the Papua New Guinea Rainforest', *Soundscape Newsletter*, no.8, June 1994, and his 'A Rainforest Acoustemology', in Michael Bull and Les Back, *The Auditory Culture Reader* (Oxford: Berg, 2003). The distinction we make between singing and speaking would seem strange to the New Zealand Maori, who see both instrumental music and song as speech (Hymes, 1972, op cit).

7 Marcel Griaule and Genevieve Calame-Griaule, cited in Frances Dyson, 'Circuits of the Voice: From Cosmology to Telephony', in Dan Lander, ed., *Radio Phonics and other Phonies* (Toronto, Musicworks, no. 43, 1992).

8 Hymes, 1972, op cit.

9 Joseph Epes Brown, *The Sacred Pipe* (University of Oklahoma Press, 1953).

10 Dwight Bolinger, *Intonation and Its Uses* (London: Edward Arnold, 1989), and Alan Cruttenden, 'Falls and Rises: Meanings and Universals', *Journal of Linguistics*, 17, 1981 although see below for instances of cross-cultural misunderstandings based on intonation, and the different ways Americans and the British ask questions.

11 Lomax, cited in Klaus R. Scherer, 'Personality Markers in Speech', in Klaus R. Scherer and Howard Giles, eds., *Social Markers in Speech* (Cambridge: Cambridge University Press, 1979).

12 Honikman, cited in John Laver and Peter Trudgill, 'Phonetic and Linguistic Markers in Speech', in Scherer and Giles, op cit.

13 Maynard cited in Nigel Ward, 'Using Prosodic Clues to Decide When to Produce Back-channel Utterances', paper given at 4th International Conference, 1996.

14 Ward, op cit.

15 Haru Yamada, *Different Games, Different Rules: Why Americans and Japanese Misunderstand Each Other* (New York: Oxford University Press, 1997).

16 B. Malinowski, 'Phatic Communion', p.151 in John Laver and Sandy Hutcheson, eds., *Communication in Face to Face Interaction* (Middlesex, England: Penguin Books, 1972).

17 ibid, p.150.

18 Hymes quoted in Douglas Biber and Edward Finegan, 'Situating Register in Sociolinguistics', p.7 in Douglas Biber and Edward Finegan, eds., *Sociolinguistic Perpsectives on Register* (New York: Oxford University Press, 1994.

19 Ethel M. Albert, 'Culture Patterning of Speech Behaviour in Burundi', in Gumperz and Hymes, op cit.

20 Judith T. Irvine, 'Registering Affect: Heteroglossia in the Linguistic Expression of Emotion', in Catherine A. Lutz and Lila Abu-Lughod, *Language and the Politics of Emotion* (Cambridge: Cambridge University Press, 1990).

21 See Pierre Bourdieu, *Disctinction: A Social Critique of the Judgement of Taste* (London: Routledge, 1984) and *The Logic of Practice* (Cambridge: Polity Press, 1990).

22 Jane Stuart-Smith, 'Glasgow: Accent and Voice Quality' in Paul Foulkes and Gerard Docherty, eds., *Urban Voices* (London: Hodder Headline, 1999). Linguists talk of speech communities, groups that share the same language,

dialect, and other linguistic norms. We could also talk of vocal communities, groups – large or small – who use their voices in similar ways.

23 Bruce L. Brown and Wallace E. Lambert, 'A Cross-cultural Study of Social Status Markers in Speech', *Canadian Journal of Behavioural Science*, 8(1), 1976.

24 Pierre Bourdieu, *Outline of a Theory of Practice*, p.93 (Cambridge: Cambridge University Press).

25 William Labov, 'The Social Stratification of (r) in New York City Department Stores', in Nikolas Coupland and Adam Jaworski, eds., *Sociolinguistics: A Reader and Coursework* (London: Macmillan, 1997).

26 Linda W.L. Young, *Crosstalk and Culture in Sino-American Communication* (Cambridge: Cambridge University Press, 1994).

27 Howard Giles, 'Ethnicity Markers in Speech', in Scherer and Giles, op cit.

28 E.T. Hall, 'A System for the Notation of Proxemic Behavior', in Laver and Hutcheson, op cit.

29 Masashi Takeuchi et al, 'Timing Detection for Realistic Dialog Systems Using Prosodic and Linguistic Information', paper given at Speech prosody 2004, Nara, Japan.

30 Yamada, op cit.

31 Tetsuya Kunihiro, 'Personality-Structure and Communicative Behaviour: A Comparison of Japanese and Americans', in Walburga von Raffler-Engel, ed., *Aspects of Nonverbal Communication* (Swets Publishing, 1983). Between speakers of the North American Indian language of Athabaskan and the English, the potential for misunderstanding is even greater. The English, because they want to establish some kind of connection and feel uncomfortable with silence, invariably initiate conversation. Athabaskans, on the other hand, if they have any doubt about a social relationship and the appropriate way to behave, avoid speaking. So an English speaker will fill in a pause in English-Athabaskan conversation earlier than the Athabaskan. (One reason for the traditional British talk about the weather is that the British generally, except with intimates, can't tolerate a silence of longer than about four seconds. Anything, even small talk about the weather, is better than silence.) 'The Athabaskans go away from the conversation thinking that English speakers are rude, dominating, superior, garrulous, smug and self-centred. The English speakers, on the other hand, find the Athabaskans rude, superior, surly, taciturn, and withdrawn' (Peter Trudgill, *Sociolinguistics: An Introduction to Language and Society*, London: Penguin Books, 1974). The Greeks regard silence even more negatively than the British. In Greece it's seen as a distancing device in a society where closeness, exuberance, and talkativeness are very highly valued. There a taciturn person is seen as unfriendly, snobbish and even dangerous (Sifanou, cited in Adam Jaworski, 'Silence and small talk', in Justine Coupland, ed., *Small Talk*, London: Longman, 2000). The potential for vocal misunderstanding exists not only between cultures but also within them. Deborah Tannen ('It's Not What You Say, It's The Way That You Say It', undated, www.surfaceonline.org) argues that New Yorkers speak fast, and start talking before other people have finished. Non-New Yorkers expect a pause before they pitch in, but it never arrives. 'People who are not from New York complain that New Yorkers interrupt them, don't listen, and don't give them a chance to talk.'

32 French and von Raffler-Engel, and Erickson, cited in Judee K. Burgoon et al, *Nonverbal Communication: The Unspoken Dialogue* (New York: McGraw-Hill, 1996). Pauses even seem to be distributed differently according to the social class of the speaker, with working-class people spending less time pausing (Bernstein 1962, cited in Aron W. Siegman, 'The Telltale Voice: Nonverbal

Messages of Verbal Communication', in Aron W. Siegman and Stanley Feldstein, *Nonverbal Behaviour and Communication*, New Jersey: Lawrence Erlbaum Associates, 1987). Discussion of Bernstein's elaborated and restricted codes is beyond the scope of this book, although he applied his distinction also to the nonverbal aspects such as intonation.

33 Ron Scollon and Suzanne B.K. Scollon, 'Interethnic Communication: How to Recognize Negative Stereotypes and Improve Communication Between Ethnic Groups' (Fairbanks, Alaska: Alaska Native Language Center, University of Alaska Fairbanks, 1980). See also Carole Douglis, 'The Beat Goes On: Social Rhythms Underlie All Speech', *Psychology Today*, November 1987.

34 'What is it like to be German in Britain?', *Guardian*, 28.6.02.

35 Alan Cruttenden, *Intonation* (Cambridge: Cambridge University Press, 1986).

36 Leo Loveday, 'Pitch, Politeness and Sexual Role: An Exploratory Investigation Into the Pitch Correlates of English and Japanese Politeness Formulae', *Language and Speech*, vol. 24, part 1, 1981.

37 Stephanie Martin and Lyn Darnley, *The Teaching Voice* (London: Whurr Publishers, 2004).

38 Deborah Tannen, 'Did You Catch That? Why They're Talking as Fast as They Can?', *Washington Post*, 5.1.03.

39 Patsy Rodenburg, *The Right to Speak*, p.168 (London: Methuen, 1992).

40 Jaakko Lehtonen and Kari Sajavaara, 'The Silent Finn', in Deborah Tannen and Muriel Saville-Troike, *Perspectives on Silence* (New Jersey: Ablex, 1985). Young Finns, it seems, can no longer tolerate silence, and like young people everywhere walk around with iPods clamped to their ears. A Finnish speech and language therapist told me that her 14- and 9-year-old sons, when they go to their family's country house, find the silence strange and hard to take.

41 Scott Allan, 'The Rise of New Zealand Intonation', in Allan Bell and Janet Holmes, eds., *New Zealand Ways of Speaking English* (Clevedon, England: Multilingual Matters, 1990).

42 Cruttenden, 1986, op cit.

43 Mark Newbrokk, cited in Matt Seaton, 'Word up', *Guardian*, 21.9.03.

44 Robin Lakoff, *Language and Woman's Place* (New York: HarperCollins, 1975).

45 McLemore, cited in Seaton, op cit.

46 K. Cave, 'What teens are saying? It's called uptalk?', Orange County Register, 1994, reprinted in Martin S. Remland, *Nonverbal Communication in Everyday Life*, p.213 (Boston: Houghton Mifflin, 2000).

47 This process, known as convergence, is discussed in chapter 17.

48 Claire Gardner, 'Accent on Mimicry from Parrot Fashion Call Centres', *Scotland on Sunday*, 2.5.04. To gain customers' trust, more than 50 per cent of staff changed their accents in order to try and sound similar to them.

49 Malavika Sangghvi, 'Indian Talkaway', *Sunday Times*, 7.12.03.

50 Peter Kingston, 'Calls to Newcastle', *Guardian*, 18.10.05.

51 Charles Haviland, 'At Your Service: Indians Heed Call of the West', *Guardian*, 18.10.03.

52 Andrew Clark, 'Catching a Train to Crewe? Call Bangalore', *Guardian*, 15.10.03.

53 Kingston, op cit. To immerse themselves in contemporary British culture, call-centre workers are shown films like *Love Actually* and *Four Weddings and a Funeral*, which themselves have been criticised for perpetuating an unrealistic view of Britain.

54 See, for instance, Laura Lehto et al, 'Voice Symptoms of Call-centre Customer Service Advisers Experienced During a Work-day and Effects of a Short Vocal

Training Course', *Logopedics Phoniatrics Vocology*, 30, 2005; David Hencke, 'Call Centres Voice Concerns', *Guardian*, 20.6.05. British trades unions have also protested about the transfer of call-centre jobs to India (Kevin Maguire, 'BT Strike Threat Over Indian Call Centres', *Guardian*, 3.6.03).

55 Jeremy Seabrook, 'Progress on Hold', *Guardian*, 24.10.03.
56 Siddhartha Deb, 'Call Me', *Guardian*, 4.4.04.
57 Amelia Gentleman, 'Indian Call Staff Quit Over Abuse on the Line', *Observer*, 29.5.05.
58 Deb, op cit.
59 Sangghvi, op cit.
60 *India Calling*, BBC Radio 4, 2002.
61 Kingston, op cit.
62 *Who Do You Think You're Talking To?*, BBC 2, 2004.
63 ibid.
64 See, for instance, George Monbiot, 'The Flight to India', *Guardian*, 21.10.03, and Seabrook, op cit.

## 13. From Oral to Literate Society

1 See J. Goody and I. Watt, 'The Consequences of Literacy', reprinted in Pier Paulo Giglioli, ed., *Language and Social Context* (Middlesex: Penguin, 1972); Marshall McLuhan, *The Gutenberg Galaxy* (London: Routledge, 1962); Walter J. Ong, *The Presence of the Word* (New Haven: Yale University Press, 1967); and Eric A. Havelock, *The Muse Learns to Write* (New Haven: Yale University Press, 1986).
2 Havelock, op cit, pp.65–66.
3 Goody and Watt, op cit; Walter J. Ong, *Orality and Literacy* (London: Methuen, 1982). Hereafter all references to Ong apply to this.
4 Bernard Hibbitts, 'Coming to Our Senses: Communication and Legal Expression in Performance Cultures', *Emory Law Journal*, 4, 1992. When Aivilik Eskimos want information, they say to themselves or others, 'Let's hear.'
5 Jane Kamensky, *Governing the Tongue: The Politics of Speech in Early New England* (New York: Oxford University Press, 1999), and, according to St Paul, '*Ex auditu fides*' – From hearing comes belief (Romans 10:17).
6 Hibbitts, op cit.
7 R. Murray Schafer, *The Soundscape: Our Sonic Environment and the Tuning of the World* (Toronto: McClelland & Stewart, 1977).
8 Ong, op cit.
9 ibid.
10 Ong, op cit, p.102.
11 McLuhan, op cit, p.232.
12 Ong, op cit.
13 Carolthers, quoted in McLuhan, op cit, p.20.
14 Ong, op cit.
15 Macluhan, op cit, p.27.
16 Although a new system has been developed for turning a telephone call into a legally binding document. The conversation is digitally recorded, encrypted, and emailed to caller, law-firm, accountant, and the manufacturers of the technology (Kane Kramer on Monicall, *Guardian*, 30.09.04).
17 McLuhan, op cit.
18 Kathleen Hall Jamieson, *Eloquence in an Electronic Age* (New York: Oxford University Press, 1988).

19  Ong, op cit, p.104.

20  ibid.

21  Ong, op cit, pp.40/1.

22  McLuhan, op cit, p.27. 'Since the ear is a hot hyperaesthetic world and the eye world is a relatively cool, neutral world, the Westerner appears to people of ear culture to be a very cold fish indeed,' remarked McLuhan (p.19). He didn't anticipate how hot visual culture could be and how cool radio.

23  David Riesman, cited in McLuhan, op cit.

24  Carothers, quoted in McLugan, op cit, p.19.

25  Leigh Eric Schmidt, 'Hearing Loss', p.46, in Michael Bull and Les Back, eds., *The Auditory Culture Reader* (Oxford: Berg, 2003).

26  Ong, op cit, p.121. In the sixteenth and seventeenth centuries punctuation was still guided by the ear and not the eye (McLuhan, op cit).

27  Ong, op cit.

28  Kamensky, op cit. Far from subscribing to the view that sticks and stones can break your bones but words can never hurt you, they believed rather that 'a soft tongue breaketh the bone'.

29  McLuhan, op cit.

30  Havelock, op cit.

31  ibid.

32  The BBC Radio panel game *Just a Minute* gets participants to pronounce on a given topic without hesitation, deviation, or repetition for a minute. It owes its long success at least partly, I feel sure, to the fact that it's a throwback to oral cultures. Trying, and often clumsily failing, to discard the written tics and spoken props on which so much broadcasting relies induces hilarity all round. This is a literate culture's comic tilt at an oral one.

33  McLuhan, op cit. Where 'primary oral' cultures were totally untouched by writing or print, these 'secondary oral' cultures had telephones, radio, and television as well as print, and shared the features of both forms (Ong, op cit).

34  Jacques Derrida, *Speech and Phenomena, and Other Essays on Husserl's Theory of Signs*, p.76 (Evanston: Northwestern University Press, 1973).

35  Ong, op cit.

36  Havelock, op cit, p.64.

37  ibid.

38  Schmidt, op cit.

39  Scott L. Montgomery and Alok Kumar, 'Telling Stories: Some Remarks on Orality in Science', p.393, *Science as Culture*, vol. 9, no. 3, 2000.

40  ibid.

41  Michel Foucault, *The Birth of the Clinic* (London: Tavistock Publications, 1976).

42  Samuel Johnson, *A Dictionary of the English Language* (London: Strahan, 1755).

43  Schmidt, op cit, p.50.

44  Jonathan Sterne, *The Audible Past: Cultural Origins of Sound Reproduction* (Durham, North Carolina: Duke University Press; 2003).

45  He conducted a series of experiments, among them one in which he produced tones from the larynx of a goose by squeezing its lungs. (Hans von Leden, 'A Cultural History of the Human Voice', in Robert Sataloff, ed., *Voice Perspectives*', San Diego: Singular Publishing, 1998). It took until 1741 for this to be repeated again by Antoine Ferrein, who coined the term vocal cords, which he pictured as comparable to the strings of a violin activated by a stream of air. In 1745 Bertin pointed out that the structures were folds and not cords and drew

attention to their elasticity (Donald S. Cooper, 'Voice: A Historical Perspective', *Journal of Voice*, vol. 3, no. 3, 1989). No one before Manual Garcia's invention of the laryngeal mirror in 1854 actually observed voice production in a living person (Cooper, op cit), although this claim has been contested. Some Americans have insisted that direct visualisation of the larynx began with Horace Green, the father of American laryngology. Green reported that, using a bent-tongue spatula to examine the throat and larynx of an 11-year-old girl in 1852, he spotted a polyp and, using direct sunlight for illumination, removed it. Laryngoscopes were refined by the Berlin physician Alfred Kirstein who, in 1895, conducted the first direct examination of the interior of larynx, and later by Gustav Killian and Chevalier Jackson. See Steven M. Zeitels, 'Universal Modular Glottiscope System: The Evolution of a Century of Design and Technique for Direct Laryngoscopy', *Annals of Otology, Rhinology and Laryngology*, – Supplement 179, vol. 108, no. 9, part 2 September 1999; Horace Green, 'Morbid Growths Within the Larynx', in Green, *On the Surgical Treatment of Polypi of the Larynx and Oedema of the Glottis* (New York: G. P. Putnam, 1852); Gayle E. Woodson, 'The History of Laryngology in the United States', *Laryngoscope*, vol.106(6), June 1996; and Richard Cooper, 'Laryngoscopy – its Past and Future', *Canadian Journal of Anesthesia*, 51, Supplement 1:R6 2004.

46  Roy Porter, *Enlightenment* (London: Penguin Books, 2000).

47  Richard Sennet, *The Fall of Public Mar*, pp.81–82 (London: Faber and Faber, 1986).

48  Markman Ellis, *The Coffee House: A Cultural History* (London: Weidenfeld & Nicholson, 2004).

49  ibid, p.365.

50  Albert Mehrabian and Susan R. Ferris, 'Inference of Attitudes from Nonverbal Communication in Two Channels', *Journal of Consulting Psychology*, vol. 31, no. 3, 1967.

51  Albert Mehrabian and Morton Wiener, 'Decoding of Inconsistent Communications', *Journal of Personality and Social Psychology*, vol. 6, no. 1, p. 113, 1967.

52  Mehrabian and Ferris, op cit.

53  Albert Mehrabian, *Nonverbal Communication*, p.182 (Chicago: Aldine, 1972).

54  Karl Popper, *The Logic of Scientific Discovery* (London: Routledge, 2002).

55  See my *Doctoring the Media: The Reporting of Health and Medicine* (London: Routledge, 1988) for a more extensive critique of the popularisation of science and medicine.

56  Carol Tavris: 'How to Publicize Science: a Case Study', in Jeffrey H. Goldstein, ed., *Reporting Science: The Case of Aggression* (Lawrence Erlbaum Associates 1986).

57  Albert Mehrabian, cited by David Lapakko, 'Three Cheers for Education', Communication Education, 1999 (quoted on d3m e-news, District 3 Toastmasters, 1 December 2000).

58  Mehrabian and Wiener, op cit.

59  www.kaaj.com, op cit, where Mehrabian also vigorously markets his personality-testing software and emotional-intelligence-testing software. In fact there's almost no aspect of emotional life for which Dr Mehrabian hasn't devised and sold software.

60  J. Lotz, *Linguistics: Symbols Make Humans* (New York: Language and Communication Research Centre, Columbia University, 1955), cited in Paul Newham, *Therapeutic Voicework* (London: Jessica Kingsley Publishers, 1998).

# 14. The Public Voice

1 Aristotle, *De Anima* (On the Soul), part two, chapter 8 (London: Gerald Duckworth, 1996).

2 And public speaking involves control over the voice: 'These are the three things – volume of sound, modulation of pitch, and rhythm – that a speaker bears in mind' (Aristotle, *Rhetoric*, Book 3, chapter 1, Loeb Classical Library, 1926).

3 R.C. Jebb, *The Attic Orators from Antiphon to Isaeos*, p.lxxviii/lxxix (London: Macmillan, 1876).

4 Browne and Behnke, cited in John Laver, 'The Analysis of Vocal Quality: from the Classical Period to the Twentieth Century', in John Laver, *The Gift of Speech* (Edinburgh: Edinburgh University Press, 1991).

5 Hans von Leden, 'A Cultural History of the Voice' in Robert Sataloff, ed., *Voice Perspectives* (San Diego: Singular, 1998).

6 ibid.

7 Ynez Viole O'Neill, *Speech and Speech Disorders in Western Thought Before 1600* (Connecticut: Greenwood Press, 1980).

8 Stanford, cited in Laver, op cit.

9 See Jacqueline Martin, *Voice in Modern Theatre* (London: Routledge, 1991).

10 Cicero, *De Oratore*, Book 3, chapter 43 (London: Heinemann, 1942).

11 Scholars were also allowed to declaim once a month (Kathleen Hall Jamieson, *Eloquence in an Electronic Age*, New York: Oxford University Press, 1988). Cicero and Quintilian's complete works were among the first books to be printed by the printing-press (Martin, op cit).

12 Patricia Bizzell and Bruce Herzberg, eds., *The Rhetorical Tradition* (New York: Bedford/St Martin's, 1990).

13 Quintilian, *Institutiones Oratoriae*, Book XI, chapter 3 (Loeb Classical Library, 1992). In men, that is. As for the female voice, 'physical robustness is essential to save the voice from dwindling to the feeble shrillness that characterises the voices of eunuchs, women and invalids.'

14 Dan Sperber and Deirdre Wilson, 'Rhetoric and Relevance' in David Wellbery and John Bender, eds., *The End of Rhetoric: History, Theory, Practice* (Stanford: Stanford University Press, 1990).

15 Jamieson, op cit.

16 Thanos Vovolis, 'The Voice and the Mask in Ancient Greek Tragedy', in Larry Sider et al, eds., *Soundscape: The School of Sound Lectures, 1998–2001*. (London: Wallflower Press, 2003).

17 Oliver Taplin, Greek Tragedy in Action (London: Methuen, 1978).

18 R. Murray Schafer, '*The Soundscape: our Sonic Environment and the Tuning of the World*' (Toronto: McClelland & Stewart, 1977). The close connection between rhetoric and acting persisted and the declamatory style was expected. The influence extended in both directions: not only did actors have to deliver their speeches in the rhetorical style, but the orator Demosthenes, famous for trying to overcome his speech impediments and achieve clear diction by putting pebbles in his mouth while reciting, also trained by using actors' methods of voice production.

19 Even though the dramatist and actor-manager Colley Cibber dismissed his 'pantomimical Manner of acting . . . his Unnatural Pauses in the middle of a Sentence' (Cibber quoted in Martin, op cit, p.8) Cibber was almost certainly driven by envy. Lampooned by Pope for his own excesses, he admitted in his autobiography that, though he longed to play the hero, 'in this Ambition I was

soon snubb'd by the Insufficiency of my Voice' (Colley Cibber, *An Apology for the Life of Colley Cibber*, Dover: 2000, originally published in 1740).

20 Tennyson's voice, for example, 'acted sometimes almost like an incantation, so that when it was a new poem that he was reading, the power of realizing its actual nature was subordinated to the wonder at the sound of the tones . . . sometimes . . . a long chant . . . sometimes a swell of sound like an organ's' (Francis Berry, *Poetry and the Physical Voice*, p.51, London: Routledge and Kegan Paul, 1962). Shelley's voice, by contrast, was described as 'intolerably shrill, harsh and discordant . . . It was perpetual, and without any remission; it excoriated my ears' (ibid, p.67).

21 Listen, for example, to the historic recordings of writers and poets from the British Library Sound Archive now issued on CD – 'The Spoken Word: Writers' and 'The Spoken Word: Poets'.

22 Tom Paulin, 'The Despotism of the Eye', p.46, in Sider, op cit.

23 This he discovered in the trenches of the First World War (Martin, op cit), through the extraordinary sounds issuing from dying soldiers. Left with aural hallucinations, he worked on his own mental state and came to realise how many different aspects of the self the voice could express (Paul Newham, *Therapeutic Voicework*, London: Jessica Kingsley, 1998).

24 The voice teacher Kristin Linklater also focused on the voice as a bridge between mind and body, trying to free it from defences and tensions through exercises and voice-work (Kristin Linklater, *Freeing the Natural Voice*, Quite Specific Media Group, 1988).

25 Angry young women, on the other hand, were nowhere on stage to be heard.

26 Brando quoted in Richard Eyre, 'Beauty is Truth, Truth Beauty', *Guardian*, 31.7.04.

27 So demotic and under-played had acting styles become that one critic described Marianne Faithfull, in the role of Irina in the 1967 London Royal Court Theatre's production of Chekhov's *Three Sisters*, as crying, 'Moscow, Moscow,' as if she were calling for her dog (Peter Noble to the author, April 1967).

28 Joan Mills, 'A Vocal Album: Snapshots from Vocal History', p.8, in Frankie Armstrong and Jenny Pearson, *Well-Tuned Women* (London: The Women's Press, 2000).

29 George Eliot, *Daniel Deronda* (London: Penguin Classics, 1995).

30 Interview, 26.11.03.

31 Interview, 30.11.03.

32 Henry Fairlie, quoted in Max Atkinson, *Our Masters' Voices: the Language and Body Language of Politics*, p.107 (London: Methuen, 1984).

33 Pasty Rodenburg, *The Need for Words* (London: Methuen, 1993).

34 Samuel Beckett, *Not I* (London: Faber and Faber, 1973).

35 For the first time those of major public figures from the past can be heard on internet sites and CDs – an expression, almost certainly, of revived public interest in the voice. See, for example the History Channel, www.historychannel.com; G. Robert Vincent Voice Library, Michigan State University, www.vvl.lib.msu.edu; American Memory: Library of Congress – American Leaders Speak, www.memory.loc.gov/ammem/nfhtm; EyeWitness to History – Voices of the 20th Century, www.eyewitnesstohistory. com; Vintage Recordings of Presidential Elections, www.edisonnj.org.menlopark/. And among the CDS, 'Voices of History – Historic Recordings from the British Library Sound Archive', www.bl.uk; Margaret Thatcher, *The Great Speeches* (Politicos Media CD).

36 Edward D.Miller, *Emergency Broadcasting and 1930s American Radio*. (Philadelphia: Temple University Press, 2003).

37 So much so that when listeners wrote to him, as they did in their thousands, they were unsure whether to address him formally or informally (ibid).

38 Robert H. Jackson, *That Man: An Insider's Portrait of Franklin D. Roosevelt.* (Oxford: Oxford University Press, 2003).

39 ibid, p.159.

40 William E. Leuchtenburg, 'The FDR Years: On Roosevelt and His Legacy', www.washingtonpost.com

41 He also changed the very texture of radio. Although his fireside chats don't sound especially intimate to today's audience, FDR's calm, measured tones, emerging as they did unsponsored from between advertisements for toothpaste and cough drops, seemed extraordinary in their day (Miller, op cit).

42 Halberstam quoted in Leuchtenburg, op cit. Roosevelt's voice, wrote the *New York Times* in 1933, 'reveals sincerity, good-will and kindliness, determination, conviction, strength, courage and abounding happiness' (*New York Times Magazine*, 18.6.33, quoted in G. W. Allport and H. Cantril, 'Judging Personality from Voice', p.155, in John Laver and Sandy Hutcheson, *Communication in Face-to-Face Interaction*, London: Penguin Books, 1972).

43 Miller, op cit. The familiarity of Roosevelt's voice made it seem as if the medium of radio had disappeared, and his voice was coming to them unmediated, out of the either, fomenting the idea that the President had superhuman powers and was omnipotent.

44 The orthodoxy that FDR's voice soothed the nation has been contested. In reality, it's been argued, it 'regulated panic', creating the conditions for his own indispensability. 'Roosevelt's voice is heard as a voice of emergency, and his presidency becomes committed not so much to ending this emergency as to transmitting it, in order to secure a place for Roosevelt's narrative voice. This voice reproduces the panic, even as it suggests that he, and his government, are well on the way to curing the ills of the country' (ibid, p.79).

45 Adolf Hitler, quoted in Ian Kershaw, *Hitler, 1889–1936*, p.147 (London: Penguin, 1998).

46 Propaganda, for Hitler, was the highest form of political activity (Kershaw, 1998, op cit) and was indistinguishable from ideology: 'The broad masses of a population are more amenable to the appeal of rhetoric than to any other force' (Hitler, *Mein Kampf*, pp. 317, 100, quoted in Alan Bullock, *Hitler: A Study in Tyranny*, London: Penguin Books, 1962).

47 William Carr, *Hitler: A Study in Personality and Politics* (London: Edward Arnold, 1978).

48 Kershaw, 1998, op cit.

49 Bullock, op cit, p.373.

50 Kershaw, 1998, op cit.

51 Carr, op cit, p.2.

52 Joachim C. Fest, *Hitler*, p.486 (London: Penguin Books, 1977).

53 Bullock, op cit, p.71.

54 Fest, op cit.

55 Konrad Heiden, *Der Führer*, p.81 (Constable and Robinson, 1999).

56 Bullock, op cit.

57 Fest, op cit, p.481. One extraordinary speech demonstrates his messianic fantasies. It's hard to know whom he was talking about – Jesus Christ or himself. 'You have once heard the voice of a man, and it has struck your hearts, and you have followed this voice. You have followed it for years, even without seeing the bearer of the voice. You only heard a voice, and you followed it. When we meet here, the miracle of this gathering fills us all. Not each of you sees me,

and I do not see each of you. But I feel you, and you feel me! (Hitler's speech to a Nuremberg Rally of 1936, quoted in Ian Kershaw, *The Hitler Myth*, pp.107–8, Oxford: Oxford University Press, 1987).

58  Traudl Junge, speaking in *Blind Spot – Hitler's Secretary* (*Im Toten Winkel – Hitlers Sekretarin*), documentary film directed by André Heller and Othmar Schmiderer, 2002. Only one recording of Hitler speaking in his normal tone of voice has been found. Recorded secretly by a Finnish radio technician in 1942, it resurfaced a few years ago (and was used by the actor Bruno Ganz to help shape his performance in the 2004 German film, *Downfall*).

59  Hitler in the early 1920s, quoted in Kershaw, 1998, p.133.

60  Fest, op cit.

61  Bullock, op cit, p.373.

62  Paul Moses, *The Voice of Neurosis* (New York: Grune and Stratton, 1954).

63  Atkinson, op cit. In his novel *The Karnau Tapes*, the German writer Marcel Beyer dubbed it 'a war of sound'. His central character, a sound engineer, described how Nazi meetings were manipulated by four microphones in front of the speaker's desk, containing batteries of loudspeakers aimed at the stadium from all angles. A fifth, designed to pick up special frequencies, would be adjusted throughout the speech to bring out special vocal effects, and a sixth, beneath the desk, could be controlled by the speaker himself. Additional microphones were installed at a radius of one metre to help create a stereophonic effect, and there was even a gigantic public address system setting up continuous vibrations in their bodies (Marcel Beyer, *The Karnau Tapes*, London: Vintage, 1998) – an echo of Vetruvius's suggestion that the acoustics of the theatre at Epidaurus could be improved by placing large bronze or earthenware 'sounding vases' in the auditorium that would pick up and reinforce certain notes (see Vovolis, op cit). Of course Beyer's is a modern fictional re-imagining of the Nazi attempt to manipulate through the voice, and yet so overwhelming was its capacity to generate paranoia that, after all these years, it still seems feasible.

64  Virginia Woolf, quoted in Gillian Beer, ' "Wireless": Popular Physics, Radio and Modernism', p.165, in Francis Spufford and Jenny Uglow, eds., *Cultural Babbage: Technology, Time and Invention* (London: Faber and Faber, 1996).

65  Susan J. Douglas, *Listening In* (New York: Times Books, 1999).

66  Joseph E. Persico, *Roosevelt's Secret War: FDR and World War 2 Espionage*. (New York: Random House, 2001).

67  Quoted in *Churchill's Roar*, BBC Radio 4, 24.01.05.

68  ibid.

69  Jean Seaton, in ibid.

70  Andrew Roberts, *Hitler and Churchill*, p.36 (London: Weidenfeld and Nicholson, 2003).

71  *Churchill's Roar*, op cit. When the programme speeded up Churchill's last sentence to Bragg's pace, it lost most of its drama. When, on the other hand, they slowed Bragg's tempo down to Churchill's, it gained enormously in power.

72  Winston Churchill speech to House of Commons, 14.07.40. The idea, floated by the revisionist historian David Irving, that the recordings of some of Churchill's most famous speeches – including the 'We shall fight them on the beaches' Dunkirk one of 4 June 1940 and the 'their finest hour' address of 18 June 1940 – were made not by Churchill but the actor Norman Shelley, has been taken up by serious historians and entered into popular mythology (see, for instance, Vanessa Thorpe, 'Finest Hour for Actor Who Was Churchill's Radio Voice', *Observer*, 29.10.00). Although recordings of Shelley speaking some of Churchill's speeches undoubtedly exist, the idea that Shelley's voice was substituted for Churchill's

seems to have been comprehensively debunked by the Winston Churchill Archives (www.winstonchurchill.org/myths).

73 Peter Preston, 'Towering He Wasn't', *Guardian*, 7.6.04.

74 Roger Rosenblatt, writing in *Time*, quoted in Paul D. Erickson, *Reagan Speaks*, p.14 (New York: New York University Press, 1985). A voice coach remarked, 'Just before he would speak he would say, "Well," and what I read into that "Well" is him giving himself permission to let go which . . . I don't think people in general do . . . [or] men in particular. "Well, here goes, what the hell . . . Now I'll just come from the heart"' (interview with Charles Michel, 28.05.03).

75 Jamieson, op cit.

76 He also suffered from significant hearing loss caused by the firing of a gun near his ear when he was making westerns in Hollywood. In 1983 he was fitted with a hearing-aid (the first US president to use one), which had to be specially 'swept' for KGB bugs at regular intervals (Brendan Bruce, *Images of Power: How the Image-Makers Shape Our Leaders*, London: Kogan Page, 1992).

77 Between 1975 and 1979, after stepping down as Governor of California and before he became President, he made more than 1,000 daily radio broadcasts, to keep himself in the public mind. He turned down television offers, on the other hand, telling friends, 'People will tire of me' (Ronald Reagan, quoted in Howard Kurtz, *Hot Air*, New York: Basic Books, 1997).

78 Atkinson, op cit.

79 Vicki Woods, 'Notebook,' *Daily Telegraph*, 18.12.01.

80 Margaret Thatcher, *The Great Speeches*, op cit.

81 Interview with Tim Bell, 19.02.03.

82 ibid.

83 ibid.

84 She disliked the Blackpool Winter Gardens especially (email to the author from Chris Collins, the Thatcher Foundation, 29.1.04).

85 Wapshott and Brock, quoted in Atkinson, op cit, p.115.

86 ibid.

87 '[For public meetings] we put pause lines in her speeches. We didn't put applause lines in – she'd get petrified if they didn't applaud' (interview with Tim Bell, 19.02.03). Yet Margaret Thatcher wasn't much given to self-doubt – self-assurance coursed through her delivery. She was good with hecklers, having learned to use the power of her office with them in an almost pantomime way. 'They've come to listen to me, not you' (Margaret Thatcher, quoted by Bell).

88 Interview with Tim Bell, 19.02.03.

89 Her statement on devolution to the House of Commons on 13 January 1976 is marked, 'Keep voice low & relaxed. Don't go too slow' (email to the author from Andrew Riley, Archivist, Thatcher Papers, Churchill Archives Centre, Cambridge, 6.2.04).

90 Edward Pearce, obituary of Sir Gordon Reece, *Guardian*, 27.09.01.

91 Obituary of Sir Gordon Reece, *Daily Telegraph*, 25.09.01.

92 Pearce, op cit.

93 Caroline Henton, 'The Abnormality of Male Speech', p.46, in George Wolf, ed., *New Departures in Linguistics* (New York: Garland Publishing, 1982).

94 The broadcaster Jon Snow said that 'she had that absolute, crisp, decisive delivery that made you feel wrong before you'd even finished the question' (interview with Jon Snow, 3.3.04).

95 Quoted in David Crystal and Hilary Crystal, *Words on Words: Quotations about Language and Languages* (London: Penguin Books, 2001).

96 Geoffrey W. Beattie, 'Turn-taking and Interruption in Political Interviews: Margaret Thatcher and Jim Callaghan Compared and Contrasted', *Semiotica*, 39–1/2, 1982.

97 ibid.

98 During the 2001 general election campaign Tony Blair declared that his goal was 'not just to win your vote. It is to win your heart and mind' (Tony Blair quoted in Sarah Hall, 'Blair Preaches Lesson of Trust and Change', *Guardian*, 9 May 2001 – not something one can imagine Harold Macmillan ever saying). This recalled Princess Diana's desire to be 'the Queen of Hearts': if Diana was the People's Princess, Blair seemed to aspire to be the People's PM.

99 Sam Wallace, 'Hague's Voice is Top Choice with Voters', *Daily Telegraph*, 12.10.99.

100 Email to Simon Mayo, BBC Radio 5 Live, 12.11.03.

101 Simon Hart, quoted in Donald Hiscock, 'The Apathy Generation', *Guardian*, 9.5.01.

102 The modern politician, wrote the playwright Arthur Miller, must learn to be an actor: 'The single most important characteristic a politician needs to display is relaxed sincerity' (Arthur Miller, 'The Final Act of Politics', *Guardian*, 21.7.01).

103 Richard Sennett, *The Fall of the Public Man* (London: Faber and Faber, 1986).

104 ibid, pp.262, 263, 265.

105 Larry Moss, cited in Dave Denison, 'Role of a Lifetime: Two Top Acting Coaches on How to Play the President', *Boston Globe*, 12.09.04.

106 Geoffrey Nunberg, quoted in Peter S. Canellos, 'In Kerry Speaking Style, Something Presidential', *Boston Globe*, 23.03.04.

107 ibid.

108 Don Aucoin, 'Kerry's Oratory Style Needs Work', *Boston Globe*, 25.03.04.

109 Renee Grant-Williams, quoted in Liza Porteus, 'Candidates Say It With Style', 6.10.04, www.foxnews.com.

110 Quoted in Porteus, op cit.

111 Arthur Miller, 'On Politics and the Art of Acting', 30th Jefferson Lecture in the Humanities, 26.03.01, www.neh.gov.

112 Denison, op cit.

113 Peggy Noonan, 'Will the Real John Kerry Please Stand Up?', *Wall Street Journal*, 22.07.04.

# 15. How Technology Has Transformed the Voice

1 Barbara Engh, 'Adorno and the Sirens: tele-phono-graphic bodies', in Leslie C. Dunn and Nancy A. Jones, eds., *Embodied Voices* (Cambridge: Cambridge University Press, 1994).

2 Henry M. Boettinger, 'Our Sixth-and-a-Half Sense', p.205, in Ithiel de Sola Pool, ed., *The Social Impact of the Telephone* (Cambridge, Mass: The MIT Press, 1977).

3 Asa Briggs, 'The Pleasure Telephone: A Chapter in the Prehistory of the Media', in ibid.

4 John Brooks, 'The First and Only Century of Telephone Literature', in ibid.

5 Marshall McLuhan, *Understanding Media*, p.283 (London: Sphere Books, 1967).

6 Boettinger, op cit.

7 Sidney H. Aronson, 'Bell's Electrical Toy: What's the Use? The Sociology of Early Telephone Usage', in de Sola Pool, op cit.

8 Sir William Preece, 1879, quoted in Ithiel de Sola Pool et al, 'Foresight and Hindsight: The Case of the Telephone', p.128, in ibid.

9 'A Reporter's Visit to the Boston Telephone Exchange', *Scientific American*, 12.02.1887, quoted in Engh, op cit.

10 ibid.

11 Brenda Maddox, 'Women and the Switchboard', in de Sola Pool, op cit.

12 C.E. McCluer, 1902, quoted in Lana F. Rakow, 'Women and the Telephone: The Gendering of a Communications Technology', pp.214–5, in Cheris Kramarae, *Technology and Women's Voice* (London: Routledge, 1988).

13 Ann Moyal, 'The Gendered Use of the Telephone: An Australian Case Study', *Media, Culture, and Society*, 1992:14.

14 Stuart Millar, 'Handheld PC Bridges Digital Divide', *Guardian*, 9.07.01.

15 Alan H. Wurtzel and Colin Turner, 'Latent Functions of the Telephone: What Missing the Extension Means', in de Sola Pool, op cit, p.256.

16 Bell quoted in Aronson, op cit, p.22.

17 'The Phonograph', *New York Times*, 7.11.1877.

18 *Harper's Weekly*, 30.03.1878.

19 Edward Johnson, 'A Wonderful Invention – Speech Capable of Indefinite Repeition from Automatic Records', *Scientific American*, 17.11.1877.

20 Gillian Beer, 'Wireless: Popular Physics, Radio and Modernism', in Francis Spufford and Jenny Uglow, eds., *Cultural Babbage: Technology, Time and Invention* (London: Faber and Faber, 1996).

21 'This machine bears a paradox: it identifies a voice, fixes the deceased (or mortal) person, registers the dead and thus perpetuates his living testimony, but also achieves his automatic reproduction *in absentia*: my self would live *without me* – horror of horrors!' (Charles Grivel, 'The Phonograph's Horned Mouth', in Douglas Kahn and Gregory Whitehead, eds., *Wireless Imagination: Sound, Radio, and the Avant-Garde*, Cambridge: The MIT Press, 1992, quoted in Michael Douglas Heumann, 'Ghost in the Machine: Sounds and Technology in Twentieth Century Literature', PhD dissertation, Faculty of English, University of California, Riverside, June 1998.) This has an interesting analysis of the role played by the phonograph in Bram Stoker's *Dracula*, a subject also discussed by John M. Picker in 'The Victorian Aura of the Recorded Voice', *New Literary History*, 32, 2001.

22 'The Talking Phonograph', reproduced from *Scientific American*, 22.12.1877, in *Engineering*, 18.01.1878. It can be found, along with many other fascinating original documents of the time, on Mike Penney's 'The Sound of a Voice' website, www.members.lycos.co.uk/MikePenney.

23 The Phonograph, *New York Times*, 7.11.1877.

24 Sousa quoted in McLuhan, op cit, p.293.

25 'The claim that sound reproduction has "alienated" the voice from the human body implies that the voice and the body existed in some prior holistic, unalienated, and self-present relation . . . But the idea of the body's phenomenological unity and sanctity gains power precisely at the moment in its history that the body is being taken apart, reconstituted, and problematized' (Jonathan Sterne, *The Audible Past: Cultural Origins of Sound Reproduction*, p.21, Durham, North Carolina: Duke University Press, 2003).

26 Amy Lawrence, *Echo and Narcissus: Women's Voices in Classical Hollywood Cinema* (Berkeley: University of California Press, 1991).

27 Sterne, op cit, p.8.

28 'Acoustic space modelled in the form of private property allow[ed] for the commodification of sound' (ibid, p.162).

29 John Belrose, 'Fessenden and Marconi: Their Differing Technologies and Trans-atlantic Experiments During the First Decade of this Century', paper given at an international conference on 100 Years of Radio, September 1995, www.anten-top.bel.ru.

30 Jeffrey Sconce, *Haunted Media: Electronic Presence from Telegraph to Television* (Durham, North Carolina: Duke University Press, 2000).

31 On the other hand Rudolf Arnheim, the Berlin-born psychologist and theorist of the arts, saw broadcasting as offering 'unity by aural means' (Rudolf Arnheim, *Radio*, p.135, London: Faber and Faber, 1936). Having fled Germany in the 1930s, radio in his utopian imagination would spread the voice of enlightenment and eliminate the boundaries between countries.

32 Kathy Newman, 'Radio-Activity: Reconsidering the History of Mass Culture in America', *Cultural Matters*, issue 2, Spring 2003; 'Changing Our Ways', www.culturalstudies.gmu,edu.

33 Susan J. Douglas, *Listening In*, p.7 (New York: Times Books, 1995).

34 Ray Lapica, quoted in Newman, op cit.

35 Arnheim, op cit, p.211.

36 Hilda Matheson, *Broadcasting*, pp.81–2 (London: Butterworth, 1933). For a discussion of the BBC radio talk in the 1930s, see Paddy Scannell and David Cardiff, *A Social History of British Broadcasting, vol. 1, 1922–1939: Serving the Nation* (Oxford: Basil Blackwell, 1991).

37 For a discussion of this voice's origins, see Frances Dyson, 'The Genealogy of the Radio Voice' in Daina Augaitis and Dan Lander, eds., *Radio Rethink* (Banff, Alberta: Walter Phillips Gallery, 1994). The fact that, until the war, BBC radio announcers were anonymous helped fortify this sense of disembodied oracle.

38 A not-too-subtle allusion to the show's producers, *Time* magazine. Van Voorhis was referred to as the 'Voice of Time', but came to be known as the 'Voice of God' (Sarah Kozloff, *Invisible Storytellers: Voice-Over Narration in American Film*, Berkeley: University of California Press, 1988).

39 Olive Shapley, *Broadcasting a Life*, p.46 (London: Scarlet Press, 1996).

40 For a discussion of ITMA's role in the war, see Barry Took, *Laughter in the Air* (London: Robson Books, 1976), and Derek Parker, *Radio: The Great Years*. (Newton Abbot, Devon: David and Charles, 1977).

41 The BBC established a news department only in 1934.

42 An extract from Morrison's report can be heard on www.eyewitnesstohistor-y.com. Morrison's tones remain so iconic of the breakdown of normal voice that their tremor and irregular breathing have even been measured by acoustic researchers (Carl E. Williams and Kenneth N. Stevens, 'Emotions and Speech: Some Acoustical Correlates', *Journal of the Acoustical Society of America*, vol. 52, no. 4, 1972).

43 Even though ironically this 'live' broadcast was actually a taped recording transmitted by NBC the day after the actual event (Edward D. Miller, '*Emergency Broadcasting and 1930s American Radio*, Philadelphia: Temple University Press, 2003).

44 Indeed, in an almost direct echo of Morrison, he exclaims, 'Ladies and gentle-men, this is the most terrifying thing I've ever witnessed' (Orson Welles, *War of the Worlds*, CBS, 30.10.38).

45 Although it was also an expression of America's economic instability and the coming war, *War of the Worlds* was transmitted just a month after the Munich crisis (Douglas, op cit).

46 ibid.

47 ibid.

48 Indeed one commentator has explained the panic over *War of the Worlds* as a reaction to broadcasting's omnipotent, omniscient presence: it was 'fascinating not so much as a story of the end of the world, but as a story of the end of the *media*. Listening to the simulated live dismantling of the network voice and the authority it implied, the "War of the Worlds" audience heard the empire of the air collapse into rubble . . . an experience both terrifying and exhilarating' (Sconce, p.16, op cit). It was as if radio, newly established font of facts, had itself been destroyed.

49 Before Pickles' first broadcast the press dusted down all its stereotypes of northerners, wondering if he'd say, 'Here is the news and ee bah gum this is Wilfred Pickles reading it' (Wilfred Pickles, *Between You and Me*, pp.136–7, London: Werner Laurie, 1949). In the event, he rounded off one midnight news bulletin with the words, 'Good night to you all – and to all northerners wherever you may be, good neet!' (ibid, p.142).

50 Although letters of abuse poured into the BBC, door-to-door interviews conducted by its Listener Research department discovered that he was more popular in the South than the North. Northerners, Pickles speculated, were probably thinking, Here we are 'avin' got on and sent t'childer to a good school to teach 'em to speak proper an' then they put a fellow on t'wireless talkin' like this (ibid, p.146).

51 Anne Karpf, *Doctoring The Media: The Reporting of Health and Medicine* (London: Routledge, 1988).

52 See Douglas for a description.

53 *Seven Ages of the Voice*, BBC World Service, 2.5.98.

54 In the same radio interview he admitted, 'When I left school, I had one of those terrifically high-pitched middle-class voices, sounding like a minor member of the Royal Family . . . [even though I spent] two years living in Dallas, Texas . . . And when I came back to Britain in 1967 I tried to get rid of that accent and thought I had been fairly successful, but recordings of me from around that time were still that kind of middle-class drawly voice' (ibid).

55 Umberto Eco, 'Independent Radio in Italy: Cultural and Ideological Diversification', *Cultures* vol. 5, no. 1, 1978.

56 Brian Hayes, quoted in Martin Shingler and Cindy Wieringa, *On Air*, p.113 (London: Arnold, 1998).

57 *Please Believe Us*, BBC Radio 3, 27.04.97. Even Miriam Margolyes, voice-over actress extraordinaire, has confessed that she doesn't get asked to do so many any more 'because my voice is of another generation. I believe in the value of the vowel and consonant' (Gerald Jacobs, 'Roles' Voice', *Jewish Chronicle*, 15.07.05)

58 Interview with Jon Snow, 3.3.04. The whole subject of prestige accents, prejudice and discrimination against less-favoured ones, and accent change is fascinating but beyond the scope of this book. For an accessible and compelling account of accent discrimination in the US, see Rosina Lippi-Green's brilliant *English With an Accent: Language, Ideology, and Discrimination* (London: Routledge, 1997). For the British experience, John Honey's *Does Accent Matter?* (London: Faber, 1989) is an excellent starting-point.

59 Interview with Dan Rather, 18.7.03.

60 See Peter Lewis and Corinne Pearlman, *Media and Power: From Marconi to Murdoch* (London: Camden Press, 1986) for a literally graphic guide to this phenomenon.

61 Lorenzo W. Milam, *Sex and Broadcasting* (California: Dildo Press, 1975).

62 Not everyone was so enthusiastic about the addition of sound to pictures. The

British writer Aldous Huxley confessed that the warbling of 'Mammy' at the film's end – 'those sodden words, that greasy sagging melody' – had made his flesh creep (quoted in Scott Eyman, *The Speed of Sound*, p.14, Baltimore: John Hopkins University Press, 1999). The director Josef von Sternberg dismissed talkies as 'a visual skeleton clattering with voices' (ibid, p.270), and the psychologist Rudolf Arnheim claimed that sound ruined film because it would be hard 'to rustle up enough fodder for the speaking machines' (Rudolf Arnheim, 'Sound Film', p.30, in Arnheim, *Film Essays and Criticism*, Madison: University of Wisconsin Press, 1997). He also argued that 'the libretti of most talkies apparently are of an utterly unbearable quality' (ibid, p.270), and believed that the first time that the actor opened his or her mouth and spoke through an amplified voice, 'film art abdicated its good old place back to the peep-show'.

63  ibid, p.182.

64  ibid.

65  ibid, p.181.

66  Edward Bernds, *Mr Bernds Goes to Hollywood: My Early Life and Career in Sound Recording at Columbia with Frank Capra and Others* (Lanham, Maryland: Scarecrow Press, 1999).

67  Few actors were given direction about what or how to speak in voice tests, other than told to say anything that came into their heads, which sometimes turned out to be nursery rhymes or non sequitur phrases. They had no idea whether to talk softly or loudly, or whether the talkies resembled the stage or were another phenomenon altogether. One production person later rued the fact that, with Clara Bow, 'We didn't train her; we should've taken six months to send her to school, teach her how to speak. [But] we were bringing in all these Broadway actors and directors' (Eyman, op cit, pp.182–4).

68  The *New York Herald Tribune* reported, 'Her voice is revealed as a deep, husky contralto that possesses every bit of that fabulous poetic glamour that has made this distant Swedish lady the outstanding actress of the motion picture world.' (Margarita Landazuri, 'Anna Christie', www.turnerclassicmovies.com).

69  Eyman, op cit, p.261. Moss Hart and George S. Kaufman's 1930 comedy, *Once in a Lifetime*, centres round a voice school for silent screen stars, set up following the advent of the talkies.

70  Even Garbo suffered from this: from her chiselled beauty, no one had imagined that her voice would be so husky, and some even tried to attribute this to microphone distortion (Michel Chion, *The Voice in Cinema*, New York: Columbia University Press, 1999).

71  ibid, p.125.

72  Eyman, op cit, p.301. By the 1930s voice had become an important element in casting. When Gregory La Cava was casting *Stage Door* in 1937, he and his female scriptwriter hung around the cafeteria listening to the voices of the RKO starlets. 'Try to get a voice', he told her, meaning a distinctive one, with individual tone and colour (Elizabeth Kendall, quoted in Sarah Kozloff, *Overhearing Film Dialogue* (Berkeley: University of California Press, 2000).

73  Rick Altman, Introduction, in Rick Altman, ed., 'Cinema/Sound', *Yale French Studies*, no. 60, 1980.

74  For a discussion of the silencing of women's voices in film, see Lawrence, op cit, and Kaja Silverman, *The Acoustic Mirror: The Female Voice in Psychoanalysis and Hollywood* (Bloomington: Indiana University Press, 1988).

75  Altman, op cit.

76  See, for instance, Christian Metz, 'Aural Objects', in Altman, op cit.

77 Alan Williams, 'Is Sound Recording Like a Language?', in ibid.

78 Chion, op cit.

79 Altman, op cit, p.6.

80 ibid.

81 See Rick Altman, '24-Track Narrative? Robert Altman's *Nashville*, *Cinema(s)*, vol. 1, no. 3, Spring 1991.

82 Interview with Charles Michel, 28.05.03.

83 Achieved with the aid of cotton wool stuffed into his mouth and designed to convey the impression, according to the notes he scrawled on his script, of 'Through the nose, high voice, nose broken early in youth to account for his difficulties' ('Godfather Script Sold for Record Price', *Guardian*, 1.07.05).

84 According to the director, McCambridge worked for three weeks on the voice. 'She was chain-smoking, swallowing raw eggs, getting me to tie her to a chair – all these painful things just to produce the sound of that demon in torment. And as she did it, the most curious things would happen in her throat. Double and triple sounds would emerge at once, wheezing sounds, very much akin to what you can imagine a person inhabited by demons would sound like. It was pure inspiration' (William Friedkin quoted in Ronald Bergen, obituary of Mercedes McCambridge, *Guardian*, 19.03.04).

85 Mary Ann Douane, 'The Voice in the Cinema: The Articulation of Body and Space', in Altman, 1980, op cit.

86 ibid. See also Chion, op cit.

87 Laura Mulvey, 'Cinema, Sync Sound and Europe 1929: Reflections on Coincidence', in Larry Sider et al, eds., *Soundscape*, p.20 (London: Wallflower Press, 2003).

88 Peter Wollen, 'Mismatches of Sound and Image', in Sider, op cit.

89 Jeff Matthews, 'Hey, You Sound Just Like Marlon Brando, Robert Redford, and Paul Newman', www.faculty.ed.umuc.edu.

90 Agnieszka Szarkowska, 'The Power of Film Translation', *Translation Journal*, vol. 9, no. 2, April 2003.

91 *Masterpiece: The Voice*, BBC World Service, 3.05.05.

92 Had television been dominant in Hitler's time, according to Marshall McLuhan, he'd either have vanished quickly or never risen to power because television is a 'cool medium' that rejects 'hot figures' (Marshall McLuhan, *Understanding Media*, London: Sphere Books, 1967).

93 Erika Tyner Allen, 'The Kennedy–Nixon Presidential Debates, 1960', The Museum of Broadcast Communications, www.museum.tv/archives.

94 A charge refuted by J.K. Chambers, 'TV Makes People Sound the Same', in Laurie Bauer and Peter Trudgill, *Language Myths* (London: Penguin, 1998).

95 In Britain at least, they're much louder than the surrounding programmes, perhaps contributing to the increasing national volume.

96 Seth Stevenson, 'The Voice-Over Gets a Make-Over', www.slate.msn.com, 28.03.05.

97 See, for instance, Rana Foroohar, 'Signal Lost', *Newsweek*, 24.01.05.

98 In one study undergraduates, tutored by computers that had been given distinct voices, responded 'as if the voices represented distinct selves. Thus, we demonstrate that users can be induced to *behave as if* computers were human, even though users know that the machines *do not* actually possess "selves" or human motivations' (Clifford Nass et al, 'Anthropomorphism, Agency, and Ethopoeia: Computers as Social Actors', SRCT Paper no. 105, presented at the INTERCHI '93 conference, Amsterdam, The Netherlands, April 1993). Other experiments from the same team found that people were willing to disclose all sorts of

personal information to computer voices, especially ones that they felt they were unlikely to encounter again (Kathleen O'Toole, 'Computers With Voices: Students Explore How People Respond', Stanford University News Service, 27.07.00www.stanford.edu).

99 Peter Norvig's splendid 'The Gettysburg PowerPoint Presentation' (19.11.1863) reconfigures the Gettysburg address into what it would have been if Abraham Lincoln had had access to PowerPoint. Slide 4 (of 6), which reviews Key Objectives and Critical Success Factors, reads:
  − What makes nation unique
  − Conceived in Liberty
  − Men are equal
  − Shared vision
  − New birth of freedom
  − Gov't of/for/by the people (www.norvig.com)
100 Sherry Turkle, quoted in ibid.


# 16. Voiceprints and Voice Theft

1 Harry Hollien, *Forensic Voice Identification*, p.19 (San Diego: Academic Press, 2001).

2 Some judges, however, have proposed a counter-argument – namely that a woman in such a terrifying situation couldn't be expected to recall the subtleties in her assailant's voice that would allow her to identify him purely on the basis of auditory memory (Harry Hollien, *The Acoustics of Crime*, New York: Plenum Press, 1990).

3 'Every language universally legible, exactly as spoken. Accomplished by means of self-interpreting symbols, based on a discovery of the exact physiological relations between sounds' (Alexander Melville Bell, *Visible Speech*, London: 1864).

4 Hollien, 1990, op cit.

5 The passage of time, however, has done little to blunt the controversies surrounding the Lindbergh case – a website, the Lindbergh Kidnapping Hoax, is still animatedly debating it (see note 2 in David Ormerod, 'Sounds Familiar? – Voice Identification Evidence', *Criminal Law Review*, pp.595–622, August 2001).

6 The rule of thumb is someone you've heard fairly regularly for about two years. The evidence also shows that we're much more accurate when recognising the voices of people our own age, and that children and old people are poorer identifiers (Hollien 2001, op cit). The distinction here is between voice recognition, where an offender's voice is already known to a witness, and voice identification, in which a witness tries to pick out a voice they hadn't heard before the offence took place (John Wilding et al, 'Sound familiar?', *Psychologist*, vol. 13, no. 11, November 2000).

7 Ormerod, op cit.

8 ibid. More than two-thirds of witnesses choose innocent voices from voice parades from which the suspect is absent (Yarmey, cited in ibid). Anthony Barron ('Speaker Identification by Earwitness: A Bigger Picture', Department of Linguistics and Phonetics, University of Leeds, May 2001) has argued that if witnesses aren't required to recognise or identify a voice but describe it, the problems are compounded – because of the lack of a shared vocabulary of voice discussed in chapter 1. In one study, neither speakers nor their friends agreed with expert and lay judges' assessments of loudness, pitch, or resonance (Scherer,

1974, cited in Ormerod, op cit). As a result, courts are reduced to techniques like asking a witness to compare a suspect's voice with their own or with that of a famous person. For the time being English law still hasn't decided whether to adopt voice parades, and in what form. An exhaustive survey of voice evidence concluded that 'it is now well recognised to be extremely unlikely that people can identify with any accuracy voices of strangers which they have heard only once before for a short period' (Ormerod, op cit).

9 Yarmey cited in Ormerod, op cit. A recent study has found that voice witnesses, at their most confident, are 'catastrophically' worse than eyewitnesses (Olson, Juslin and Winman, cited in ibid).

10 Hollien, 2001, op cit.

11 Hollien, 2001, op cit.

12 ibid.

13 Alexander Solzhenitsyn, *The First Circle*, p. 33 (London: Fontana Books, 1970).

14 Clifford Irving had secured a lucrative publishing deal with McGraw-Hill to turn in what he claimed was Hughes's ghosted autobiography, but which Hughes in the call pronounced a blatant fake. NBC hired Lawrence Kersta to make a voice-print analysis. Kersta, comparing pitch, tone, and volume in the call with those in a recording of Hughes from 1947, concluded that the phone-caller was indeed the authentic Hughes. Irving was arrested, convicted of forgery, obliged to reimburse the publisher and sent to jail.

15 The opposite argument, though, has also been advanced: that 'all visual projections of sounds are arbitrary and fictitious', and that the more that sound has been pictorialised, the worse our listening skills have become. Many modern acousticians, goes this argument, have exchanged an ear for an eye, and simply read sound from sight, although R. Murray Schafer (*The Soundscape: Our Sonic Environment and the Tuning of the World*, p. 127, Toronto: McClelland and Stewart, 1977), was tilting more at engineers and audiologists than phoneticians.

16 Hollien, 2001, op cit.

17 For these reasons the term spectrographic voice recognition is preferred to voiceprint: though it hardly trips off the tongue, and is far less likely to capture the popular imagination, it's much more accurate.

18 Hollien, 1990, op cit, p.10.

19 Judges, though, are expected to give the usual cautionary advice required of them in identification cases, particularly if prosecution is based solely on a witness's identification of a suspect's voice, or if there's any risk that voice evidence might jeopardise fairness and lead to an 'unsafe conviction'. See Ormerod, op cit, for a thorough discussion of the problems associated with voice identification evidence.

20 Bizarrely, British juries may sometimes be called upon to become speaker-recognition experts themselves: they can be invited to listen to a recording and then compare it with another voice.

21 Other methods being developed or already in use include iris scanning, signature analysis, even body-odour matching. Biometrics is one of the oldest forms of identification – the Egyptians identified people by their scars, complexion, eye colour, and height – but speaker verification developed originally in the early 1980s to support communications and intelligence projects funded by the US government (Steven F. Boll, 'Testimony before the Subcommittee on Domestic and Internal Monetary Policy, Commerce, Banking and Financial Services', House of Representatives, Washington, DC, May 20 1998).

22 Richard Mammone, 'Your Voiceprint Will Be Your Key', *Speech Technology Magazine*, Jan/Feb 1998 (www.speechtechmag.com).

23 www.authentiz.com.
24 Clive Summerfield, 'The Future Is Hear: Securing Government Services Using Speaker Verification', VeCommerce Ltd (www.biometricsinstitute.org).
25 And 768 million ATM transactions annually require authentication (ibid).
26 Mammone, op cit.
27 It works like this. A customer pre-records, either over the phone or in person, a two to six second example of their voice, by speaking their name or a predetermined phrase. If the background noise is excessive, they may be told, 'Please move away from the dishwasher' (Orla O'Sullivan, 'Biometrics comes to life', *ABA Banking Journal*, January 1997, www.banking.com). This is then digitally encoded. When they call in, the sound of their voice, cleaned up, is set beside the pre-recorded sample. By checking the formants or dominant frequencies over segments of sound in the caller's voice and comparing them with all the other samples in the database, those with differing elements can be eliminated, leaving the matched one. A voice can be accepted or rejected in half a second, perhaps even one-tenth of a second.
28 Summerfield, op cit.
29 O'Sullivan, op cit.
30 'Passwords hold key to IT Help Desk problems', 14.1.03 (www.axiossytems.-com).
31 Sheryl P Simons, 'Voice Recognition Market Trends', Faulkner Information Services 2002 (www.stanford.edu).
32 'Voice Authentication: Datamonitor Consumer Survey Results (www.vocent.-com).
33 'Large Scale Evaluation of Automatic Speaker Verification Technology', The Centre for Communication Interface Research, University of Edinburgh, May 2000.
34 'At the present time,' according to one group of researchers, 'there is no scientific process that enables us to uniquely characterise a person's voice or to identify with absolute certainty an individual from his or her voice . . . Given the current state of knowledge, there are no methods, either automatic or based on human experience, that enable one to state with certainty that a person is (or is not) the speaker in a particular recording' (Jean-François Bonastre et al, 'Person Authentication by Voice: A Need for Caution', *Proceedings of the 8th European Conference on Speech Communication and Technology, Geneva*, Switzerland: Eurospeech, 2003).
35 'Large Scale Evaluation of Automatic Speaker Verification Technology', op cit.
36 Mizuko Ito, 'Mobile Phones, Japanese Youth, and the Replacement of Social Contact', www.itofisher.com.
37 Denis Campbell, 'Mobiles, DVDs and MP3s Send Under Eights to Techno Heaven', *Observer*, 13.02.05.
38 Ling and Yttri, quoted in Ito, op cit.
39 McLuhan, op cit.
40 Sadie Plant, 'On the mobile', 2002, www.receiver.vodaphone.com.
41 ibid.
42 See Jeannette Hyde, 'Hello! I'm on the beach', *Observer*, 6.05.01.
43 Ito, op cit. There's even a website, www.cellmanners.com, trying to develop courtesy in mobile-phone use.
44 Rachel Cusk, 'Finding Words', *Guardian*, 14.09.01.
45 James Meek, 'Hi, I'm in G2', *Guardian*, 11.11.02.
46 Interview with Jeremy Green, 16.07.04.
47 Meek, op cit.

48 Rebound Voice Verification Service (www.reboundecd.com).

49 Judith Markowitz, 'Speaker Verification for Community Release', *Speech Technology Magazine*, Nov/Dec 2003 (www.speechtechmag.com).

50 Each morning a judge gets a full printed readout of calls (admissible in court). Says one, 'If there was a violation I know the type of violation it was: no answer, hang up, unrecognised voice . . . Usually we get little violations at the beginning. It's kind of a training process. Once you train them how to use the system, you'll see "voiceprint successful" over and over. With the bad ones you still see "No answer, no answer, no answer, hang up, hang up, no answer". Not every child is a candidate for voice' (ibid).

51 Michel Foucault, *Discipline and Punish*, p.206 (New York: Vintage Books, 1979).

52 Indeed Bentham also envisaged, as a crucial part of the Panopticon, a vast system of eavesdropping through speaking tubes (Leigh Eric Schmidt, 'Hearing Loss', in Michael Bull and Les Back, eds., *The Auditory Culture Reader*, Oxford: Berg, 2003).

53 Robert Winnett and David Leppard, 'Yobs face losing their mobile phones', *Sunday Times*, 21.3.04. Since many of the young people who'd be penalised in this way are the same ones who mug other young people for their mobiles, one can't help thinking that this proposal would simply increase the number of mobile thefts.

54 This automatically synchronises our voice with the lip movements of the character we've chosen to play, stretching or shrinking the recording of our voice to match them without affecting our pitch (Barry Fox, 'Dub Your Own Voice to *Shrek* Characters', *New Scientist*, 10.12.01, www.newscientist.com).

55 You can hear it on www.ai.mit.edu.

56 Yudhijit Bhattacharjee, 'Making Robots More Like Us', *New York Times*, 06.03.03.

57 It leaves the recipient with a voice that's neither their own nor that of the donor (James Meek, 'Voice Box Transplants May Restore Lost Speech Within Decade', *Guardian*, 16.06.01; Anthony Browne, 'New Op Gives Living Voice of the Dead', *Observer*, 26.11.00).

58 Some argue that these produce synthetic voices that no longer resemble Stephen Hawking's but, on the contrary, sound increasingly natural since they've been pumped full of the acoustic of human emotion (Iain R. Murray et al, 'Emotional Stress in Synthetic Speech: Progress and Future Directions', *Speech Communication*, 20, 1996; Clifford Nass et al, 'The Effects of Emotion of Voice in Synthesized and Recorded Speech', Proceedings of the AAAI Symposium Emotional and Intelligent II: The Tangled Knot of Social Cognition, North Falmouth, Massachusetts, 2001; Iain R. Murray and John L. Arnott, 'Synthesizing Emotions in Speech: Is it Time to Get Excited?', Proceedings of ICSLP 96 the 4th International Conference on Spoken Language Processing, Philadelphia, PA, 1996. Others argue that, even after twenty years of speech synthesis, you still know that you're listening to a computer (David Howard, PEVOC 6, Royal Academy of Music, London, 2.09.05).

59 Marc Moens, quoted in www.pentechvc.com.

60 From audio artists like Disinformation, Scanner (Anne Karpf, 'Scanner in the Works', *Guardian*, 15.09.99), Hildegard Westerkamp (www.sfu.ca//westerka/bio.html), David Toop (David Toop, *Haunted Weather*, London: Serpent's Tail, 2004). Bruce Naumann's 2004 sound installation, 'Raw Materials', in the Turbine Hall at Tate Modern, enveloped visitors in cajoling, caressing, and melancholic voices. Gregory Whitehead's work uses the voice and is also about

it: pieces like 'Principia Schizophonica' engage playfully with the unearthly delights of the voice in the era of electronic media (on 'The Pleasure of Ruins', 1993, released by Staalplaat. See also Gregory Whitehead, 'Radio Play Is No Place', *Drama Review*, 40, 3 (T151), Fall 1996, and 'Who's There? Notes on the Materiality of Radio', *Art & Text*, December–February, 1989).

61 John Arlidge, 'Tyranny@work' (*Observer*, 24 February 2002).

62 Lucy Ward, 'Email Could Replace Talks with Teacher' (*Guardian*, 10 December 2001).

63 'Y TEXTING MAYBE BAD 4 U', www.textually.org, 8.3.04.

64 Will Woodward, 'Parents Not Preparing Children for School, Ofsted Head Warns' (*Guardian*, August 1 2003).

65 Rebecca Smithers, 'Teachers to Work on Pupils Lost for Words' (*Guardian*, 13 November 2003).

66 Oliver Pritchett quoted in 'Grunting', *Guardian*, 18.1.02.

67 Smithers, op cit.

68 BBC TV 6 o'clock news, 17 November 2003.

69 John L. Locke, *The De-Voicing of Society* (New York: Simon and Schuster, 1998).

70 Even the inspiring social analyst Ivan Illich evoked an egalitarian utopia when he recalled arriving in 1926 on the Dalmatian island of Brac at the same time as the first loudspeaker. 'Up to that day, all men and women had spoken with more or less equally powerful voices. Henceforth this would change. Henceforth the access to the microphone would determine whose voice shall be magnified. Silence now ceased to be in the commons; it became a resource for which loudspeakers compete . . . the encroachment of the loudspeaker has destroyed that silence which so far had given each man and woman his or her proper and equal voice. Unless you have access to a loudspeaker, you now are silenced. Ivan Illich, 'Silence is a Commons' (www.preservenet.com/theory/Illich.html).

71 Kathleen Edgerton Kendall, 'Do Real People Ever Give Speeches?' (*The Central States Speech Journal*, Fall vol.xxv, no.3, 1974.

72 Daniel Goleman, *Emotional Intelligence* (London: Bloomsbury, 1996).

73 Richard Sennett, *The Fall of Public Man* (London: Faber & Faber, 1986).

74 Nancy Cartwright, *My Life as a 10-Year-Old Boy*, p.175 (London: Bloomsbury, 2000).

75 Duncan Campbell, 'Homer Banned from Public Speaking', *Guardian*, July 2002.

76 Nick Campbell, 'Ask the Scientists: Alan 2.0', *Scientific American Frontiers* archives, PBS, Fall 1990 to Spring 2000 (www.pbs.org/safarchives). A new voice-processing system can help a voice 'evolve' to sound the way the speaker wants it to through a microphone – less weedy, say, or more joyful, calm, or even manly. A genetic algorithm analyses the voice signal to work out which aspects of it need to be enhanced or suppressed to produce a better sound (Ian Sample, 'Genetic Algorithm Tunes up Public Speakers', *New Scientist*, 17.07.02).

77 Vocaloid, Yamaha's new software, has synthesised the human singing voice. A male soul voice called Leon, and a female soul voice called Lola, have been generated after recording two real singers for a week and then processing the sounds into a database of snippets, which can then be reassembled in a different form and the pitch altered to fit a new tune. Leon and Lola, it hardly needs saying, will cost a lot less to hire than their real-life counterparts.

78 Will Knight, 'Computer Program Raises Possibility of Voice Theft', *New Scientist*, 13.8.01.

79 Sigmund Freud, 'The Uncanny', in *Standard Edition, vol. XVII*, trans. James Strachey (London: Hogarth Press, 1955).

80 For an invigorating discussion of ventriloquism, see Steven Connor, *Dumb-struck: A Cultural History of Ventriloquism* (Oxford: Oxford University Press, 2000).

## 17. How People and Corporations are Trying to Change the Voice

1 'How Does Chanting Improve Our Health and Wellbeing?', www.russilpaul.com; Vandana Mohata, 'Mantras: An Interview with Jonathan Goldman', www.healingsounds.com.

2 Jill Purce, pioneer of overtone or harmonic singing or chanting, a version of Tibetan and Mongolian monks' traditional throat singing, calls it 'sonic massage', or 'musical medicine' (Jill Purce, 'The Healing Voice' flyer, 2005, and www.healingvoice.com). Tibetan Buddhists explain its origin through the story of the Lama Je Tzong Sherab Senge who, one day in 1433, awoke from a dream in which he heard two voices – one deep as the growling of a wild bull, and the other high, pure, and sweet like a child's. Both were combined into one voice and emanated from him. The dream instructed him to take this 'one voice chord' and use it for tantric chanting, uniting male and female aspects of divine energy.

3 A more novel form of sound healing has recently emerged – the reading aloud of Homer's hexameters. Used by many Ancient Greek epic poems, the hexameter has six metres to the line: according to a German study, this helps synchronise the heart and breathing, slow the breathing down, and raise oxygen in the blood. Dirk Cysarz et al, 'Oscillations of Heart Rate and Respiration Synchronize During Poetry Recitation', *American Journal of Physiology – Heart and Circulatory Physiology*, 287, 2004. *The Iliad*, in other words, may be good for the heart.

4 Daniel R. Boone, *Is Your Voice Telling on You? How to Find and Use Your Natural Voice*, p.7 (San Diego: Singular, 1997).

5 One of the best-known books of this kind for actors is called *Freeing the Natural Voice* by Kristin Linklater, first published in 1976 (Quite Specific Media Group, 1998).

6 See, for instance, Paul Newham, *Therapeutic Voicework*, (London: Jessica Kingsley, 1998), a broad survey of ideas about the voice whose author is a leading practitioner of therapy through voicework.

7 Eugene Rontal et al, 'Vocal Cord Dysfunction – An Industrial Health Hazard', *Annals of Otology, Rhinology, and Laryngology*, 88: 1979, Ann-Christine Ohlsson et al, 'Vocal Behaviour in Welders A Preliminary Study', *Folia Phoniatrica et Logopaedica*, 39, 1987; Joanne Long et al, 'Voice Problems and Risk Factors Among Aerobics Instructors', *Journal of Voice*, vol. 12, no. 1. For the side-effects of medication like inhaled corticosteroids see Eva Ihre et al, 'Voice Problems as Side Effects of Inhaled Corticosteroids in Asthma Patients – a Prevalence Study', *Journal of Voice*, 18(3), September, 2004.

8 Heavier workloads brought about by the national curriculum have aggravated the problem (Stephanie Martin and Lyn Darnley, *The Teaching Voice*, London: Whurr, 2004). The idea that the demands of a job can impinge upon the voice isn't a modern one, however. In 1897 Mrs Emil Behnke wrote that 'there are two sections of professional voice users in whom the effects of want of training for the physical side of their work are increasingly apparent – namely, the clergy and school teachers. In many of these cases vocal power becomes greatly diminished

and its quality injured by wrong voice use, even when the speaker is not incapacitated from all work by entire breakdown' (Mrs Emil Behnke, *The Speaking Voice: Its Development and Preservation*, p.3, London: J. Curwen & Sons, 1897)).

9  'Not training teachers about voice skills would be like training a surgeon how to do an operation without explaining about the tools or instruments they have to use' (James Williams, quoted in Janet Murray, 'A Quiet Word of Advice', *Guardian*, 13.01.04).

10  The Voice Care Network UK, www.voicecare.org.uk.

11  M. Robin DiMatteo, 'Nonverbal Skill and the Physician–Patient Relationship', in Robert Rosenthal, ed., *Skill in Nonverbal Communication: Individual Differences* (Cambridge, Mass: Oelgeschlager, Gunn and Hain, 1979). See also Lesley Fallowfield et al, 'Efficacy of a Cancer Research UK Communication Skills Training Model for Oncologists: a Randomised Controlled Trial', *Lancet*, 359, 2002; Adam Jones, 'Hearing the worst', *Guardian*, 25.04.01; Jo Carlowe, 'And the Good News Is', *Observer*, 21.07.02; and Jo Revill, 'How Doctors Deliver the Curt Words that Mean Life or Death', *Observer*, 10.11.02.

12  Sir Donald Irvine, quoted in Sarah Boseley, 'Doctors Failing 3m Patients', *Guardian*, 18.12.04.

13  DiMatteo, op cit.

14  Nalini Ambady, 'Surgeons' Tone of Voice', *Surgery*, 132, 2002. 'If physicians improve their communication skills,' said an attorney elsewhere, 'they can reduce their legal risk' (Heidi Foster, quoted in Judith Kapuscinski, 'Who Sues, Who Gets Sued and Why?' *Insights into Risk Management*, vol. 3, no.4, Fall 2002).

15  Chris Arnot, 'Let's Put Theory into Practice', *Guardian*, BT School Awards supplement, 27.09.05.

16  Sue Horner, quoted in Stephen Hoare, 'It Takes Two to Communicate', *Guardian* BT Schools Award Supplement, 27.09.05.

17  ibid.

18  Standford Gregory et al, 'Voice Pitch and Amplitude Convergence as a metric of Quality in Dyadic Interviews', *Language and Communications*, vol. 13, no. 3, 1993.

19  Howard Giles et al, 'Speech Accommodation Theory: The First Decade and Beyond', in M. McLaughlin, ed., *Communication Yearbook*, vol.10 (Newbury Park: Sage, 1987).

20  Howard Giles and Angie Williams, 'Accommodating Hypercorrection: A Communication Model', *Language and Communication*, vol. 12, no. 3/4, 1992.

21  ibid.

22  For example, a speaker talking at 50 words per minute can move to match exactly another speaker's rate of 100 words per minute (total convergence) or 75 words per minute (partial convergence) (Giles et al 1987, op cit).

23  Giles and Williams, op cit.

24  ibid, p.349.

25  Joseph N. Capella, 'Management of Conversational Interaction', in Mark L. Knapp and Gerald R. Miller, eds., *Handbook of Interpersonal Communication* (Newbury Park: Sage, 1994).

26  Stanley Feldstein et al, 'Gender and Speech Rate in the Perception of Competence and Social Attractiveness', *Journal of Social Psychology*, 141(6), 2001, although the ratings of competence were also influenced by the gender of both listeners and speakers.

27 Beatrice Beebe et al, 'Systems Models in Development and Psychoanalysis: The Case of Vocal Rhythm, Coordination, and Attachment', *Infant Mental Health Journal* vol. 21(1–2), 2000.

28 Norbert Freedman and Joan Lavender, 'Receiving the Patient's Transference: the Symbolising and Desymbolising of Counter-transference', *Journal of American Psychoanalytic Association*, vol. 45, no. 1, 1997. Schizophrenics and learning-disabled people, on the other hand, seem to display less convergence (Condon and Ogston, and Condon, cited in Judee K. Burgoon, *Interpersonal Adaptation: Dyadic Interaction Patterns*, Cambridge: Cambridge University Press, 1995).

29 Clifford Nass and Kwan Min Lee, 'Does Computer-Synthesised Speech Manifest Personality? Experimental Tests of Recognition, Similarity-Attraction, and Consistency-Attraction', *Journal of Experimental Psychology: Applied*, vol. 7, no. 3, September 2001.

30 Noriko Suzuki et al, 'Effects of Echoic Mimicry Using Hummed Sounds on Human-Computer Interaction', *Speech Communication*, 40 p.569, 2003.

31 Patrick Boylan, 'Accomodation Theory Revisited', paper given at Sietar European Congress, Brussels, 16.3.00. Converging works with words as well as voice. A waitress in a restaurant got larger tips when she repeated back the customer's order using exactly the same words than when she paraphrased it (van Baaren et al, cited in Rick B. van Baaren, 'Mimicry and Prosocial Behaviour', in *Psychological Science*, vol. 15, no. 1, January 2004).

32 Stanford W. Gregory Jr. et al, 'Verifying the Primacy of Voice Fundamental Frequency in Social Status Accomodation', *Language and Communication*, 21, 2001.

33 Berg, cited in Giles et al 1987. When interviewers speak louder or more softly, one study found, the interviewee changes their volume to match them (Michael Natale, 'Convergence of Mean Vocal Intensity in Dyadic Communication as a Function of Social Desirability', *Journal of Personality and Social Psychology*, vol. 32, no. 5, 1975). Another interviewer, by halving the length of time of his own questions, influenced the length of the interviewee's answers (Matarazzo et al, cited in Robert G. Harper et al, eds., *Nonverbal Communication: The State of the Art*, New York: John Wiley, 1978), while in a different experiment, the interviewee's interruption rate regularly matched that of the interviewer (Wiens et al, cited in Harper et al, op cit).

34 Stanford W. Gregory, Jr. and Stephen Webster, 'A Nonverbal Signal in Voices of Interview Partners Effectively Predicts Communication Accomodation and Social Status Perceptions', pp.1232, 1239, *Journal of Personality and Social Psychology*, vol. 70, no. 6, 1996. 'It is quite obvious that the lower-frequency signal of the human voice communicates much more than just pitch . . . the lower frequency of the voice communicates social status relations between partners.'

35 Stanford W. Gregory, Jr. and Timothy J. Gallagher, 'Spectral Analysis of Candidates' Nonverbal Vocal Communication: Predicting US Presidential Election Outcomes', *Social Psychology Quarterly*, vol. 65, no. 3, p.305, 2002. Acoustic analysis of the candidates' pitch produced 'a nonverbal, unconscious measure of social dominance . . . [which] may be communicated to observers of the debate, and the resulting perception of one candidate's social dominance over the other ultimately may be expressed through the observers' voting behaviour.' The single exception was 2000, when the more dominant speaker – Al Gore – won the popular vote but lost the election.

36 Boylan, op cit. So accommodation 'can include being the kind of foreigner that one's interlocutors wish one to be: that is, one can accommodate optimally by

accommodating minimally just as, in other cases, by accommodating to the hilt' (ibid).

37 Howard Giles and Peter Powesland, 'Accomodation Theory', in Nikolas Coupland and Adam Jaworski, eds., *Sociolinguistics: A Reader and Coursework* (London: Macmillan, 1997).

38 Interview, 31.10.03.

39 Mulac et al, cited in Giles et al, 1987.

40 Capella, op cit.

41 Leslie M. Beebe and Howard Giles, 'Speech Accommodation Theories: a Discussion in Terms of Second Language Acquisition', *International Journal of the Sociology of Language*, 46, 1984.

42 'Mimicry Makes Computers the User's Friend', *New Scientist*, 2.6.03.

43 Renee Grant-Williams, *Voice Power: Using Your Voice to Captivate, Persuade, and Command Attention*, p.106 (New York: Amacom American Management Association, 2002). She urges not just converging but also some canny diverging, reminding readers that 'you have a part to play in this, too. Use whatever you know about colour and variety to steer the duet in the direction you want to take. Your partner may soon be adjusting their voice to yours' (p.108).

44 Joseph O'Connor and John Seymour, *Introducing NLP*, p.20 (HarperCollins, 2002)

45 Davis, cited in Burgoon et al, 1995, op cit.

46 Interview, 28.5.03.

47 ibid.

48 Interview, 10.10.04.

49 Interview, 10.10.04.

50 Erving Goffman, *The Presentation of Self in Everyday Life*, Preface, p.15 (London: Pelican Books, 1971).

51 www.thewinningvoice.com.

52 Boyd Clarke, 'The Leader's Voice' (Select Books, 2002).

53 Rene Grant-Williams, *Voice Power: Using Your Voice to Captivate, Persuade, and Command Attention* (New York: Amacom American Management Association, 2002).

54 Khalid Aziz, *Presenting to Win* (London: Oak Tree Press, 2000).

55 See Rosanna Lippi-Green's coruscating attack on these in her perceptive analysis, *English with an Accent: Language, Ideology and Discrimination in the United States* (London: Routledge, 1997).

56 See 'Add a "Voice Lift" to Your Tummy Tuck', 19.04.04, www.CNN.com; Toby Moore, 'Tune In, Stay Young', *The Times*, 21.04.04; Claire Coleman, 'Voice Lift', *Daily Mail*, 3.10.05.

57 Peter Jaret, 'My Voice Has Got to Go', *New York Times*, 21.7.05.

58 Advice to call-centre workers quoted in Deborah Cameron, *Good to Talk?*, pp.105–6 (London: Sage, 2000).

59 Arlie Russell Hochschild, *The Managed Heart*, p.108 (Berkeley: University of California Press, 1983).

60 David J. Lieberman, *Never Be Lied to Again* (New York: St Martin's Press, 1998).

61 Lillian Glass, *I Know What You're Thinking* (Hoboken, New Jersey: John Wiley, 2002).

62 Jo-Ellan Dimitrius, *Reading People* (London: Vermillion, 1999).

63 Elisabeth Zetterholm, 'Intonation Patterns and Duration Differences in Imitated Speech', in Bernard Bel and Isabelle Marlien, eds., *Proceedings of Speech Prosody 2002* (Aix-en-Provence: Laboratoire Parole et Langage, 2002).

64 David Howard in *I'd Know That Voice Anywhere*, BBC Radio 4, 14.6.03.
65 Interview with Rory Bremner, 12.05.03.
66 Interview with Rory Bremner, 12.05.03.
67 Deso A. Weiss, 'The Psychological Relations to One's Own Voice', *Folia Phoniatrica*, vol.7, no. 4, p.213, 1955.
68 ibid.
69 ibid, p.214. In this specialist's view, 'A successful therapy of the voice might be equivalent to the effects of a thoroughgoing psychotherapy.'
70 Robert F. Coleman and Ira W. Markham, 'Normal Variations in Habitual Pitch', *Journal of Voice*, vol. 5, no. 2, pp.176–7, 1991. See also John A. Haskell, 'Vocal Self-Perception: The Other Side of the Equation', *Journal of Voice*, vol. 1, no. 2, p177, 1987, and John A. Haskell, 'Adjusting Adolescents' Vocal Self-Perception', *Language, Speech, and Hearing Services in Schools*, vol. 22, July 1991.
71 Interview with Charles Michel, 30.05.03.
72 Interview with Andrea Haring and Elena McGee, 28.05.03.
73 See Newham, op cit.
74 Interview, 30.10.03.

# Conclusion

1 Andrew Gumbel, Obituary of Sidney Morgenbesser, *Independent*, 06.08.04. Thanks to Stan Cohen for drawing my attention to this.
2 Anita McAllister and Svante Granqvist, 'Child Voice and Noise: The Effects of a Day at the Daycare on Vocal Parameters in 10 Five-year-old Children', paper given at PEVOC 6, Royal Academy of Music, September 2005.
3 See R. Murray Schafer's pioneering analysis of legislation against the human voice and changes to the soundscape, *The Soundscape: Our Sonic Environment and the Tuning of the World* (Toronto: McClelland & Stewart, 1977).
4 Michael Bull, 'Introduction: Into Sound', in Michael Bull and Les Back, *The Auditory Culture Reader* (Oxford: Berg, 2003).

# Select Bibliography

## BOOKS

Abercrombie, David, *Problems and Principles* (London: Longman, 1956)

Aikin, W.A., *The Voice* (London: Longmans, Green and Co., 1951)

Andersen, Elaine Slosberg, *Speaking With Style: The Sociolinguistic Skills of Children* (London: Routledge, 1992)

Argyle, Michael, *Bodily Communication* (London: Routledge, 1975)

Aristotle, *De Anima* (London: Routledge, 1993)

Aristotle, *Rhetoric*, Book 3, chapter 1 (Cambridge, Mass: Loeb Classical Library, 1926)

Armstrong, Frankie and Jenny Pearson, *Well-Tuned Women* (London: The Women's Press, 2000)

Atkinson, Max, *Our Masters' Voices: The Language and Body Language of Politics* (London: Methuen, 1984)

Beck, Alan, *Radio Acting* (London: A&C Black, 1997)

Behnke, Mrs Emil, *The Speaking Voice: Its Development and Preservation* (London: J. Curwen & Sons, 1897)

Bell, Alexander Graham, *Lectures upon the Mechanism of Speech* (New York: Funk and Wagnalls, 1906)

Bernds, Edward, *Mr Bernds Goes to Hollywood: My Early Life and Career in Sound Recording at Columbia with Frank Capra and Others* (Lanham, Maryland: Scarecrow Press, 1999)

Bernheimer, Charles and Claire Kahane, *In Dora's Case: Freud, Hysteria, Feminism* (London: Virago, 1985)

Bolinger, Dwight, *Intonation and Its Uses* (London: Edward Arnold, 1989)

Boone, Daniel, *Is Your Voice Telling on You?* (San Diego: Singular, 1997)

Bourdieu, Pierre, *Dictinction: A Social Critique of the Judgement of Taste* (London: Routledge, 1984)

Bowlby, John, *Attachment* (London: Penguin Books, 1984)

Boysson-Bardies, Benedicte De, *How Language Comes to Children: From Birth to Two Years* (Cambridge, Mass: MIT Press, 1999)

Brown, Lyn Mikel and Carol Gilligan, *Meeting at the Crossroads: Women's Psychology and Girls' Development* (Massachusetts: Harvard University Press, 1992)

Bull, Michael and Les Back, *The Auditory Culture Reader* (Oxford: Berg, 2003)

Bullock, Alan, *Hitler: A Study in Tyranny* (London: Penguin Books, 1962)

Burgoon, Judee K. et al., *Nonverbal Communication: The Unspoken Dialogue* (New York: McGraw-Hill, 1996)

Cameron, Deborah, ed., *The Feminist Critique of Language: A Reader* (London: Routledge, 1998)

Cameron, Deborah, *Good to Talk?* (London: Sage, 2000)

Chion, Michel, *The Voice in Cinema* (New York: Columbia University Press, 1999)

Cicero, *De Oratore*, Book 3 (London: Heinemann, 1942)

Coates, Jennifer and Deborah Cameron, eds., *Women in Their Speech Communities* (London: Longman, 1988)

Connor, Steven, *Dumbstruck: A Cultural History of Ventriloquism* (Oxford: Oxford University Press, 2000)

Cook, Norman D., *Tone of Voice and Mind* (Amsterdam: John Benjamins, 2002)

Cruttenden, Alan, *Intonation* (Cambridge: Cambridge University Press, 1986)

Crystal, David, *A Dictionary of Linguistics and Phonetics* (Oxford: Blackwell, 2003)

Crystal, David, *Prosodic Systems and Intonation in English* (Cambridge: Cambridge University Press, 1969)

Crystal, David, *The English Tone of Voice* (London: Edward Arnold, 1975)

Dal Vera, Rocco, ed., *The Voice in Violence* (Cincinnati: Voice and Speech Trainers Association, 2001)

Darby, John K., ed., *Speech Evaluation in Psychiatry* (New York: Grune and Stratton, 1981)

Darwin, Charles, *The Descent of Man* (Loughton, Essex: Prometheus Books, 1991)

Darwin, Charles, *The Expression of the Emotions in Man and Animals* (London: Fontana Press, 1999)

Deacon, Terrence, *The Symbolic Species* (London: Penguin Books, 1998)

Denes, Peter B. and Elliot N. Pinson, *The Speech Chain* (New York: W.H. Freeman, 1993)

Derrida, Jacques, *Speech and Phenomena, and Other Essays on Husserl's Theory of Signs* (Evanston: Northwestern University Press, 1973)

DiBattista, Maria, *Fast-Talking Dames* (New Haven: Yale University Press, 2003)

Douglas, Susan J., *Listening In* (New York: Times Books, 1999)

Dunn, Leslie C. and Nancy A. Jones, *Embodied Voices* (Cambridge: Cambridge University Press, 1994)

Eyman, Scott, *The Speed of Sound* (Baltimore: Johns Hopkins University Press, 1999)

Fest, Joachim C., *Hitler* (London: Penguin Books, 1977)

Foucault, Michel, *The Birth of the Clinic* (London: Tavistock Publications, 1976)

Foucault, Michel, *Discipline and Punish* (New York: Vintage Books, 1979)

Fry, D. B., *The Physics of Speech* (Cambridge: Cambridge University Press, 1979)

Goffman, Erving, *The Presentation of Self in Everyday Life* (London: Pelican Books, 1971)

Graddol, David and Joan Swann, *Gender Voices* (Oxford: Blackwell, 1989)

Greene, Margaret C. L., *Disorders of the Voice* (Austin, Texas: PRO-ED, 1986)

Greene, M. C. L. and Lesley Mathieson, *The Voice and Its Disorders* (London: Whurr Publishers, 1989)

Gumperz, John J. and Dell Hymes, *Directions in Sociolinguistics* (New York: Holt, Rinehart and Winston, 1972)

Gumperz, John J., *Discourse Strategies* (Cambridge: Cambridge University Press, 1982)

Harper, Robert G., ed., *Nonverbal Communication: The State of the Art* (New York: John Wiley, 1978)

Hart, Betty and Todd Risley, *Meaningful Differences in the Everyday Experience of Young American Children* (Baltimore, Maryland: Paul H. Brookes Publishing, 1995)

Havelock, Eric A., *The Muse Learns to Write* (New Haven: Yale University Press, 1986)

Hochschild, Arlie Russell, *The Managed Heart* (Berkeley: University of California Press, 1983)

Hollien, Harry, *The Acoustics of Crime* (New York: Plenum Press, 1990)

Hollien, Harry, *Forensic Voice Identification* (San Diego: Academic Press, 2002)

Honey, John, *Does Accent Matter?* (London: Faber, 1989)

Hymes, Dell, *Foundations in Sociolinguistics* (London: Tavistock, 1977)

Jaffe, Joseph and Stanley Feldstein, *Rhythms of Dialogue* (New York: Academic Press, 1970)

Jamieson, Kathleen Hall, *Eloquence in an Electronic Age* (New York: Oxford University Press, 1988)

Kamensky, Jane, *Governing the Tongue: The Politics of Speech in Early New England* (New York: Oxford University Press, 1999)

Karpf, Anne, *Doctoring the Media: The Reporting of Health and Medicine* (London: Routledge, 1988)

Kendon, Adam et al., eds., *Organization of Behaviour in Face-to-Face Interaction* (The Hague: Mouton Publishers, 1975)

Kershaw, Ian, *Hitler, 1889–1936* (London: Penguin, 1998)

Knapp, Mark L., *Nonverbal Communication in Human Interaction* (New York: Holt, Rinehart & Winston, 1972)

Kozloff, Sarah, *Invisible Storytellers: Voice-Over Narration in American Film* (Berkeley: University of California Press, 1988)

Kramarae, Cheris, *Technology and Women's Voices* (London: Routledge, 1988)

Labov, W., *The Social Stratification of English in New York City* (Washington, DC: Center for Applied Linguistics, 1966)

Lakoff, Robin, *Language and Woman's Place* (New York: HarperCollins, 1975)

Laver, John, *The Gift of Speech* (Edinburgh: Edinburgh University Press, 1991).

Laver, John, *The Phonetic Description of Voice Quality* (Cambridge: Cambridge University Press, 1980)

Laver, John and Sandy Hutcheson, eds., *Communication in Face to Face Interaction* (Middlesex, England: Penguin Books, 1972)

Lawrence, Amy, *Echo and Narcissus: Women's Voices in Classical Hollywood Cinema* (Berkeley: University of California Press, 1991)

Levelt, William J.M., *Speaking: From Intention to Articulation* (Cambridge, Mass: The MIT Press, 1989)

Lieberman, Philip, *Eve Spoke* (London: Picador, 1998)

Lippi-Green, Rosina, *English with an Accent: Language, Ideology, and Discrimination* (London: Routledge, 1997)

Locke, John L., *The De-Voicing of Society* (New York: Simon and Schuster, 1998)

Luchsonger, Richard and Godfrey Arnold, *Voice–Speech–Language* (California: Wadsworth, 1965)

Malinowski, Bronislaw, *Coral Gardens and Their Magic*, vol. 2, 'The Language of Magic and Gardening' (London: Routledge, 2001)

Martin, Jacqueline, *Voice in Modern Theatre* (London: Routledge, 1991)

Martin, Stephanie and Lyn Darnley, *The Teacher's Voice* (London: Whurr Publishers, second edition, 2004)

Matheson, Hilda, *Broadcasting* (London: Butterworth, 1933)

McLuhan, Marshall, *The Gutenberg Galaxy* (London: Routledge, 1962)

McLuhan, Marshall, *Understanding Media* (London: Sphere Books, 1967)

Mehrabian, Albert, *Nonverbal Communication* (Chicago: Aldine, 1972)

Miller, Edward D., *Emergency Broadcasting and* 1930s American Radio (Philadelphia: Temple University Press, 2003)

Moses, Paul J., *The Voice of Neurosis* (New York: Grune and Stratton, 1954)

Newham, Paul, *Therapeutic Voicework* (London: Jessica Kingsley Publishers, 1998)

Niederland, William G., 'Early Auditory Experiences, Beating Fantasies, and Primal Scene', *Psychoanalytic Study of the Child*, 1958

Ong, Walter J., *Orality and Literacy: The Technologizing of the Word* (London: Methuen, 1982)

Pear, T.H., *Voice and Personality* (London: Chapman and Hall, 1931)

Pike, Kenneth, *The Intonation of American English* (Ann Arbor: University of Michigan Press, 1945)

Pinker, Steven, *The Language Instinct* (London: Penguin Books, 1994)

Quintilian, *Institutio Oratorio*, Book XI, chapter 3 (Cambridge, Mass: Loeb Classical Library, 1920)

Ree, Jonathan, *I See a Voice* (London: HarperCollins, 1999)

Reik, Theodor, *Listening with the Third Ear* (New York: Farrar, Strauss and Company, 1949)

Rodenburg, Patsy, *The Right to Speak* (London: Methuen, 1992)

Rush, James, *The Philosophy of the Human Voice* (Philadelphia: Grigg and Elliott, 1833)

Sapir, Edward, *Language: An Introduction to the Study of Speech* (Thomson Learning, 1955)

Sataloff, Robert, ed., *Voice Perspectives* (San Diego: Singular Publishing, 1998)

Schafer, R. Murray, *The Soundscape: Our Sonic Environment and the Tuning of the World* (Toronto: McClelland and Stewart, 1977)

Scherer, Klaus R. and Howard Giles, eds., *Social Markers in Speech* (Cambridge: Cambridge University Press, 1979)

Scherer, Klaus R. and Paul Ekman, eds., *Approaches to Emotion* (Hillsdale, New Jersey: Lawrence Erlbaum Associates, 1984)

Schore, Allan N., *Affect Regulation and the Repair of the Self* (New York: W.W. Norton, 2003)

Sconce, Jeffrey, *Haunted Media: Electronic Presence from Telegraph to Television* (Durham, North Carolina: Duke University Press, 2000)

Sennet, Richard, *The Fall of Public Man* (London: Faber and Faber, 1986)

Siegman, Aron W. and Stanley Feldstein, *Nonverbal Behaviour and Communication* (New Jersey: Lawrence Erlbaum Associates, 1987)

Silverman, Kaja, *The Acoustic Mirror* (Bloomington: Indiana University Press, 1988)

Snow, Catherine E. and Charles A. Ferguson, *Talking to Children: Language Input and Acquisition* (Cambridge: Cambridge University Press, 1977)

Sola Pool, Ithiel de, ed., *The Social Impact of the Telephone* (Cambridge, Mass: The MIT Press, 1977)

Solzhenitsyn, Alexander, *The First Circle* (London: Fontana Books, 1970)

Spufford, Francis and Jenny Uglow, eds., *Cultural Babbage: Technology, Time and Invention* (London: Faber and Faber, 1996)

Stern, Daniel, *The Interpersonal World of the Infant* (London: Karnac Books, 1998)

Stern, Daniel N., *The First Relationship: Infant and Mother* (Cambridge, Mass: Harvard University Press, 2002)

Sterne, Jonathan, *The Audible Past: Cultural Origins of Sound Reproduction* (Durham, North Carolina: Duke University Press, 2003)

Tannen, Deborah, ed., *Gender and Conversational Interaction* (New York: Oxford University Press, 1993)

Thorne, Barrie and Nancy Henley, *Language and Sex: Difference and Dominance* (Massachusetts: Newbury House, 1975)

Thorne, Barrie et al., eds., *Language, Gender, and Society* (Boston: Heinle and Heinle, 1983)

Tomatis, Alfred A., *The Conscious Ear* (New York: Station Hill Press, 1991)

Tomatis, Alfred, *The Ear and Language* (Ontario: Moulin Publishing, 1996)

Trudgill, Peter, *Sociolinguistics* (London: Penguin Books, 2000)

Vasse, Denis, *L'Ombilic et la Voix: Deux Enfants en Analyse* (Paris: Editions du Seuil, 1974)

# ARTICLES

Abitbol, Jean et al., 'Does a Hormonal Vocal Cycle Exist in Women? Study of Vocal Premenstrual Syndrome in Voice Performers by Videostroboscopy-Glottography and Cytology on 38 Women', *Journal of Voice*, vol. 3, no. 2, 1989

Abitbol, Jean et al., 'Sex Hormones and the Female Voice', *Journal of Voice*, vol. 13, no. 3, 1999

Addington, David W., 'The Relationship of Selected Vocal Characteristics to Personality Perception', *Speech Monographs*, vol.35, 1968

Allan, Scott, 'The Rise of New Zealand Intonation', in Allan Bell and Janet Holmes, eds., *New Zealand Ways of Speaking English* (Clevedon, England: Multilingual Matters, 1990)

Alpert, Murray, 'Encoding of Feelings in Voice', in P.I. Clayton and J.E. Barrett, *Treatment of Depression: Old Controversies and New Approaches* (New York: Raven Press, 1983)

Altman, Rick, ed., 'Cinema/Sound', *Yale French Studies*, no. 60, 1980

Ambady, Nalini et al., 'Surgeons' Tone of Voice: A Clue to Malpractice History', *Surgery*, 2002

Anderson, A. K. and E. A. Phelps, 'Intact Recognition of Vocal Expressions of Fear Following Bilateral Lesions of the Human Amygdala', *Neuroreport*, 9, 16 November 1998

Andrews, Moya L. and Charles P. Schmidt, 'Gender Presentations: Perceptual and Acoustical Analyses of Voice', *Journal of Voice*, vol.11, no.3, 1997

Anolli, Luigi and Rita Ciceri, 'The Voice of Deception: Vocal Strategies of Naïve and Able Liars', *Journal of Nonverbal Behaviour*, 21 (4), Winter 1997

Anzieu, Didier, 'The Sound Image of the Self', *International Review of Psychoanalysis*, 6, 1979

Apple, William et al., 'Effects of Pitch and Speech Rate on Personal Attributes', *Journal of Personality and Social Psychology*, vol. 37, no. 5, 1979

Arensburg, B., et al., 'A Reappraisal of the Anatomical Basis for Speech in Middle Palaeolithic Hominids', *American Journal of Physical Anthropology*, 83, 1980

Arensburg, Baruch and Anne-Marie Tillier, 'Speech and the Neanderthals', *Endeavour*, vol. 15, no. 1, 1991

Bady, Susan Lee, 'The Voice as a Curative Factor in Psychotherapy', *Psychoanalytic Review*, 72(3), Fall 1985

Baltaxe, Christiane, 'Vocal Communication of Affect and Its Perception in Three to Four-Year-Old Children', *Perceptual and Motor Skills*, 72, 1991

Banse, Rainer and Klaus R. Scherer, 'Acoustic Profiles in Vocal Emotion Expression', *Journal of Personality and Social Psychology*, vol.70, no.3, 1996

Barron, Anthony, 'Speaker Identification by Earwitness: A Bigger Picture', Department of Linguistics and Phonetics, University of Leeds, May 2001

Bateson, Mary Catherine, 'Mother–Infant Exchanges: the Epigenesis of Conversation Interaction', *Annals of the New York Academy of Sciences*, 263, 1975

Beattie, Geoffrey W., 'Turn-taking and Interruption in Political Interviews: Margaret Thatcher and Jim Callaghan Compared and Contrasted', *Semiotica*, 39-1/2, 1982

Beebe, Leslie M. and Howard Giles, 'Speech Accommodation Theories: a Discussion in Terms of Second Language Acquisition', *International Journal of the Sociology of Language*, 46, 1984

Beebe, Beatrice et al., 'Systems Models in Development and Psychoanalysis: the Case of Vocal Rhythm, Coordination, and Attachment', *Infant Mental Health Journal*, vol.21(1-2), pp.105-6, 2000

Belin, Pascal et al., 'Voice-selective Areas in Human Auditory Cortex', *Nature*, vol. 403, 20 January 2000

Blair, R. James, 'Turning a Deaf Ear to Fear: Impaired Recognition of Vocal Affect in Psychopathic Individuals', *Journal of Abnormal Psychology*, vol.111, 2002

Blount, Ben G. and Elise J. Padgug, 'Prosodic, Paralinguistic, and Interactional Features in Parent–Child Speech: English and Spanish', *Journal of Child Language*, 4, 1976

Bolinger, Dwight, 'Intonation Across Languages', in Joseph H. Greenberg (ed.), *Universals of Human Language* (California: Stanford University Press, 1978)

Bouhuys, Antoinette L. and Wilhemina E.H. Mulder-Hajonides Van Der Meulen, 'Speech Timing Measures of the Severity, Psychomotor Retardation, and Agitation in Endogenously Depressed Patients', *Journal of Communication Disorders*, 17, 1984

Breuer, Josef and Sigmund Freud, 'Studies on Hysteria', in *The Standard Edition of the Complete Psychological Works of Sigmund Freud*, vol. 2 (London: The Hogarth Press, 1955)

Brody, Morris W., 'Neurotic Manifestations of the Voice', *Psychoanalytic Quarterly*, vol. 12, 1943

Brown, Bruce L. and Wallace E. Lambert, 'A Cross-Cultural Study of Social Status Markers in Speech', *Canadian Journal of Behavioural Science*, 8(1), 1976

Burnham, Denis et al., 'What's New, Pussycat? On Talking to Babies and Animals', *Science*, vol. 296, issue 5572, May 2002

Butcher, Peter, 'Psychological Processes in Psychogenic Voice Disorder', *European Journal of Disorders of Communication*, 30, 1995

Chen, Xin et al., 'Auditory-Oral Matching Behaviour in Newborns', *Developmental Science*, 7:1, 2004

Cheour-Luhtanen, M. et al., 'The Ontogenetically Earliest Discriminative Response of the Human Brain', *Psychophysiology*, 33, July 1996

Clark, A.G. et al., 'Inferring Nonneural Evolution From Human-Chimp-Mouse Orthologous Gene Trios', *Science* 302, 2003

Coleman, Robert F. and Ira W. Markham, 'Normal Variations in Habitual Pitch', *Journal of Voice*, vol. 5, no. 2, pp.176-7, 1991

Condon, William S. and Louis W. Sander, 'Neonate Movement Is Synchronised with Adult Speech: Interactional Participation and Language Acquisition', *Science, New Series*, vol. 183, No. 4120, 11 January 1974

Crelin, Edmund S., 'The Skulls of Our Ancestors: Implications Regarding Speech, Language, and Conceptual Thought Evolution', *Journal of Voice*, vol. 3, no. 1, 1989

Crown, Cynthia L. et al., 'The Cross-Modal Coordination of Interpersonal Timing: Six-Week-Old Infants' Gaze with Adults' Vocal Behaviour', *Journal of Psycholinguistic Research*, vol. 31, no. 1, January 2002

Cruttenden, Alan, 'Falls and Rises: Meanings and Universals', *Journal of Linguistics*, 17, 1981

Daly, Nicola and Paul Warren, 'Pitching It Differently in New Zealand English: Speaker Sex and Intonation Patterns', *Journal of Sociolinguistics*, 5/1, 2001

Daniel, Paul J., 'Voice Change Surgery in the Transsexual', *Head and Neck Surgery*, May-June 1982

Darby, John K. and Harry Hollien, 'Vocal and Speech Patterns of Depressed Patients', *Folia Phoniatrica et Logopaedica*, 1977, 29

Davis, Matthew H. and Ingrid S. Johnsrude, 'Hierarchical Processing in Spoken Language Comprehension', *The Journal of Neuroscience*, 15 April 2003

Debruyne, Frans et al., 'Speaking Fundamental Frequency in Monozygotic and Dizygotic Twins', *Journal of Voice*, vol. 16, no. 4, 2002

DeCasper, Anthony J. and William P. Fifer, 'Of Human Bonding: Newborns Prefer Their Mothers' Voices', *Science*, vol. 208, 6 June 1980

DeCasper, Anthony J. and Phyllis A. Prescott, 'Human Newborns' Perception of Male Voices: Preference, Discrimination, and Reinforcing Value', *Developmental Psychobiology*, 17(5), 1984

DeGusta, David et al., 'Hypoglossal Canal Size and Hominid Speech', *Proceedings of the National Academy of Sciences, USA, Anthropology*, vol. 96, February 1999

DePaulo, Bella M. and Lerita M. Coleman, 'Evidence for the Specialness of the "Baby Talk" Register', *Language and Speech*, vol. 24, part 3, 1981

DePaulo, Bella M. and Lerita M. Coleman, 'Verbal and Nonverbal Communication of Warmth to Children, Foreigners, and Retarded Adults', *Journal of Nonverbal Behaviour*, 11(2), 1987

Devereux, George, 'Mohave Voice and Speech Mannerisms', in Dell Hymes, *Language in Culture and Society* (New York: Harper and Row, 1967)

DiMatteo, M. Robin, 'Nonverbal Skill and the Physician–Patient Relationship', in Robert Rosenthal, ed., *Skill in Nonverbal Communication: Individual Differences* (Cambridge, Mass: Oelgeschlager, Gunn and Hain, 1979)

Dondi, Marco et al., 'Can Newborns Discriminate Between Their Own Cry and the Cry of Another Newborn Infant?' *Developmental Psychology*, vol.35(2), March 1999

Doupe, Allison J. and Patricia K. Kuhl, 'Birdsong and Human Speech: Common Themes and Mechanisms', *Annual Review of Neuroscience*, vol. 22, 1999

Edelsky, Carole, 'Question Intonation and Sex Roles', *Language in Society*, 8, 1979

Eimas, Peter D. et al., 'Speech Perception in Infants', *Science, New Series*, vol. 171, no. 3968, 22 January 1971

Erickson, Frederick, 'Timing and Context in Everyday Discourse: Implications for the Study of Referential and Social Meaning', *Sociolinguistic Working Paper* no. 67 (Austin, Texas: Southwest Educational Development Laboratory, 1980)

Falk, Dean, 'Prelinguistic Evolution in Early Hominins: Whence Motherese?', *Behavioural and Brain Science*, 27:6, 2004

Feldstein, Stanley et al., 'Gender and Speech Rate in the Perception of Competence and Social Attractiveness', *Journal of Social Psychology*, 141, 2001

Ferguson, Charles A., 'Baby Talk in Six Languages', *American Anthropologist, New Series*, vol. 66, no. 6, part 2: 'The Ethnography of Communication', December 1964

Fernald, Anne, 'Meaningful Melodies in Mothers' Speech to Infants', in Hanus Papousek et al., *Nonverbal Vocal Communication: Comparative and Developmental Approaches* (Cambridge: Cambridge University Press, 1992)

Fernald, Anne et al., 'A Cross-language Study of Prosodic Modifications in Mothers' and Fathers' Speech to Preverbal Infants', *Journal of Child Language*, 16, 1989

Ferrand, Carole T. and Ronald L. Bloom, 'Gender Differences in Children's Intonational Patterns', *Journal of Voice*, vol. 10, no. 3

Fifer, William P. and C.M. Moon, 'The Role of the Mother's Voice in the Organization of Brain Function in the Newborn', *Acta Paediatric Supplement*, 397, 1994

Fifer, William P. and Chris M. Moon, 'The Effects of Fetal Experience with Sound', in Jean-Pierre Lecanuet et al, eds., *Fetal Development : A Psychobiological Perspective* (New Jersey : Lawrence Erlbaum Associates, 1995)

Fitch, W. Tecumseh and David Reby, 'The Descended Larynx Is Not Uniquely Human', *Proceedings of the Royal Society, London*, B, vol 268, 2001

Floccia, Caroline et al., 'Unfamiliar Voice Discrimination for Short Stimuli in Newborns', *Developmental Science*, 2000, 3:3

Fraisse, Paul, 'Rhythm and Tempo', in Diana Deutsch, ed., *The Psychology of Music* (London: Academic Press, 1982)

France, D.J. et al., 'Acoustical Properties of Speech as Indicators of Depression and Suicidal Risk', *Transactions on Biomedical Engineering*, 47, July 2000

Freedman, Norbert and Joan Lavender, 'Receiving the Patient's Transference: the Symbolising and Desymbolising Counter-transference', *Journal of American Psychoanalytic Association*, vol. 45, no. 1, 1997

Freud, Sigmund, 'Fragment of an Analysis of a Case of Hysteria', in *The Standard Edition of the Complete Psychological Works of Sigmund Freud*, vol. 7 (London: The Hogarth Press, 1953)

Freud, Sigmund, 'The Uncanny', in *The Standard Edition of the Complete Works of Sigmund Freud*, vol. 17, (London: Hogarth Press, 1955)

Freud, Sigmund, 'Recommendations to Physicians Practising Psychoanalysis', in *The Standard Edition of the Complete Works of Sigmund Freud*, vol. 12 (London: Hogarth, 1958)

Garnica, Olga K., 'Some Prosodic and Paralinguistic Features of Speech to Young Children', in Catherine E. Snow and Charles A. Ferguson, *Talking to Children: Language Input and Acquisition* (Cambridge: Cambridge University Press, 1977)

Garrett, Kathryn L. and E. Charles Healey, 'An Acoustic Analysis of Fluctuations in the Voices of Normal Adult Speakers across Three Times of Day', *Journal of the Acoustical Society of America*, vol. 82, No. 1, July 1987

Giles, Howard et al., 'Speech Accommodation Theory: the First Decade and Beyond', in M. McLaughlin, ed., *Communication Yearbook*, vol. 10 (Newbury Park: Sage, 1987)

Giles, Howard and Angie Williams, 'Accommodating Hypercorrection: a Communication Model', *Language and Communication*, vol. 12, no. 3/4, 1992

Giles, Howard and Peter Powesland, 'Accommodation Theory', in Nikolas Coupland and Adam Jaworski, eds., *Sociolinguistics: A Reader and Coursework* (London: Macmillan, 1997)

Gill, Rosalind, 'Justifying Injustice: Broadcasters' Accounts of Inequality in Radio', in Caroline Mitchell, ed., *Women and Radio: Airing Differences* (London: Routledge, 2000)

Gilligan, Carol, 'Remembering Iphigenia: Voice, Resonance, and the Talking Cure', in Edward R. Shapiro, ed., *The Inner World in the Outer World* (New Haven: Yale University Press, 1997)

Goody, J. and I. Watt, 'The Consequences of Literacy', reprinted in Pier Paulo Giglioli, ed., *Language and Social Context* (Middlesex: Penguin, 1972)

Gratier, Maya, 'Expressive Timing and Interactional Synchrony between Mothers and Infants: Cultural Similarities, Cultural Differences, and the Immigration Experience', *Cognitive Development*, 18, 2003

Gray, Steven et al., 'Witnessing a Revolution in Voice Research: Genomics, Tissue Engineering, Biochips and What's Next!', *Logopedics, Phoniatrics, Vocology*, vol. 28:1, 2003

Greden, John F. et al., 'Speech Pause Time: a Marker of Psychomotor Retardation Among Endogenous Depressives', *Biological Psychiatry*, vol. 16, no 9, 1981

Gregory, Jr., Stanford W. et al., 'Voice Pitch and Amplitude Convergence as a Metric of Quality in Dyadic Interviews', *Language and Communications*, vol. 13, no. 3, 1993

Gregory Jr., Stanford W. and Stephen Webster, 'A Nonverbal Signal in Voices of Interview Partners Effectively Predicts Communication Accommodation and Social Status Perceptions', in *Journal of Personality and Social Psychology*, vol. 70, no. 6, 1996

Gregory, Jr., Stanford W. et al., 'Verifying the Primacy of Voice Fundamental Frequency in Social Status Accommodation', *Language and Communication*, 2001, 21

Gregory, Jr., Stanford W. and Timothy J. Gallagher, 'Spectral Analysis of Candidates' Nonverbal Vocal Communication: Predicting US Presidential Election Outcomes', *Social Psychology Quarterly*, vol. 65, no. 3, 2002

Grieser, DiAnne L. and Patricia Kuhl, 'Maternal Speech to Infants in a Tonal Language: Support for Universal Prosodic Features in Motherese', *Developmental Psychology*, vol. 24, no. 1, 1988

Haskell, John A., 'Vocal Self-perception: the Other Side of the Equation', *Journal of Voice*, vol. 1, no. 2, 1987

Henton, C.G. and R.A.W. Bladon, 'Breathiness in Normal Female Speech: Inefficiency Versus Desirability', *Language and Communication*, vol. 5, no. 3, 1985

Henton, Caroline G., 'Fact and Fiction in the Description of Female and Male Pitch', *Language and Communication*, vol. 9, no. 4, 1989

Henton, Caroline G., 'The Abnormality of Male Speech', in George Wolf, ed., *New Departures in Linguistics* (New York: Garland Publishing, 1992)

Hepper, P.G. et al., 'Newborn and Fetal Response to Maternal Voice', *Journal of Reproductive and Infant Psychology*, vol. 11, 1993

Hibbitts, Bernard, 'Coming to Our Senses: Communication and Legal Expression in Performance Cultures', *Emory Law Journal*, 4, 1992

Hirsh-Pasek, K. and R. Treiman, 'Doggerel: Motherese in a New Context', *Journal of Child Language*, 9(1), 1982

Holzman, Philip S. and Clyde Rousey, 'The Voice as Precept', *Journal of Personality and Social Psychology*, vol. 4, no. 1, 1966

Houston, Derek M. and Peter W. Jusczyk, 'The Role of Talker-specific Information in Word Segmentation by Infants', *Journal of Experimental Psychology*, vol. 26, no. 5, 2000

Houston, Derek M. and Peter W. Jusczyk, 'Infants' Long-term Memory for the Sound Patterns of Words and Voices', *Journal of Experimental Psychology: Human Perception and Performance*, vol. 29(6), December 2003

Irvine, Judith T., 'Registering Affect: Heteroglossia in the Linguistic Expression of Emotion', in Catherine A. Lutz and Lila Abu-Lughod, *Language and the Politics of Emotion* (Cambridge: Cambridge University Press, 1990)

Isakower, Otto, 'On the Exceptional Position of the Auditory Sphere', *International Journal of Psychoanalysis*, 1939

Jaffe, Joseph et al., 'Rhythms of Dialogue in Infancy', *Monographs of the Society for Research in Child Development*, vol. 66, no. 2, serial no. 265, April 2001

Jarvis, Erich D. et al., 'Behaviourally Driven Gene Expression Reveals Song Nuclei in Hummingbird Brain', *Nature*, 406, 10 August 2000

Johnstone, Tom and Klaus R. Scherer, 'Vocal Communication of Emotion', in Michael Lewis and Jeanette M. Haviland-Jones, *Handbook of Emotions*, second edition (New York: The Guilford Press, 2000)

Jurgens, Uwe, 'On the Neurobiology of Vocal Communication', in Hanus Papousek et al., eds., *Nonverbal Vocal Communication: Comparative and Developmental Approaches* (Cambridge: Cambridge University Press, 1992)

Kahane, Claire, 'Freud and the Passions of the Voice', in John O'Neill, ed., *Freud and the Passions* (Philadelphia: Pennsylvania State University Press, 1996)

Kay, Richard F. et al., 'The Hypoglossal Canal and the Origin of Human Vocal Behaviour', *Proceedings of the National Academy of Sciences of the United States of America*, vol. 95, Issue 9, 28 April 1998

Kessen, William et al., 'The Imitation of Pitch in Infants', *Infant Behaviour and Development*, vol. 2, no. 1, January 1979

Kisilevsky, B.S. and J.A. Low, 'Human Fetal Behaviour: 100 Years of Study', *Developmental Review*, 18, 1998

Kisilevsky, Barbara S. et al., 'Effects of Experience on Fetal Voice Recognition', *Psychological Science*, vol. 14, No. 3, May 2003

Kitamura, C. et al., 'Universality and Specificity in Infant-directed Speech: Pitch Modifications as a Function of Infant Age and Sex in a Tonal and Non-tonal Language', *Infant Behaviour and Development*, 2002, 24

Kitamura, Christine and Denis Burnham, 'Pitch and Communicative Intent in Mothers' Speech: Adjustments for Age and Sex in the First Year', *Infancy*, 4(1), 2003

Kramarae, Cheris, 'Women's Speech: Separate But Unequal?', in Barrie Thorne and Nancy Henley, *Language and Sex: Difference and Dominance* (Massachusetts: Newbury House, 1975)

Kramarae, Cheris, 'Resistance to Women's Public Speaking', in Senta Tromel-Plotz, ed., *Gewalt durch Sprache* (Frankfurt: Fisher Taschenbuch Verlag, 1984)

Kuhl, Patricia K. and Andrew N. Meltzoff, 'Evolution, Nativism and Learning in the Development of Language and Speech', in M. Gopnik, ed., *The Inheritance and Innateness of Grammars* (New York: Oxford University Press, 1997)

Kuhl, Patricia K. et al., 'Linguistic Experience Alters Phonetic Perception in Infants by 6 Months of Age', *Science, New Series*, vol. 255, no. 5044, 31 January 1992

Kuhl, Patricia K. and Andrew N. Meltzoff, 'Infant Vocalizations in Response to speech: Vocal Imitation and Developmental Change', *Journal of the Acoustical Society of America*, 100: 4, Part 1, October 1996

Kuhl, Patricia K. et al., 'Cross-Language Analysis of Phonetic Units in Language Addressed to Infants', *Science, New Series*, vol. 277, no. 5326, 1 August 1997

Kuhl, Patricia K., 'A New View of Language Acquisition', *Proceedings of the National Academy of Sciences of the United States of America*, vol.97, no.22, 24 October 2000

Labov, William, 'The Social Stratification of (r) in New York City Department Stores', in Nikolas Coupland and Adam Jaworski, eds., *Sociolinguistics: A Reader and Coursework* (London: Macmillan, 1997)

Lecanuet, J.P. et al., 'Prenatal Discrimination of a Male and Female Voice Uttering the Same Sentence', *Early Development and Parenting*, vol. 2(4), 1993

Lecanuet, Jean-Pierre et al., 'Human Fetal Auditory Perception', in Jean-Pierre Lecanuet et al., eds., *Fetal Development: A Psychobiological Perspective* (New Jersey : Lawrence Erlbaum Associates, 1995)

Lehtonen, Jaakko and Kari Sajavaara, 'The Silent Finn', in Deborah Tannen and Muriel Saville-Troike, *Perspectives on Silence* (New Jersey: Ablex, 1985)

Lehto, Laura et al., 'Voice Symptoms of Call-centre Customer Service Advisers Experienced during a Workday and Effects of a Short Vocal Training Course', *Logopedics, Phoniatrics, Vocology*, 30, 2005

Leinonen, Lea et al., 'Shared Means and Meanings in Vocal Expression of Man and Macaque' *Logopedics, Phoniatrics, Vocology*, Vol.28: 2, 2003

Lester, Barry M. and C.F. Zachariah Boukydis, 'No Language but a Cry', in Hanus Papousek et al., eds., *Nonverbal Vocal Communication: Comparative and Developmental Approaches*, (Cambridge: Cambridge University Press, 1992)

Leuchtenburg, William E., 'The FDR Years: On Roosevelt and His Legacy', www.washingtonpost.com

Linke, C.E., 'A Study of Pitch Characteristics of Female Voices and Their Relationship to Vocal Effectiveness', *Folia Phoniatrica et Logopaedica*, 25, 1973

Linville, Sue Ellen, 'Acoustic Correlates of Perceived versus Actual Sexual Orientation in Men's Speech', *Folia Phoniatrica et Logopaedica*, 1998

Loveday, Leo, 'Pitch, Politeness and Sexual Role: an Exploratory Investigation into the Pitch Correlates of English and Japanese Politeness Formulae', *Language and Speech*, vol. 24, part 1, 1981

Lynch, Michael P. et al., 'Phrasing in Prelinguistic Vocalizations', *Developmental Psychobiology*, 28(1), 1995

Mahl, George, 'Disturbances and Silences in the Patient's Speech in Psychotherapy', *Journal of Abnormal and Social Psychology*, vol. 53, 1956

Maiello, Suzanne, 'The Sound-Object: a Hypothesis about Prenatal Auditory Experience and Memory', *Journal of Child Psychotherapy*, vol. 21, no. 1, 1995

Maiello, Suzanne, 'Prenatal Trauma and Autism', *Journal of Child Psychotherapy*, vol. 27, no. 2, 2001

Maiello, Suzanne, 'The Rhythmical Dimension of the Mother-infant – Transcultural Considerations', *Journal of Child and Adolescent Mental Health*, 15(2), 2003

Manning, J. T. et al., 'Ear Asymmetry and Left-Side Cradling', *Evolution and Human Behaviour*, 18, 1997

Martin, G.B. and R.D. Clarke, 'Distress Crying in Neonates: Species and Peer Specificity', *Developmental Psychology*, 18, 1982

Mastropieri, Diane and Gerald Turkewitz, 'Prenatal Experience and Neonatal Responsiveness to Vocal Expressions of Emotion', *Developmental Psychology*, 35(3), November 1999

McClure, Erin B. and Stephen Nowicki, Jr., 'Associations between Social Anxiety and Nonverbal Processing Skill in Preadolescent Boys and Girls', *Journal of Nonverbal Behaviour*, 25(1), Spring 2001

McHugh-Munier, Caitriona et al., 'Coping Strategies, Personality, and Voice Quality in Patients with Vocal Fold Nodules and Polyps', *Journal of Voice*, vol.11, no.4, 1997

McKay, Anne, 'Speaking Up: Voice Amplification and Women's Struggle for Public Expression', in Cheris Kramarae, 'Technology and Women's Voices (London: Routledge and Kegan Paul, 1988)

Mehrabian, Albert and Susan R. Ferris, 'Inference of Attitudes from Nonverbal Communication in Two Channels', *Journal of Consulting Psychology*, vol. 31, no. 3, 1967

Mehrabian, Albert and Morton Wiener, 'Decoding of Inconsistent Communications', *Journal of Personality and Social Psychology*, vol. 6, no. 1, 1967

Mendoza, Elvira and Gloria Carballo, 'Vocal Tremor and Psychological Stress', *Journal of Voice*, vol. 13. no. 1

Michelsson, Katarina et al., 'Cry Characteristics of 172 Healthy 1 to 7-Day-Old Infants', *Folia Phoniatrica et Logopaedica*, 54, 2002

Michelsson, K. et al., 'Cry Score – an Aid in Infant Diagnosis', *Folia Phoniatrica et Logopaedica*, 36, 1984

Milmoe, Susan et al., 'The Mother's Voice: Postdictor of Aspects of Her Baby's Behaviour', *Proceedings of the 76th Conference of the American Psychological Association*, 1968

Miller, Juliet, 'The Crashed Voice – a Potential for Change : a Psychotherapeutic View', *Logopedics, Phoniatrics, Vocology*, 28, 2003

Mitchell, Robert W., 'Americans' Talk to Dogs: Similarities and Differences With Talk to Infants', *Research on Language and Social Interaction*, vol. 34, no. 2, 2001

Montepare, Joann M. and Cynthia Vega, 'Women's Vocal Reactions to Intimate and Casual Male Friends', *Personality and Social Psychology Bulletin*, vol. 14, no. 1, March 1988

Montgomery, Scott L. and Alok Kumar, 'Telling Stories: Some Remarks on Orality in Science', *Science as Culture*, vol. 9, no. 3, 2000

Moon, Christine et al., 'Two-Day-Olds Prefer Their Native Language', *Infant Behaviour and Development*, 16, 1993

Morris, J.S. et al., 'Saying it with Feeling: Neural Responses to Emotional Vocalizations', *Neuropsychologia*, 1999, 37

Morton, J. Bruce and Sandra E. Trehub, 'Children's Understanding of Emotion in Speech', *Child Development*, vol. 72, no. 3, May/June 2001

Moyal, Ann, 'The Gendered Use of the Telephone: an Australian Case Study', *Media, Culture, and Society*, 1992:14

Murray, Iain R. and John L. Arnott, 'Toward the Simulation of Emotion in Synthetic Speech: a Review of the Literature on Human Vocal Emotion', *Journal of the Acoustic Society of America*, 93 (2), February 1993

Murray, Iain R. et al., 'Emotional Stress in Synthetic Speech: Progress and Future Directions', *Speech Communication*, 20, 1996

Nass, Clifford and Kwan Min Lee, 'Does Computer-synthesised Speech Manifest Personality? Experimental Tests of Recognition, Similarity-Attraction, and Consistency-Attraction', *Journal of Experimental Psychology: Applied*, vol. 7, no. 3, September 2001

Natale, Michael, 'Convergence of Mean Vocal Intensity in Dyadic Communication as a Function of Social Desirability', *Journal of Personality and Social Psychology*, vol. 32, no. 5, 1975

Nazzi, Thierry et al., 'Language Discrimination by Newborns : Towards an Understanding of the Role of Rhythm', *Journal of Experimental Psychology: Human Perception and Performance*, vol 24, no. 3, 1998

Nazzi, Thierry et al., 'Language Discrimination by English-Learning 5-month-olds: Effects of Rhythm and Familiarity', *Journal of Memory and Language*, 43, 2000

Niedzielska, Grazyna et al., 'Acoustic Evaluation of Voice in Individuals with Alcohol Addiction', *Folia Phoniatrica et Logopaedica* 46, 1994

Ockleford, Elizabeth M. et al., 'Responses of Neonates to Parents' and Others' Voices', *Early Human Development*. 18, 1988

Ohala, John J., 'Cross-language Use of Pitch: An Ethological View', *Phonetica*, 1983

Oller, D. Kimbrough and Rebecca E. Eilers, 'Development of Vocal Signalling in Human Infants: Towards a Methodology for Cross-species Vocalization Comparisons', in Hanus Papousek et al., eds., *Nonverbal Vocal Communication: Comparative and Developmental Approaches* (Cambridge: Cambridge University Press, 1992)

Ormerod, David, 'Sounds Familiar? – Voice Identification Evidence', *Criminal Law Review*, August 2001

Papousek, Mechthild, 'Early Ontogeny of Vocal Communication in Parent-Infant Interactions', in Hanus Papousek et al., eds., *Nonverbal Vocal Communication: Comparative and Developmental Approaches* (Cambridge: Cambridge University Press, 1992)

Papousek, Mechthild et al., 'Infant Responses to Prototypical Melodic Contours in Parental Speech', *Infant Behaviour and Development* 1990, 13

Patterson, Michelle L. and Janet F. Werker, 'Two-month-old Infants Match Phonetic Information in Lips and Voice', *Developmental Science* 6:2, 2003

Pemberton, Cecilia et al., 'Have Women's Voices Lowered Across Time? A Cross-sectional Study of Australian Women's Voices', *Journal of Voice*, vol. 12, no. 2, 1998

Phillips, M. L. et al., 'Neural Responses to Facial and Vocal Expressions of Fear and Disgust', Proceedings of the Royal Society, London, *Biological Science*, 7 Oct 1998

Plant, Sadie, 'On the Mobile', www.receiver.vodaphone.com

Pye, Clifton, 'Quiché Mayan Speech to Children', *Journal of Child Language*, 13, 1986

Querleu, D. et al., 'Reaction of a Newborn Infant less than 2 Hours after Birth to the Maternal Voice', *Journal de gynecologie, obstetrique, et biologie de la reproduction*, Paris, 13(2), 1984

Querleu, D. et al., 'Intra-amniotic Transmission of the Human Voice', *Revue Française de Gynecologie et d'Obstetrique*, 1988, January 83

Reissland, Nadja et al., 'The Pitch of Maternal Voice: a Comparison of Mothers Suffering from Depressed Mood and Non-depressed Mothers Reading Books to their Infants', *Journal of Child Psychology and Psychiatry*, 44:2, 2003

Robinson, Christopher W. and Vladimir M. Sloutsky, 'Auditory Dominance and its Change in the Course of Development', *Child Development*, vol. 75, no. 5, September/October 2004

Rosolato, Guy, 'La Voix: Entre Corps et Langage', *Revue Française de Psychanalyse* 37, no. 1, 1974

Ross, Elliott D., 'How the Brain Integrates Affective and Propositional Language into a Unified Behavioural Function', *Archives of Neurology*, vol. 38, December 1981

Ross, Mileva, 'Radio', in Josephine King and Mary Stott, *Is This Your Life?: Images of Women in the Media* (London: Virago, 1977)

Russell, Alison et al., 'Speaking Fundamental Frequency Changes Over Time in Women: A Longitudinal Study', *Journal of Speech and Hearing Research*, vol. 38, February 1995

Sachs, Jacqueline et al., 'Anatomical and Cultural Determinants of Male and Female Speech' in Roger W. Shuy and Ralph W. Fasold, eds., *Language Attitudes: Current Trends and Prospects* (Washington, DC: Georgetown University Press, 1973)

Sachs, Jacqueline, 'The Adaptive Significance of Linguistic Input to Prelinguistic Infants', in Catherine E. Snow and Charles A. Ferguson, *Talking to Children: Language Input and Acquisition* (Cambridge: Cambridge University Press, 1977)

Sagi, Abraham and Martin L Hoffman, 'Empathetic Distress in the Newborn', *Developmental Psychology*, 12: 2, 1976

Sansavini, Alessandra, 'Neonatal Perception of the Rhythmical Nature of Speech: the Role of Stress Patterns', *Early Development and Parenting*, vol. 6 (1), 1997

Sapir, E., 'Speech as a Personality Trait', in John Laver and Sandy Hutcheson, eds., *Communication in Face-to-Face Interaction* (Harmondsworth: Penguin Books, 1972).

Sataloff, Robert Thayer, 'Genetics of the Voice', *Journal of Voice*, vol. 9, no.1

Scherer, Klaus R., 'Personality Inference from Voice Quality: The Loud Voice of Extroversion', *European Journal of Social Psychology*, vol. 8, 1978

Scherer, Klaus R., 'Vocal Affect Expression: a Review and a Model for Future Research', *Psychological Bulletin*, vol. 99, no. 2, 1986

Scherer, Klaus R., 'Expression of Emotion in Voice and Music', *Journal of Voice*, vol. 9, no. 3, 1995

Schmitt, Jorg J. et al., 'Hemispheric Asymmetry in the Recognition of Emotional Attitude Conveyed by Facial Expression, Prosody and Propositional Speech', *Cortex*, 33, 1977

Schucker, Beth and David R. Jacobs, 'Assessment of Behavioral Risk for Coronary Disease by Voice Characteristics', *Psychosomatic Medicine*, vol. 39, no. 4, July–August 1977

Scott, S. K. et al., 'Impaired Auditory Recognition of Fear and Anger After Bilateral Amygdala Lesions', *Nature*, 16 January1997

Scott, Sophie K. et al., 'Identification of a Pathway for Intelligible Speech in the Left Temporal Lobe', *Brain*, 2000

Shahidullah, S. and P.G. Hepper, 'Frequency Discrimination by the Fetus', *Early Human Development*, 36, January 1994

Simner, M. L., 'Newborns' Response to the Cry of Another Infant', *Developmental Psychology*, 5, 1971

Singh, Leher et al., 'Infants' Listening Preferences : Baby Talk or Happy Talk', *Infancy*, vol. 3, no. 3, 2002

Sloutsky, Vladimir M. and Amanda C. Napolitano, 'Is a Picture Worth a Thousand Words? Preferences for Auditory Modality in Young Children', *Child Development*, vol. 74, no. 3, May/June 2003

Spence, Melanie J. and Anthony J. DeCasper, 'Prenatal Experience with Low-Frequency Maternal-Voice Sounds Influence Neonatal Perception of Maternal Voice Samples', *Infant Behaviour and Development*, 1987, 10

Stemple, Joseph C., 'Voice Research: So What? A Clearer View of Voice Production, 25 Years of Progress; the Speaking Voice', *Journal of Voice*, vol. 7, No. 4, 1993

Stern, Daniel N., 'Putting Time Back into Our Considerations of Infant Experience: a Microdiachronic View', *Infant Mental Health Journal*, vol. 21(1–2), 2000

Stross, Brian, 'Speaking of Speaking: Tenejapa Tzeltal Metalinguistics', in Richard Bauman and Joel Sherzer, eds., *Explorations in the Ethnography of Speaking* (Cambridge: Cambridge University Press, 1989)

Stuart-Smith, Jane, 'Glasgow: Accent and Voice Quality', in Paul Foulkes and Gerard Docherty, eds., *Urban Voices* (London: Hodder Headline, 1999)

Svec, Jan G. et al., 'Vocal Dosimetry: Theoretical and Practical Issues', in G. Schade et al., eds., *Proceeding Papers for the Conference: Advances in Qualitative Laryngology, Voice and Speech Research* (Stuttgart: IRB Verlag, 2003)

Szarkowska, Agnieszka, 'The Power of Film Translation', *Translation Journal*, vol. 9, no. 2, April 2003

Tanford, J. Alexander et al., 'Novel Scientific Evidence of Intoxication: Acoustic Analysis of Voice Recordings from the Exxon Valdez', *Journal of Criminal Law and Criminology*, vol. 82, no. 3, 1991

Tartter, V.C., 'Happy Talk: Perceptual and Acoustic Effects of Smiling on Speech', *Perception and Psychophysics*, vol. 27, 1980

Thiessen, Erik D. and Jenny R. Saffran, 'When Cues Collide: Use of Stress and Statistical Cues to Word Boundaries by 7 to 9-month-old Infants', *Developmental Psychology*, vol. 39(4), July 2003

Thorpe, W.H., 'Vocal Communication in Birds', in R.A. Hinde, ed., *Non-Verbal Communication* (Cambridge: Cambridge University Press, 1972)

Titze, Ingo R., 'Physiologic and Acoustic Differences between Male and Female Voices', *Journal of the Acoustical Society of America*, 85(4), April 1989

Titze, Ingo R. et al., 'Populations in the US Workforce who Rely on Voice as a Primary Tool of Trade: a Preliminary Report', *Journal of Voice*, vol. 11, no. 3, 1997

Trager, George L., 'Paralanguage: a First Approximation', reprinted in Dell Hymes, *Language in Culture and Society* (New York: Harper-Row, 1964)

Trainor, Laurel J. et al., 'Is Infant-directed Speech Prosody a Result of the Vocal Expansion of Emotion?', *Psychological Science*, vol. 11, no. 3, May 2000

Trevarthen, Colwyn, 'Intrinsic Motives for Companionship in Understanding: Their Origin, Development, and Significance for Infant Mental Health', *Infant Mental Health Journal*, vol.22(1-2), 2001

Trevarthen, Colwyn and Kenneth J. Aitken, 'Infant Intersubjectivity: Research, Theory, and Clinical Applications', *Journal of Child Psychology and Psychiatry*, vol. 42: no. 1, 2001

Trevarthen, Colwyn and Stephen Malloch, 'Musicality and Music before Three: Human Vitality and Invention Shared with Pride', *Zero to Three*, vol. 23, no. 1, September 2002

Tuomi, Seppo K. and James E. Fisher, 'Characteristics of Simulated Sexy Voice', *Folia Phoniatrica*, 31, 1979

Valentine, Carol Ann and Banisa Saint Damian, 'Gender and Culture as Determinants of the "Ideal Voice"', *Semiotica*, 71-3/4, 1988

van Bezooijen, Renee, 'Sociocultural Aspects of Pitch Differences between Japanese and Dutch Women', *Language and Speech*, 38(3), 1995

Van Lancker, Diana, 'Speech Behaviour as a Communication Process', in John K. Darby, *Speech Evaluation in Psychiatry* (New York: Grune & Stratton, 1981)

Vouloumanos, Athena and Janet F. Werker, 'Tuned to the Signal: the Privileged Status of Speech for Young Infants', *Developmental Science*, 2004

Vovolis, Thanos, 'The Voice and the Mask in Ancient Greek Tragedy', in Larry Sider et al., eds., *Soundscape: The School of Sound Lectures, 1998-2001* (London: Wallflower Press, 2003)

Vrij, Aldert et al., 'People's Insight into Their Own Behaviour and Speech Content when Lying', *British Journal of Psychology*, 92, 2001

Ward, Cynthia D. and Robin Panneton Cooper, 'A Lack of Evidence in 4-month-old Human Infants for Paternal Voice Preference', *Developmental Psychobiology*, 35, 1999

Warren-Leubecker, Amye and John Neil Bohannon, 'Intonation Patterns in Child-directed Speech: Mother–Father Differences', *Child Development*, 55, 1984

Weinberg, Bernd and Suzanne Bennett, 'Speaker Sex Recognition of 5- and 6-year-old Children's Voices', *Journal of the Acoustical Society of America*, vol. 50, no. 4 (pt 2), 1971

Weintraub, Sandra et al., 'Disturbances in Prosody: A Right-Hemisphere Contribution to Language', *Archives of Neurology*, vol. 38, December 1981

Weiss, Deso A., 'The Pubertal Change of the Human Voice', *Folia Phoniatrica*, vol. 2, 1950

Weiss, Deso A., 'The Psychological Relations to One's Own Voice', *Folia Phoniatrica*, vol. 7, no. 4, 1955

Wells, Rulon, 'The Pitch Components of English', *Language*, 21, 1943

Werker, Janet F. and Renée N. Desjardins, 'Listening to Speech in the First Year of Life: Experiential Influences on Phoneme Perception', *Current Directions in Psychological Science*, vol. 4, no. 3, June 1995

Werker, Janet F. and Richard C. Tees, 'Influences on Infant Speech Processing: Towards a New Synthesis', *Annual Review of Psychology*, 50, 1999

Werker, Janet F. and Richard C. Tees, 'Speech Perception as a Window for Understanding Plasticity and Commitment in Language Systems of the Brain', *Developmental Psychobiology*, 46, 2005

Whitehead, Gregory, 'Radio Play Is No Place', *The Drama Review*, 40, 3 (T151), Fall 1996

Wilding, John et al., 'Sound Familiar?', *Psychologist*, vol.13, no 11, November 2000

Williams, Carl E. and Kenneth N. Stevens, 'Emotions and Speech: Some Acoustical Correlates', *Journal of the Acoustical Society of America*, vol.52, no.4, Part 2, 1972

Wolff, P.H., 'The Natural History of Crying and Other Vocalizations in Early Infancy', in B.M. Foss ed., *Determinants of Infant Behaviour*, vol. 4 (London: Methuen, 1969)

Yamazawa, Hideko and Harry Hollien, 'Speaking Fundamental Frequency Patterns of Japanese Women', *Phonetica*, 49, 1992

Zalusky, Sharon, 'Telephone Analysis: Out of Sight, But Not Out of Mind', www.psychomedia.it

Zeskind, Philip Sanford and Victoria Collins, 'Pitch of Infant Crying and Caregiver Responses in a Natural Setting', *Infant Behaviour and Development*, 10, 1987

Zlochower, Adena J. and Jeffrey F. Cohn, 'Vocal Timing in Face-to-face Interaction of Clinically Depressed and Nondepressed Mothers and Their 4-month-old Infants', *Infant Behaviour and Development*, 19, 1996

# DOCUMENTS, SPEECHES AND UNPUBLISHED PAPERS

Bonastre, Jean-François et al., 'Person Authentication by Voice: a Need for Caution', *Proceedings of the 8th European Conference on Speech Communication and Technology* (Geneva, Switzerland: Eurospeech, 2003)

Chapman, Janice, 'What Is Primal Scream?', paper given at the First International Conference of the Physiology and Acoustics of Singing, Groningen, October 305, 2002

.Davis, Pamela, 'Emotional Influences in Singing', paper given at the Fourth International Congress of Voice Teachers, London, 1997

Lovett, L.M. and B. Richardson, 'Talk: Rate, Tone and Loudness – How They Change in Depression', paper given to the Autumn Quarterly Meeting, Royal College of Psychiatrists, 1992

Malloch, Stephen N. et al., 'Measuring the Human Voice: Analysing Pitch, Timing Loudness and Voice Quality in Mother–Infant Communication', paper presented at the International Symposium of Musical Acoustics, Edinburgh, August 1997 (reprinted in *Proceedings of the Institute of Acoustics*, vol. 19, part 5).

Martin, Stephanie, 'An Exploration of Factors Which Have an Impact on the Vocal Performance and Vocal Effectiveness of Newly Qualified Teachers and Lecturers', paper given at the Pan European Voice Conference, Graz, Austria, 31 August 2003.

McAllister, Anita and Svante Granqvist, 'Child Voice and Noise: the Effects of a Day at the Daycare on Vocal Parameters in 10 Five-year-old Children', paper given at the Pan European Voice Conference, Royal Academy of Music, September 2005

Svec, Jan G. et al., 'Measurement with Vocal Dosimeter: How Many Meters do the Vocal Folds Travel During a Day?', paper given at the Pan European Voice Conference, Graz, Austria, 28 September 2003.

Thorn, Richard 'The Anthropology of Sound', paper given at 'Hearing is Believing' conference, University of Sunderland, 2 March 1996.

Titze, Ingo, 'How Far Can the Vocal Folds Travel", paper given at the American Speech–Language–Hearing Association Annual Conference, Atlanta, Georgia, 2002

Titze, Ingo, 'The Human Voice at the Intersection of Art and Science', paper given at the Pan European Voice Conference, Graz, Austria, 28 August 2003.

Vilkman, Erkki, 'The Role of Voice (Quality) Therapy in Occupation Safety and Health Context', paper given at the Pan European Voice Conference, Graz, Austria, 29 August 2003

# Index

Anne Karpf is a writer, sociologist, and award-winning journalist. For seven years she was radio critic of the *Guardian*, where she now writes a weekly column. Her last book, *The War After: Living with the Holocaust*, a memoir of growing up the child of Jewish Holocaust survivors, was published to acclaim in 1996. She broadcasts on radio and television, and teaches at London Metropolitan University.

## A NOTE ON THE TYPE

The text of this book is set in Linotype Sabon,
named after the type founder, Jacques Sabon.
It was designed by Jan Tschichold and jointly
developed by Linotype, Monotype and Stempel,
in response to a need for a typeface to be
available in identical form for mechanical hot
metal composition and hand composition
using foundry type.

Tschichold based his design for Sabon roman
on a font engraved by Garamond, and Sabon
italic on a font by Granjon. It was first used
in 1966 and has proved an enduring
modern classic.